The Reticuloendothelial System in Health and Disease

Immunologic and Pathologic Aspects

ADVANCES IN EXPERIMENTAL MEDICINE AND BIOLOGY

Recent Volumes in this Series

The Reticuloendothelial System in Health and Disease

Immunologic and Pathologic Aspects

Edited by

Herman Friedman
Albert Einstein Medical Center
Philadelphia, Pennsylvania

Mario R. Escobar
Medical College of Virginia
Richmond, Virginia

and

Sherwood M. Reichard
Medical College of Georgia
Augusta, Georgia

PLENUM PRESS • NEW YORK AND LONDON

Library of Congress Cataloging in Publication Data

Main entry under title:

The Reticuloendothelial system in health and disease.

(Advances in experimental medicine and biology; 73A-B)
"Proceedings of the seventh international congress of the Reticuloendothelial Society and the first scientific meeting of the European Reticuloendothelial Society held in Pamplona, Spain, September 15-20, 1975."
Includes bibliographical references and index.
CONTENTS: pt. A. Functions and characteristics.—pt. B. Immunologic and pathologic aspects.
1. Reticulo-endothelial system—Diseases—Congresses. 2. Reticulo-endothelial system—Congresses. I. Reichard, Sherwood M. II. Escobar, Mario R. III. Friedman, Herman, 1931-   IV. Reticuloendothelial Society. V. European Reticuloendothelial Society. VI. Series. [DNLM: 1. Reticuloendothelial system—Congresses. W1 AD559 v. 73 1975 / [WH650 R438 1975]]
RC645.5.R48   616.4'1   76-24833

ISBN-13: 978-1-4684-3302-9   e-ISBN-13: 978-1-4684-3300-5
DOI:10.1007/ 978-1-4684-3300-5

Proceedings of the second half of the Seventh International Congress of the Reticuloendothelial Society and the First Scientific Meeting of the European Reticuloendothelial Society held in Pamplona, Spain, September 15-20, 1975

PROCEEDINGS OF THE

7TH INTERNATIONAL CONGRESS OF THE RETICULOENDOTHELIAL SOCIETY

and

FIRST SCIENTIFIC MEETING OF THE EUROPEAN RETICULOENDOTHELIAL SOCIETY

Pamplona, Spain   September 15-20, 1975

EXECUTIVE COMMITTEE: S. M. Reichard, *President*; R. A. Good, *Vice-President* and M. R. Escobar, *Secretary General*

ADVISORY COMMITTEE: A. C. Allison, G. Astaldi, J. A. Bellanti, L. A. Chedid, E. L. Dobson, N. B. Everett, M. Feldman, K. E. Hellstrom, M. Kojima, M. Landy, G. B. Mackaness, S. J. Normann, G. J. V. Nossal, J. W. Rebuck, F. Rossi, W. G. Spector and G. J. Thorbecke

SCIENTIFIC PROGRAM COMMITTEE: H. Friedman, *Co-Chairman*; W. Th. Daems, *Co-Chairman*; L. J. Berry; I. Carr; N. R. Di Luzio; Q. N. Myrvik; P. Patriarca; A. J. Sbarra; B. Schildt; M. M. Sigel; D. J. Wilkins

This volume is dedicated in deference to the following pioneer investigators, honored by the Congress, whose significant contributions have triggered the explosive development of our knowledge of the reticuloendothelial system:

K. K. Y. Akazaki

R. E. Billingham

G. Biozzi

F. M. Burnet

K. B. P. Flemming

P. Garzon

J. L. Gowans

J. E. Harris

J. Da Silva Harris

P. J. Jacques

N. K. Jerne

R. Paoletti

R. T. Prehn

K. Stern

A. E. Stuart

M. Timar

I. Toro

Preface

An extraordinary development of the knowledge, concepts and
biomedical applications has occurred during the past two decades
in the biological sciences, including the Reticuloendothelial System (RES). For example, it is now widely recognized that distinct
classes of cells are involved in the recognition of "exogenous"
invaders of the body such as microorganisms and other foreign antigens, as well as of "endogenous" parasites represented by transformed neoplastic cells and altered "self" antigens. Prominent
among cell populations involved in such recognition of antigens
and subsequent immune responses are those constituting the RES.
In recent years, there has been much discussion and indeed controversy as to what constitutes such a system and even whether the
term "RES" is appropriate. Some investigators feel that the phagocytes are the most important cells of the RES as they play a major
role in the defense mechanism of the host. Mononuclear phagocytes
include tissue macrophages as well as circulating monocytes and
their precursors. Although phagocytosis is a major functional activity, it is only one of their several functions. The important
role of phagocytes and other mononuclear cells in antibody formation, cell-mediated immunity, specific and non-specific resistance
to microorganisms and tumor cells, as well as homeostatic adjust-
ments in general has become the focus of attention for many investigators.

The Reticuloendothelial Society, which is concerned with the
advancement of knowledge concerning the many diverse functions of
RE cells, organizes national and international meetings and pub-
lishes a scientific journal. The Seventh International Congress
of the RES Society was held in Pamplona, Spain, together with the
First Scientific Meeting of the European RES Society on September
15-20, 1975. The Congress had as its scientific objective a wide
range of subjects concerning the RES in health and disease. Special reference was given to macrophage structure, ontogeny and
classification of cell populations. Emphasis was placed on the
function and biochemistry of macrophages and other cells of the
RES in the modulation of the immune response, their role in tumor

and transplantation immunology, and their influence in regulating the growth and function of other cell types. The secretion of soluble materials by macrophages and lymphocytes and the mode of action of chemotherapeutic drugs and other agents on RE cell function and immune reactivity were also stressed.

A symposium was presented each morning during the Congress with five to six internationally renowned biomedical and basic scientists who discussed specific aspects of the RES in detail. Each afternoon was devoted to two simultaneous scientific sessions where short papers were presented on subjects of current interest and relevance to the RES. Workshops devoted to particularly important phases of the RES were also held. These volumes constitute the published record of the proceedings of the symposia, the scientific paper sessions, and the workshops. However, the written contributions are not arranged exactly as presented at the Congress, but rather these two volumes intersperse papers selected for their current relevance. The first volume includes papers given during the first half of the Congress and is entitled "Functions and Characteristics of the RES". The second volume comprises papers given during the second half of the Congress and is entitled "Immunologic and Pathologic Aspects of the RES".

The setting for this meeting was in the magnificent Basque country of northern Spain. Pamplona, founded in 75 B. C. by Pompey and captured in 778 by Charlemagne, is a city of great tradition and charm. The nine-day Fiesta de San Fermin is still celebrated by running the bulls through the streets crowded with enthusiastic spectators. The sidewalk cafes on Plaza de Mayor are festive and reminiscent of the vivid descriptions left by Hemingway. In contrast, the Universidad de Navarra is a modern, beautifully furnished and equipped facility, which added the final touch of perfection to a wonderful Congress.

April, 1976

M. R. Escobar
H. Friedman
S. M. Reichard

Acknowledgments

The editors are indebted to Drs. J. A. Bellanti, N. R. Di Luzio, R. Evans, A. Oehling, F. Rossi, M. M. Sigel, G. J. Thorbecke and D. J. Wilkins who chaired the symposia. Many thanks to Drs. G. W. Gander, N. Harris, A. M. Kaplan, M. Kojima, G. Lazar, S. S. Lefkowitz, T. J. Linna, E. Ouchi, R. E. Ritts, H. Schorn and E. Wisse who chaired the short paper sessions. Gratitude is expressed to Drs. W. Th. Daems, F. J. Di Carlo, N. B. Everett, K. B. P. Flemming, P. J. Jacques, P. Patriarca and A. J. Sbarra for their valuable help in preparing the workshops. In addition, Drs. J. A. Astruc, E. Bresnick, H. P. Dalton, G. E. Rodriguez and H. M. Syrop assisted in reviewing some of the manuscripts. This Congress could not have succeeded without the financial assistance from pharmaceutical companies, members of the RE Society and, especially, the generous donation from the United States Energy Research and Development Administration. Recognition is extended to Ms. M. A. Dearing, Ms. G. Browder and Ms. I. Friedman who shared in the monumental task of typing all the manuscripts and to Ms. Phyllis Straw of Plenum Publishing Corporation for her expert editorial advice.

Contents of Part B

SECTION 1
IMMUNOLOGIC AND PATHOLOGIC ASPECTS OF THE RES

SECTION 2
THE RES IN IMMUNOPATHOLOGY AND AUTOIMMUNITY

SECTION 3
THE RES IN TRANSPLANTATION IMMUNOLOGY

WORKSHOP

    K. Flemming, S. Reichard and M. Escobar, *Co-Chairmen*

POST—CONGRESS WORKSHOP SUMMARY

Contents of Part A

SECTION 1
MORPHOLOGY AND SURFACE PROPERTIES OF MACROPHAGES

SECTION 2
ONTOGENY AND DEVELOPMENT OF RE CELLS

SECTION 3
BIOCHEMISTRY AND PHYSIOLOGY OF THE RES

SECTION 4
THE PHAGOCYTIC EVENT

Introduction

This volume reviews the immunologic and pathologic aspects
of the RES and it is divided into four areas: The first section
is concerned mainly with newer aspects of the immune response
mechanism as related to RES cells. The role of T lymphocytes in
immune responses and the multiplicity of these cells are described
in detail. Recent basic findings which have altered the understanding of how T lymphocytes enhance and suppress immune responses
are presented. Various aspects of immunologic tolerance from a
number of viewpoints are also included. For example, the role of
T lymphocytes in immunologic tolerance, including the unexpected
discovery that two cells may be involved in immune induction of
T-independent antigens, comprising both T cells and a cooperating
A cell , is discussed. The relationship of "classic" RES cell
"blockade" and how this relates to modern concepts of antibody
formation and immunologic tolerance at the cellular level are viewed in regard to a model system whereby colloidal carbon depressed
antibody formation. The role of macrophages in such blockade and
induction of tolerance is included. Modulation of the expression
of secondary Ig and an anti-hapten response _in vitro_, Ig as a control mechanism for the shift in antibody synthesis and the role
of antigen and localization of immunologically specific B lymphocytes and immunoglobulins in immune tissue are covered in this
section. The second section deals with the RES in immunopathology
and autoimmunity. A description of the immunopathology of the
nervous system is followed by discussions regarding immunodeficiency
diseases, immune complex diseases, autoimmune thyroiditis in an experimental animal model and the role of viruses in antibody-dependent cell-mediated cytotoxicity. Section three is involved with
the RES in transplantation immunology. This section includes topics such as the immunogenetics of classic allografting, the role
of alloantibodies in relation to the rejection of skin grafts, the
effect of immunosuppression on rosette forming cells, the biological and pathologic characteristics of mice with different levels of
leukocytes and the development of cellular immunity to transplanted
kidneys. In section four, the RES in tumor immunology is reviewed
including chapters on the role of leukemia viruses on immune func-

tion, both cellular and humoral, the immunotherapy of leukemia in
experimental mice and the effects of a leukemia virus infection on
the morphology of splenocytes. The host response to tumor specific
antigens and a description of a polypeptide antigen associated
with certain malignancies in man, the role of cell-mediated and
humoral immunity in protection of experimental animals from malignancy, the role of macrophages in modulation of tumor growth and
their participation in spontaneously regressing syngeneic tumors
follow. A discussion of the interaction of macrophages with tumor
cells at the cellular level completes this section.

A summary of a workshop on radiobiological and environmental
effects on the RES is also included in this volume. This workshop
reviewed some experimental work dealing with radiation effects on
cells of the RES, physiologic mechanisms in radiation resistance
and immunologic aspects of malnutrition. The final portion of this
volume summarizes briefly the proceedings of the post-congress workshop which was devoted to certain areas of RES methodology and
which was very informative and of great interest to participants
within a relaxed and informal setting. It is hoped that a publication of this session summary will be the spark for future workshops
on new procedures which are constantly being developed and evaluated.

Immunologic and Pathologic

Aspects of the RES

THE ROLE OF THE T CELL IN THE IMMUNE RESPONSE

R. K. GERSHON

Yale University School of Medicine, New Haven,
Connecticut (USA)

T cells are a heterogenous group of cells and thus have multiple functions in the immune response. This multiplicity of functions has led to a multiplicity of names for T cell subsets; there
are "killer cells," "helper cells," and "suppressor cells" implying
that T cells kill foreign cells as well as help and suppress syngeneic effector lymphoid cells to perform their immunological functions. There have been a number of exciting basic discoveries made
recently, which have significantly altered our understanding of
how T cells perform some of these functions. These new insights
should be helpful in designing experimental protocols to use the
immune response to control cancer growth.

Host-Tumor Interactions

In almost all cases which have been adequately studied, tumor
cells have been known to possess antigens on their surface which
are recognized as foreign by the host in which the tumors arise
(reviewed in 35). Under certain circumstances, these antigens can
be shown to provoke immune responses which inhibit tumor growth,
while in other instances tumors may thrive and progress in spite
of the existence of foreign transplantation antigens on their surfaces. There are two basic mechanisms by which antigenic tumors
might avoid destruction:
  1. Tumor cells may be "non-immunogenic." Antigens may
exist in a nonstimulatory form, and thus behave like haptens.
Free hapten fails to evoke an immune response but does so quite
well when conjugated to a carrier which then gives the hapten
"immunogenicity." Some antigens may be stimulatory but have a

3

propensity to stimulate suppressor mechanisms. For example; the
polymer of glutamic acid alanine and tyrosine ($G_{60}A_{30}T_{10}$) fails
to evoke an immune response in some strains of mice because it
preferentially stimulates suppressor T cells (26,5). Other antigens may preferentially stimulate the production of antibodies,
which act to feedback and suppress the development of cell mediated immune responses (55). For example, it has been shown that pre-
treatment of animals with lyophilized tumor cells leads to the
production of enhancing antibodies and a more rapid growth of
subsequent tumor transplants, while pretreatment of the animals
with irradiated intact tumor cells leads to the production of cell
mediated immunity and a decreased ability of subsequent grafts to
grow (35). Another mechanism which might affect the immunogenicity of an antigen is simply antigen dose. Excess antigen may inhibit the response by activating suppressor mechanisms or by
direct paralysis of responding cells.
    2. Additionally tumor cells may escape destruction by being
immunogenic but resistant to the immune response they evoke.
Resistance to host killing mechanisms might be conferred upon a
tumor cell by several attributes. The tumor might not be affected by the host killer mechanism. For example: the tubercle
bacillus evokes a potent immune response which sometimes may be
adequate to contain its growth but is usually inadequate to kill
all of the organisms. Resistance could also be conferred upon
tumor cells which were susceptible to host killing mechanisms but
which grew so rapidly that they would generate more cells than the
host could kill in a given time. This seems to be the basis for
concomitant immunity, a situation in which the host mounts a very
potent immune response but fails to contain the original tumor
inoculum (which is quite large by the time the response is generated) but is capable of rejecting new grafts of the same tumor (27).
Highly immunogenic tumors such as choriocarcinoma and Burkitt's
lymphoma probably escape host defenses by such a mechanism.

KILLER CELL MECHANISMS

    The host has almost as many mechanisms for killing tumor
cells as tumor cells have in avoiding the killers. Thus, some
tumors are killed by cytotoxic T cells (16), some by antibodies
(39), some by K cells in antibody dependent cell mediated cytolysis (38,42) and some by macrophages (40). Macrophages may
phagocytize tumor cells when cytophilic antibody is present, but
they also can kill tumor cells by non-phagocytic means (6).
In some cases, the killing is specific and in those cases it
seems that the macrophages are "armed" with specific T cell
products (24). In addition macrophages may be nonspecifically activated by T cell products (37). This mechanism seems

to be the basis for increased resistance of hosts immunized
with BCG (1).

It was recognized in the early 1960's that lymphoid cells of
animals sensitized in vivo against allografts were cytotoxic for
donor cells in culture (45). Since then a number of different
methods have been applied to assay the cell mediated cytotoxic
effector phase, the most commonly used of which is the one
described initially by Brunner (9) and modified by Canty and
Wunderlich (14): the short term release of chromium 51 from the
target cells. In this system it has clearly been shown that the
cytotoxicity is affected by T cells which have differentiated
into specific cytotoxic lymphocytes. By the use of this assay,
Hayry and Defendi (30) as well as Wunderlich and Canty (57) were
able to show that killer T cells were generated in mouse mixed
lymphocyte reactions occurring in vitro while Solliday and Bach
reached similar conclusions in their studies of human lymphocytes (53). More recently it has been shown that allogeneic tumor
cells can also induce killer T cell activity in vitro (56); and,
in some circumstances, weak but definite tumor specific killer
T cell activity can be induced in culture by syngeneic tumor
cells (44).

IMMUNOGENETICS

Recent advances in immunogenetics and in methods to fractionate T cell subpopulations have produced further insight into the
mechanism of generation of killer T cells. Shreffler and David
have recently published a magnificent review on the makeup of the
major histocompatibility complex (MHC) of the mouse (50). A brief
summary of the points relative to this discussion is as follows:
the MHC is coded for by a tightly linked series of genes to the
right of the centromere of the seventeenth mouse chromosome and
has been divided into 4 major regions, K I S and D, reading from
left to right.

The S region controls the Ss and Slp serum protein traits and
more recently it has been suggested to control complement C4 levels.
This region serves a good marker for genetic studies, but is not
particularly relevant to this discussion and thus will not be
considered further.

The antigens coded for by the K and D regions are serologically determined. It is generally felt that K and D arose from a
common ancestral gene through duplication and mutational diversions.
Thus, certain anti-H-2 K and D sera show cross reactions and these
identify "public antigens." In addition, it has been clearly
demonstrated that each H-2 haplotype has two private specificities
which can be consistently arranged into separate H-2K and H-2D

sets. Biochemical evidence also supports this two region model.
Both K and D are highly polymorphic and the lymphocytes of
heterozygotes express both parental alleles. These loci have been
called SD (serologically determined) by Fritz Bach (2) because
mice which differ from one another only at these loci generally do
not make a good mixed lymphocyte reaction (MLR). There are some
exceptions to this generalization, the most notable being the B6
(Hzl) mutant of Bailey (3). This is supposed to be a point muta-
tion which occurred in the K region but induces a strong MLR.
The basis for this apparent exception to the generalization is not
fully resolved. Antigens coded for by the K and D regions of the
MHC are expressed on all lymphocytes.

Antigens coded for by the I region of the locus are only
expressed on subpopulations of lymphocytes. They are found on
most B cells and on a minority of T cells. Immune response
genes are coded for in this region. The antigens coded for by the
I region of the MHC have been called LD loci because differences
between cells in this region leads to strong mixed lymphocyte
reactions. However, antisera can also be raised against these
regions. Although little killer cell activity is generated
against the I (L.D.) region, in some instances it acts to elicit a
helper function in that its presence increases the killer activity
generated against K or D. The mechanism by which this works will
be considered below. Most people consider it likely that the
killer activity is directed against K or D but, there is the report
of Edidin and Henney that cells with their K and D antigens
capped are fully susceptible to killer cell activity (22). This
may mean that the actual loci by which the killer cell recognizes
the target are closely linked to, but distinct from, the SD
region. However, no recombinants or mutants have been found to
define such loci.

It is generally felt that killer cells react only with the
products of the K and D regions, even when killer cells are
generated against non MHC coded antigens such as viral-antigen
(19) haptenated cells (49), antigens determined by the Y locus
(52), or other non-H-2 antigens (7). In these cases killing is
restricted to cells which bear the same K or D antigens as the
sensitizing antigens, suggesting that the killer cells are
reacting with K and D antigens which have been altered by the
virus, the hapten, or the other non-H-2 antigens. Unpublished
evidence from several laboratories suggests however, that under
appropriate conditions killer cells can react with I region
antigens (10). It is not clear whether these differences can be
accounted for by killing by different T cell subpopulations.

T CELL SUBSETS

In this regard, it would be important to review what is
known about the lymphocyte subsets participating in these reactions. Interactions between subsets of T cells were first described by Cantor and Asofsky (11,43) and from their work as well
as others, two types of T cells were defined: T-1) a short-lived
cell which does not recirculate, is found predominantly in the
spleen and thymus, and acts as an amplifier cell; T-2) is a long-
lived recirculating, more mature T cell which acts as an effector
cell. Suppressor activity has often been associated with cells
with T-1-like qualities.

More recently, T cell subsets have been further refined using
antisera gainst the T cell differentiation antigens of the Ly
series discovered and characterized by Boyse (12). Basically 4
antigenically distinct Ly markers have been found on T cells; they
are: Ly1, 2, 3, and 5 (Ly4 identifies a B cell differentiation
antigen). The Ly series of antigens like the theta antigen has
two alleles. Some of the functions of the cells bearing the Ly1
2, and 3 antigens have been defined while Ly5 remains to be
characterized. Approximately 80-90% of T cells can be killed by
sequential lysis with the Ly 1,2,3 sera. The cells containing
Ly1 represent approximately 30% of the adult pool. These cells
have shown to be helper cells both for B cells as well as for
killer T cells. They also mediate reactions of delayed hypersensitivity (28). Another subset of lymphocytes has both Ly 2
and Ly 3 antigens on it. This subset is the effector cell in the
killer assay and represents somewhere between 5-10% of the T
cell pool. It adheres to K and D (SD) antigens on fibroblast
monolayers and thus specific antigen recognizing clones can be
deleted. In addition, this cell has been shown to be responsible
for suppression in two different suppressor T cell systems; the
chronic allotype suppression of Herzenberg (31), and the Con A
suppression system described by Dutton (20) and by Pierce (46).
It has been shown to have Fc receptors on it (54) and also
antigens coded for by the I region of the MHC (13). The Ly 1
cell on the other hand, is the cell which responds in the mixed
lymphocyte reaction to I (L.D.) antigens and does not contain I
antigens on it and does not adhere to monolayers. About 30% of
the T cells of adult mice contain all three Ly antigens while
greater than 90% of the peripheral T cells of mice aged one to
two weeks have all three Ly antigens on them. No specific
function as yet has been found for this subset.

CHRONIC ALLOTYPE SUPPRESSION

At this point I would like to digress slightly and review

the chronic allotype suppression story (31) because it is so
well characterized and can perhaps give us some clues concerning
the interactions which generate killer cells. Allotypes are
antigenic determinants on immunoglobulins and unlike the other
antigens I have been discussing, there is allelic exclusion by
allotype bearing cells. Anti-allotype antibody given in utero
to heterozygotes results in the weanling animals not producing any
allotype with the antigenic constitution of the target for the anti-
allotype serum. This acute suppression usually wanes towards the
end of the second month of life. However, in SJL x Balb/c mice
around the seventh or eighth week of life, after some of the
suppressed allotype has started to reappear, a chronic suppression
sets in and the allotype redisappears, usually never to appear
again. It has been shown that the chronic allotype suppression
is mediated by a T cell and can be adoptively transferred with
T cells. Most interestingly, and important for our purposes, is
that it has been shown recently that the "allotype suppressor T
cell" has as its target another T cell, not a B cell, and that the
suppressor cell has Ly 2, 3, antigens on it (32). Thus, when
"allotype suppressor T cells" are killed with an anti-Ly 2, 3
serum and complement, the remaining cells no longer have suppressor activity. However, when the remaining cells are mixed
with normal B cells and immunized, the antibody made is essentially devoid of the suppressed allotype. When normal T cells are
added to the Ly 2, 3 depleted T cells taken from suppressed
animals and normal B cells, suppressed allotype is made.

It would thus appear that there are suppressor T cells which
specifically block the activity of helper T cells which in turn
may interact specifically with B cells bearing the appropriate
allotype. It would seem most likely, in order for this to occur
that a) the allotype helper T cell (Ly 1) has specificity for the
B cell Ig allotype and b) either the suppressor T cell has
specificity for the helper T cell or the helper T cell recognizes
the suppressor and becomes suppressed by "back recognition"
mechanisms. The suppressor T cell might recognize the idiotype
(see following section for review of idiotypes) of the helper
T cell receptor which recognizes the B cell Ig allotype. The
helper T cell might then function by interacting with B cells
expressing the allotype, causing transformation of these cells in
much the same fashion as anti-allotype sera induce blast transformation of lymphocytes (48). This hypothesis finds support in
the recent observation that a) T cells bear idiotypic determinants and b) can recognize idiotypic determinants as discussed
below. Moreover the proposed idiotypic specificity of the
suppressor T cell would explain why the allotype suppressor cell is
not inhibited by free allotype. In this model such a cell does
not recognize allotype but rather anti-idiotype allotype. The
objection that circulating idiotype would interfere with allotype

helper function might be overcome if in fact the allotype
helper T cell recognized an allotype which has been slightly modified by interaction with another membrane moiety. A possible
candidate for this component could be the membrane Fc receptor.
Thus, the allotype bearing B cell might bind antigen and the
aggregation of receptors would thus trigger interaction with the
Fc receptor. The allotype helper cell would then recognize the
complex of Fc receptor and the region of the receptor heavy chain
bearing the allotype. One important consequence of this would be
the relationship of I region antigens and Fc receptors (18).
Thus histocompatibility requirements for efficient cell collaboration would be due to recognition by the allotype-helper cell of
the MHC product-allotype complex.

The alternate model, "back stimulation," is that the Ly 1
helper cell recognizes and reacts with the membrane bound allotype
on the suppressor cell (Ly 2, 3, I positive) activates it and
becomes suppressed. It would be important to distinguish between
these models as the mode of activation of suppressor cells may have
important bearing on the generation of killer cells by the MHC
locus.

Regulation of Responder Cells by Stimulator Cells

This brings me to the next major point, that is regulation of
the MLR and the generation of killer cells by the target cells
in a mixed lymphocyte reaction. In in vivo situations it is well
established, but not widely recognized, that host cells regulate
GVH reactions (51) (as well as the adoptive transfer of memory
cells (15). This because irradiation of recipients markedly
augments the activity of cells transferred into adult recipients
(51,29,4). Early workers thought that the host cells were
preventing the added cells (either the parental cells in the
GVH or the syngeneic cells in the adoptive transfer experiments)
from localizing in appropriate places in the lymphoid tissue.
Recent work has adequately ruled that out as a possibility by
showing that the added cells enter the host pool in equal numbers
in irradiated and normal mice (17). Chromosome marker studies have
shown that not only do they enter the pool in normal animals but
they remain there for the life of the animal and circulate and
recirculate similarly to the way they did in the host from which
they were taken (21). Although they enter the host circulation
and seem to behave normally in terms of circulatory capacities,
their activity is greatly diminished in hosts which have not been
irradiated. The enhaced activity they display in irradiated hosts
can be abrogated by returning normal T cells to the irradiated
host (29,4,21). The addition of the normal T cells does not
affect their lodgement and subsequent recirculating capacities

but does diminish their immunological capacities. Thus it is
quite clear in the in vivo counterpart of the MLR (i.e., the
GVHR) simulator T cells can regulate the reponse of responder
cells.

We have recently shown similar effects (regulation of responder by stimulator) in in vitro MLR's and also in the generation of cytotoxic cells (Metzler and Gershon unpublished observations). It has been shown by several workers that stimulator
cells respond in a MLR due to the recognition of them by the
responder T cell. However, the consequences of this back-stimula-
tion in terms of regulation have not been studied at any length.
Possible mechanisms include:
     1.   Role of I antigens. It may be that recognition by the
helper cell (i.e., Ly 1) of the allotype and/or the I antigen of a
stimulator T cell might lead to activation of suppressor T cells
which would lead to suppression of the helper. One of the key
points in this analogy between I antigens and allotype is the
fact that the suppressor cell (i.e., the 2, 3, cell) has I (L.D.)
antigen on its surface. It could well be that the recognition of
the I antigen by the Ly 1 responder cell might directly activate the
suppressor and thus ablate the helper effect.
     2.   Possible Role of Idiotype Recognition by Stimulator Cells.
There is another possible way by which stimulator cells in an MLR
could regulate the response of the responder cells; that is by
recognition of idiotypes on the parental cells. Idiotype,
(36,41) refers to the antigens on immunoglobulin molecules which
are uniquely derived from the amino acid sequences that give the
immunoglobulin molecules antigen recognizing characteristics.
Although these antigens are not necessarily part of the combining
site or coded for by the hypervariable region, they do identify
molecules within an immunoglobulin pool that have antigen specificity. For example: antibodies made against a myeloma protein
which has antigen binding characteristics, can inhibit the ability of normally made antibodies to bind the specific antigen
(reviewed in 47). Of course in a normal immune response antigen
binding antibodies with different idiotypes are also made but the
antigen recognizing capacities of some of them can be blocked
by the anti-idiotype serum.
     Recent evidence has shown that T cells have idiotype on their
surface (8,23) and also that T cells are capable of recognizing
idiotype on other cells (33). Thus in an MLR it is possible
that the stimulator cell can recognize, via anti-idiotype receptors, those T cells in the responder population which have the
unique determinants (idiotype) that give them the capacity to
recognize the reciprocal parents' antigens on the Fl cells.
This because the Fl animal will not contain cells capable of
recognizing itself and therefore will not be tolerant to those
idiotypes. (Again note the possible homology with the anti-

allotype suppression system in that acute suppression leads to loss of tolerance, antigen returns, activates the allotype recognizing cell (Ly 1), which then responds and activates the allotype anti-idiotype cell (Ly 2, 3) which leads to suppression (see allotype discussion above). Idiotype recognizing T cells have been shown to be able to regulate (both suppress and increase) the response of idiotype bearing cells (8,23) even in the face of large amounts of circulating idiotype, again suggesting membrane bound idiotype is preferentially reacted with, which seems to be the case with allotypes. In fact there is some evidence, although not fully convincing at this stage, that F1 T cells do recognize antigens on parental cells (25). Since F1 animals are tolerant of all the histocompatibility antigens of both parents, antigen they are recognizing may well be an idiotype.

3. <u>The Possible Role of Lymphokines</u>. It has been shown that when T cells repond to histocompatibility antigens they release blastogenic factors (34). It is possible that these blastogenic factors nonspecifically activate stimulator cells which in turn may make lymphokine-like products in response to the nonspecific activating signal. These secondarily made lymphokines may in turn feedback and alter the activity of the responder cell.

4. <u>The Possible Role of Antigen Dose</u>. As discussed above, excess antigen can be immunosuppressive either directly by paralyzing the antigen recognizing cell or indirectly by activating suppressor mechanisms. Stimulator cells could easily affect the activity of responder cells by making and shedding large amounts of antigen and in so doing regulate the response to them. The making and shedding of membrane determinants by different subsets of lymphocytes is an area that is currently under intense investigation.

In this brief review I have discussed several of the new developments in cellular immunology which I find particularly intriguing; dissection of the MHC, idiotype and allotype recognition, T cell differentiation antigens and suppressor cells. The prediction is that studies interrelating or assimilating the new findings will result, in the very near future, in some important conceptual breakthroughs in our understanding of the role of the T cell in the immune response.

REFERENCES

1. Ariyan, S. and Gershon, R.K., J. Nat. Cancer Inst. 51 (1973) 1145.
2. Bach, F.H., Bach, M.L., Sondel, P.M. and Sundharadas, G., Transpl. Rev., 12 (1972) 30.
3. Baily, D.W., Transpl. 11 (1971) 426.
4. Bell, E. B. and Shand, F.L., Eur. J. Immunol., 5. (1975) 1.
5. Benacerraf, B., Kapp, J.A., and Pierce, C.W., In: Immunolo-

gical Tolerance: Mechanisms and Potential Therapeutic

Applications, (Eds. D.H. Katz and B. Benacerraf) Academic
Press, New York, New York (1974) 507.
6. Bennett, B., Old, L. J. and Boyse, E.A., Transpl. 2 (1964)
    183.
7. Bevan, M.J., Nature, 256 (1975) 419.
8. Binz, H. and Wigzell, H., J. Exp. Med. 142 (1975) 197.
9. Brunner, R.T., Mauel, J., Rudolf, H. and Chapius, B.,
    Immunology, 18 (1970) 501.
10. Cantor, H., personal communication.
11. Cantor, H. and Asofsky, R., J. Exp. Med., 131 (1971) 235.
12. Cantor, H. and Boyse, E.A., J. Exp. Med., 141 (1975) 1376.
13. Cantor, H., personal communication.
14. Canty, T.G. and Wunderlich, J.R., J. Nat. Cancer Inst., 45
    (1970) 761.
15. Celada, F., J. Exp. Med., 124 (1966) 1.
16. Cerottini, J.C., Nordin, A.A. and Brunner, K.T., Nature,
    228 (1970) 1308.
17. Davies, A.J.S., personal communication.
18. Dickler, H.B. and Sachs, D.H., J. Exp. Med., 140 (1974) 779.
19. Doherty, P.C. and Zinkernagel, R.M., Lancet, I (1975) 1406.
20. Dutton, R. W., J. Exp. Med., 136 (1972) 1445.
21. Eardley, D.D. and Gershon, R.K., J. Exp. Med., 142 (1975) 524.
22. Edidin, M. and Henney, C.S., Nature New Biol., 246 (1973) 47.
23. Eichmann, K. and Rajewsky, K., personal communization.
24. Evans, R. and Alexander, P., Immunology, 23 (1972) 677.
25. Gebhardt, B.M., Nakao, Y. and Smith, R.T., J. Exp. Med.
    140 (1974) 370.
26. Gershon, R.K., Maurer, P.H. and Merryman, C.F., Proc. Nat.
    Acad. Sci., 70 (1973) 250.
27. Gershon, R.K., Carter, R.L., and Kondo, K., Nature, 213
    (1967) 674.
28. Gershon, R.K. and Cantor, H., unpublished observation.
29. Gershon, R.K., Cohen, P., Hencin, R.S. and Liebhaber, S.A.,
    J. Immunol., 108 (1972) 586.
30. Hayry, P. and Defendi, V., Science, 168 (1970) 133.
31. Herzenberg, L.A. and Herzenberg, L.A., Contemp. Topics in
    Immunobiol., 3 (1974) 41.
32. Herzenberg, L.A., Personal communication.
33. Janeway, C.A., Sakato, N. and Eisen, H., Proc. Nat. Acad. of
    of Sci., (in press).
34. Kasakura, S. and Lowenstein, L., Nature 208 (1965) 795.
35. Kirkwood, J.M. and Gershon, R.K., Progress Exp. Tumor Res.,
    19 (1974) 757.
36. Kunkel, H.G., Mannik, M. and Williams, R.G., Science, 140
    (1963) 1218.
37. Mackeness, G.B., Infectious Agents and Host Reactions (Ed.
    S. Mudd) W.B. Saunders Co., Philadelphia, Pennsylvania
    (1970) 62.

38. Mac Lennon, I.C.M., Loewi, G. and Howard, A., Immunology, 17 (1969) 897.
39. Old, L.J. and Boyse, E.A., Annu. Rev. Med., 15 (1964) 167.
40. Old, L.J., Boyse, E.A., Bennett, B. and Lilly, F., In: Cell Bound Antibodies, (Eds. B. Amos and H. Koprowski) Wistar Institute Press, Philadelphia, Pennsylvania (1963).
41. Oudin, J., Proc. Roy. Soc. 166 (1966) 207.
42. Perlmann, P. and Perlmann, H., Cell. Immunol., 1 (1970) 300.
43. Raff, M.C. and Cantor, H., In: Progress in Immunology, (Ed. B. Amos) Academic Press, New York, New York (1971) 83.
44. Rollinghoff, M., Wagner, H., Warner, N.L. and Nossal, G.J.V., Israel J. Med. Sci., 10 (1974) 1001.
45. Rosenau, W. and Moon, H.D., J. Nat. Cancer Inst., 27 (1961) 471.
46. Rich, R.R. and Pierce, C.W., J. Exp. Med., 137 (1973) 649.
47. Sakato, N. and Eisen, H., J. Exp. Med., 141 (1975) 1411.
48. Sell, S. and Gell, P.G.H., J. Exp. Med., 122 (1965) 423.
49. Shearer, G.M., Eur. J. Immunol., 4 (1974) 257.
50. Shreffler, D.C. and David, C.S., Adv. Immunol., 20 (1975) 125.
51. Simmonsen, M., Prog. Allergy, 6 (1962) 349.
52. Simpson, E., personal communication.
53. Solliday, S. and Bach, F.H., Science, 107 (1970) 1406.
54. Stout, R., personal communication.
55. Uhr, J.W. and Moller, G., Adv. Immunol., 8 (1968) 81.
56. Wagner, H., Rollinghoff, M. and Nossal, G.J.V., Transpl. Rev., 17 (1973) 3.
57. Wunderlich, J.R. and Canty, T.G., Nature, 228 (1970) 63.

IMMUNOLOGICAL TOLERANCE: NEW PARAMETERS FOR B-CELL TOLERANCE

IN VITRO

E. DIENER

University of Alberta, Edmonton, Alberta (CANADA)

Before expanding on the ramifications of the mechanisms of immune tolerance, I will briefly describe the meaning of the term, in the context of this presentation. According to its original meaning tolerance defines the state of specific unresponsiveness to a secondary antigenic challenge. More than ever before have we become aware that the above definition of tolerance is an operational one and in no way reflects the basic mechanisms which underlie the phenomenon of unresponsiveness. Although the evidence is strong that one form of tolerance exists as a result of specific cell mediated suppression of immunocompetence (38), there is still sufficient justification to have us consider the more conservative concept of antigen induced functional deficiency at the level of the immunocompetent cell per se. Admittedly, the variety of current explanatory concepts concerning this type of tolerance may reflect the absence of solid analytical data on the subject rather than confirm a comprehensive body of experimental evidence. In spite of this deficiency, it has nevertheless been possible to define some of the conditions under which tolerance may be induced at the cellular and the molecular level in vitro (20). If analyzed in depth, our informations concerning the minimal requirements for tolerance induction in terms of cell dynamics, antigen structure and concentration may ultimately disclose the more fundamental mechanisms involved. Some of these parameters have previously been discussed in detail (20,21,22,23, 33).

Conditions for Tolerance Induction

Antigen Structure

A remarkable correlation exists between the degree of immunogenicity of an antigen and its capacity to induce tolerance in
vitro. Detailed experimental analysis has shown that these two
properties are common for antigens which share certain structural
similarities independent of individual antigenic specificities.
This was found to be the case when testing immunological properties in vitro of Salmonella H-antigens of different molecular
forms but of the same serological specificities. It was shown
that the degree of both immunogenicity and tolerogeneity increased
with increasing degrees of polymerization (20). This means that
antigenic determinants have to be present on the same carrier
molecule in sequences of repeating specificities. An investigation of this principle to explain certain forms of antibody
induced unresponsiveness led to the discovery that complexes of
monomeric antigen with appropriate amounts of antibody could also
act as tolerogene when bound to the surface of immunocompetent
cells (24). This concept may explain the well known phenomenon of
tumor or allograft enhancement which is thought, in part, to be due
to central inhibition by antigen-antibody complexes (41). The
general validity of the above conclusions concerning the depen-
dance between the structure of an antigen and its immunogenic and
tolerogenic properties has elegantly been confirmed by the work
of Feldmann and colleagues (34,17). Thus DNP on various carriers
such as polymerized flagellin (POL), KLH, levan, SIII and
dextran has been found to induce hapten specific tolerance in
vitro, whereby the degree of tolerogenicity as well as immunogenicity could be influenced by the degree of epitope density. At
the time this work was published it was difficult to accomodate its
interpretation with the fact that non polymeric antigens such as
ultracentrifuged serum proteins are capable of exerting tolerogenicity in vivo but not in vitro. This problem has largely been
solved, however, with the discovery of suppressor cells with
respect to tolerance towards serum proteins (5); a phenomenon which
is not the subject of this discussion.

A striking feature of polymeric antigens is their apparent in-
dependence at least in vitro of thymus derived helper cells (35,
25). Furthermore, it appears that such antigens may even express
immunogenicity in cell populations which are depleted of accessory
(A) cells to the extent that an immune response to sheep erythrocytes may no longer occur (18,55,51,31). These observations -
have led to the conclusion that an in vitro response to polymers
results from a direct interaction of immunocompetent B-cells with
antigen. The association therefore between the requirement for

A-cells and the dependence in the case of non polymers on T-cells,
has provided the basis for one of the most plausible theories on B-
cell triggering, known as the "matrix theory" (36). According to
this hypothesis, monomeric antigens are presented to B-cells on the
surface of A-cells in the form of a matrix of antigenic epitopes.
This matrix is thought to be bound to the A-cell surface by
properly spaced T-cell derived helper factor, thereby mimicking
the conditions which for polymeric antigens have proven optimally
immunogenic. This model has, however, lost its attractiveness in
view of the recent finding that contrary to earlier claims, even
T-cell independent antigens require A-cells to express immuno-
geneity in vitro, although the optimal number of these cells is
considerably less for T-cell independent (POL, DNP-Ficoll, soluble
sheep erythrocyte antigens) than for T-cell dependent antigens
(sheep erythrocytes) (43). On the basis of what has been discussed
above, the dependence of both, immunity and tolerance on the same
structural characteristics of the antigen, suggests that the two
phenomena may involve similar pathways for induction. Consequently
when discussing underlying mechanisms of tolerance the possible
role of both, B-cells and A-cells as participants should be
considered.

                          Antigen Concentration

     Experimental evidence shows that besides antigen structure, a
second key parameter for tolerance induction is antigen concentration (22). Generally speaking, polymeric antigens induce tolerance at supraoptimal concentrations whereby the difference between
an immunogenic and a tolerogenic dose may be as low as five-
fold as in the case for POL. With the availability of POL,
intrinsically labeled with tritium, it has been possible by means
of autoradiographic techniques to very accurately quantitate the
amount of antigen bound to immunocompetent B cells under tolerogenic conditions (26). Grain count analysis was carried out
after incubation of normal mouse spleen cells for one hour at 4 C
in the presence of antigen concentrations ranging from immunogenic
(200 ng/ml) to tolerogenic (1 µg/ml and more) amounts. The number
of antigen binding cells per 10^6 nucleated spleen cells reached
a definite plateau with increasing antigen concentrations, suggesting that POL binding cells comprise a distinct population.
There also occurred a shift in the grain count distribution profile from low grain counts at immunogenic to high grain counts at
tolerogenic antigen concentrations (27). This confirms the postulate that at least one of the conditions to render an immunocompetent cell tolerant is reflected in the amount of antigen
bound to its surface. This quantitative parameter combined with
the structural requirement for in vitro tolerogenicity of an
antigen resulted in the following definition of the conditions at

the cell surface which leads to either immunity or tolerance: A
T-cell independent antigen, in order to trigger an immunocompetent
B-cell is required to cause a certain degree of receptor cross-
binding. A degree of cross-binding which exceeds a given threshhold
may cause tolerance (20,22,28). If this definition reflects
reality, one would predict that the higher the epitope density on
an antigen, the lower would be its concentration required to induce
tolerance. However, as shown by other investigators, this prediction cannot be verified (18,19) for reasons which become obvious in
the next paragraph.

New Parameters

Accessory Cells

Recent advances in the analysis of triggering mechanisms as
exerted by T-independent antigens have rendered it likely that the
process of discrimination between a tolerogenic and an immunogenic
signal does not simply occur as the result of a direct interaction
between the immunocompetent B-cell and antigen. This conclusion
derives from the fact that contrary to earlier work (31,55,51,36),
T-independent antigens do require the presence of A-cells for immune induction (43). It has been suggested (50) that one of the
functions of A-cells might be the removal of antigen from the environment of immunocompetent cells in order to minimize its tolerogeneity due to high concentrations. As for T-cell independent but
A-cell dependent POL this theory predicts that the shape of the
antigen dose response curve as expressed by normal cells should
remain the same for cell populations depleted of A-cells but should
shift in its entirety to lower antigen concentrations. Initial
tests have yielded results which are partially in support of this
hypothesis in so far as tolerance with high antigen concentrations
is more readily achieved (by a factor of 100 x) with A-cell depleted than with normal spleen cell populations. On the other hand,
the failure of A-cell depleted cell cultures, regardless of antigen concentration, to give rise to immune responses comparable in
magnitude with normal control cultures is in disagreement with the
above hypothesis (A-cell depleted cultures would be fully restored
to normal immune performance by reconstitution with heavily irradiated A-cells). Furthermore, the addition of A-cells to normal
spleen cell cultures failed to alter the normal antigen dose response curve (unpublished data). Thus, other possible functions
for A-cells in T-cell independent responses must be considered;
e.g., the presentation or processing of antigen (56) or secretion of B-cell stimulating factors (53,7,8). The problem becomes
even more complex in view of recent findings which brought a dissociation between stimulatory and inhibitory properties of accessory

cells (44). Stimulatory activity as defined by the ability to
reconstitute the in vitro immune response of non-adherent spleen
cells was found in fractions rich in medium sized mononuclear phagocytes. These cells could be separated by physical means from large
phagocytic cells with immunosuppressive properties. These regulatory functions of A-cells could act via surface bound mechanisms of
antigen presentation. However, this assumption is unlikely in view
of the failure to find evidence for tritum labelled POL to bind to
macrophage surfaces at an incubation temperature of 4 C (26).
The only association of antigen with such cells was by ingestion
which occurred at 37 C. Thus the question as to how A-cells regulate the discrimination between immunogenic and tolerogenic signals
must remain open at the present time. The requirement for A-cells
in the immune responses to polymeric antigens may even serve as
an argument in favor of the idea that all antigens require T-cell
helper factor; its relative efficiency would, however, depend on
the type of antigen. Should this be so, the A-cell dependent
antigen dose response curve discussed above would likely be
accounted for by the classical interaction scheme between B-, T-
and A-cells.

The Effect of Glucocorticosteroids on Tolerance

Glucocorticosteroids have been widely used clinically as
immunosuppressants and their inhibitory effect on different cell
classes has been investigated at the experimental level (11,13,47,2,14
45). In humoral immunity, recent work has demonstrated that cortisone causes a disfunction in accessory cells and in thymus derived
helper cells without seriously affecting precursors of antibody
forming B cells (45). Furthermore, the observation was made that
cortisone at certain concentrations causes a marked shift in the
dose response curve in vitro to POL, whereby a potentially immunogenic concentration of the antigen is reduced by up to 200-fold.
For example, spleen cells from mice injected 24 hrs previously with
10 mg of cortisone acetate would respond optimally to one ng/ml but
not to 250 ng/ml of POL. To test for the possibility that an otherwise immunogenic concentration of the antigen had in fact induced
unresponsiveness in cortisone-treated lymphocytes, spleen cells
from mice injected with cortisone 24 hrs earlier were preincubated
with one ng/ml or 250 ng/ml of POL for six hours at 37 C, washed
and cultured for four days in the presence of one ng/ml of POL
(note that one ng/ml of the antigen was found to be optimally
immunogenic for cortisone treated cells). These experiments
showed that unresponsiveness had indeed been induced in cortisone
treated cells with a concentration of antigen which in normal
lymphocytes induces immunity (29). It is of particular interest
that the state of cortisone mediated unresponsiveness could be
reversed by the addition of heavily irradiated A-cells from normal

mice. In view of the previously mentioned phenomenon concerning
the regulatory influence of A-cells on tolerance, it appears likely that these cells are affected in their function by cortisone.
While high concentrations of glucocorticosteroid facilitate the
induction of tolerance, the opposite has been found to occur in
the presence of the steroid at close to physiological concentrations in vitro (29). In these experiments, normal spleen cells
were cultured in vitro in the presence of 10 ng/ml of the steroid.
Under these conditions tolerance was no longer induced at
concentrations of POL which in normal cells induced unresponsiveness. It is likely that this effect also reflects an immuno-
regulatory role of A-cells which may be under the control of
glucocorticosteroids. On these grounds, one is tempted to
speculate that the well documented differences in the susceptibility of certain mouse strains to tolerance induction (49,37) may
be caused by differences either in the blood concentration of
glucocorticosteroids or by differences in susceptibility of A
cells to the regulatory influence of these hormones.

Cell Surface Dynamics in Immunity and Tolerance

Whatever the intermediary steps between the first contact of
tolerogen and the operational manifestation of unresponsiveness,
the medium which finally transmits the signal for unresponsiveness is the B cell surface membrane. Regardless of the means by
which A-cells (in the case of T-independent antigens) regulate
the antigen dose response curve, it is sensible (but not proven
experimentally) to assume that the final step in discrimination
between an immunogenic and a tolerogenic signal is at least in
part determined by the antigen bound to the immunocompetent B
cell. For this reason, the kinetics of antigen redistribution
on antigen binding cells under conditions of immunity or tolerance
induction have been studied by means of autoradiographic techniques
using biosynthetically tritium labelled POL (3H-POL) (26,30).
This assay detects a distinct population of antigen binding
lymphocytes (on the average 25 cells/10^6 nucleated cells regardless of whether labelling was carried out in the presence of
immunogenic or tolerogenic antigen concentration). Using auto-
radiography together with velocity sedimentation techniques, it
was shown that 3H-POL binding cells from normal mouse spleen in
the presence of immunogenic amounts of the antigen undergo blastogenesis upon an as yet undetermined number of cycles of antigen
capping and receptor reformation (30). Under conditions of
tolerance induction in vitro, however, with supraoptimal concentrations of POL, capping and loss of antigen from the surface
membrane was inhibited and the cells remained uniformly labelled
for a period of at least 12 hrs (30). There has been considerable
interest concerning the question as to whether the tolerant state

of a B-cell is maintained by the continuous presence of antigen on
its surface (receptor blockade) (6) or whether its initial
encounter with the cell mediates a tolerogenic signal to the cytoplasm. The latter of these two possible events is the most likely
one since the inhibition of antigen capping by mitogens occurs
independently of immune induction and capping, and therefore is not
a prerequisite for immune triggering (46). Work by others has
shown that high concentrations of concanavalin-A inhibit the
redistribution of surface Ig-receptors and that this inhibitory
effect could be reversed by agents such as colchicine, colcemide,
vinblastine and vincristine (32). In view of this it was decided
to test the possibility that colchicine would also reverse antigen
induced inhibition of receptor capping and, if so, whether such
an effect would coincide with reversibility of the tolerant state
(30). It was indeed possible to show that colchicine not only
caused reversibility of capping inhibition which had been induced
by tolerogenic amounts of POL but also enabled the cells to form
new receptors. It was therefore of major interest to ascertain
whether these cells had also regained the capability to participate
in a normal immune response upon antigenic challenge in vitro.
However, it was found that cells after treatment with colchicine
subsequent to the inhibition of capping by antigen failed to
become immunized upon exposure to an immunogenic concentration of
POL. That this unresponsiveness was not due to the presence of
colchicine during the short period of tolerance induction is
indicated by the fact that the control group which was treated
with immunogenic amounts of POL together with colchicine subsequently gave a normal immune response. These data suggest that
the induction of tolerance under the conditions described is not
due to "receptor blockade" in the physical sense but represents
an antigen induced signal which renders the cell unresponsive via
metabolic pathways, the nature of which is yet to be discovered.
Alternatively or in addition to such an effect, tolerance induction may also have affected A-cells in their capacity to cooperate
with B-cells.

SUMMARY AND CONCLUSION

It is evident from a long history of in vivo experimentation
that both, antigen structure and concentration are the main parameters which determine tolerance induction. In vitro analysis has
established these parameters to be also valid at the level of the
immunocompetent B-cell. On its surface, antigen dose and structure both create different degrees of immunoglobulin-receptor
cross-linking. It has been reasoned that the extent of such cross-
linking may at least in an operational sense determine whether
immunity or tolerance is induced (20). Functionally, these conditions have been shown to cause receptor aggregation followed by

capping at antigen concentrations which induce immunity (26) and
inhibition at tolerogenic concentrations (26,27). These findings
show not only that antigen capping is not a prerequisite of immune
induction (46), but also that inhibition of capping by antigen is
not a prerequisite for tolerance induction (30). Thus the failure
of colchicine to prevent the induction of tolerance in vitro,
under conditions at which inhibition of receptor capping is
reversed, supports the notion that tolerance induction does not
require "receptor freezing" by antigen.

Most investigators believe that tolerance results from the
direct interaction of antigen with immunocompetent cells; a
concept which has received enforcement from the tolerogenic properties of polymeric antigens and the apparent fact that such antigens require neither T-cells nor A-cells to stimulate B-cells
(35,55,31). Consequently, a common pathway was thought to exist
for both, the triggering of immunity and tolerance to T-independent antigens. Furthermore, the dramatic shift in the antigen
dose response curve for tolerance which has been observed to occur
upon A-cell depletion has suggested to some investigators that A-
cells mediate this effect by removing high concentrations of antigen which would otherwise induce tolerance (50). However, several
recent findings suggest that these speculations oversimplify the
mechanisms of tolerogenicity and immunogenicity. Firstly, in
contrast to previous findings, we have made the unexpected discovery of a two cell model for immune induction by T-independent
antigens, comprising B-cells and cooperating A-cells (43). Secondly, the suggestion that A-cells prevent tolerance by removing
excess antigen is unlikely since it has not been possible to reduce
the antigen concentration required to induce tolerance in normal
cells by adding additional A-cells (unpublished data). Furthermore, partial A-cell depletion, although causing reduction in the
antigen dose required to induce tolerance and generally lowering
the immune response, failed to exert an antigen dose shift in
the immunogenic dose range (43).

The effect of A-cells on tolerance induction may also be
consistent with the effect of glucocorticosteroids on tolerance
(29), provided one postulates that these agents act via A-cells.
In support of this is the fact that cortisone mediated tolerance
to otherwise immunogenic antigen concentrations is abrogated by
the addition of A-cells. The similarity of structure between
T-independent antigens and B-cell mitogens, including polyclonal
activators such as lipopolysaccharide (LPS) has been emphasized
in the theory of B-cell triggering by T-independent antigens
(15). The hypothesis presents an alternative to the earlier
view that B-cell receptor molecules are involved in transmission
of the triggering stimulus. Instead, it is suggested that Ig
receptors serve only to focus antigen on to the specific B-cell,

whereas triggering of the B-cell, in the case of T-independent antigens, is accomplished by interaction of the cell with non-specific mitogenic determinants on the antigen molecule (16). The response to T-dependent antigens is thought to be mediated by nonspecific mitogenic factors, produced by T-cells, which are focused onto the relevant B-cell as a result of antigen binding between B and T cells. If one disregards the existing controversy of whether LPS acts directly on B-cells or whether T-cells and A-cells are involved (3,4) then, according to the above model, LPS should break B-cell tolerance. That this is indeed the case has been shown in a number of examples of tolerance to T-dependent antigens (4,12,9,39,10,48). However, recent data establish that tolerance to POL cannot be reversed by LPS (54) in situations where it is reversible by other means such as by cortisone or by Trypsin (23,39,54); from these data one would predict that with respect to B-cell activation, both synergy at low concentrations and suppression at high concentrations of LPS and POL would occur. However, only suppression by LPS of the POL response was seen, regardless of dose (54). These data suggest that either the structural similarity between LPS and other polyclonal activators is coincidental and does not imply functional analogy with respect to triggering mechanisms or, that the structural similarity of these agents reflects a functional analogy in the sense that they all are at least A-cell dependent. Whatever is the case it appears that different mechanisms may be involved in signal discrimination at the B-cell level for T-dependent antigens.

It is evident from this discussion that a model to explain B-cell immunity or tolerance to T-independent antigens on the basis of direct interaction alone of immunocompetent cells with antigen can no longer be maintained. Clearly more experimental insight into the nature of cell cooperation in T-independent responses is required to permit the formulation of concepts of B-cell triggering which explain current data.

REFERENCES

1. Allison, A.C. and Davies, A.J.S., Nature, 233, (1971) 330.
2. Anderson, B. and Blomgren, H., Cell Immunol., 1 (1970) 362.
3. Andersson, J., Sjöberg, O. and Möller, G., Transplant Rev., 11 (1972) 131.
4. Armeding, D. and Katz, D.H., J. Exp. Med., 139 (1974) 24.
5. Basten, A., Miller, J.F.A.P. and Johnson, P., Transpl. Rev., 26 (in press).
6. Borel, Y. and Aldo-Bersson, M., Immunol. Tol. (Eds. D.H. Katz and B. Benacerraf) Academic Press, New York, New York (1974)
7. Calderon, J. and Unanue, E.R., Nature, 253 (1975) 359.

8. Calderon, J., Kiely, M. M., Lefko, J. and Unanue, E.R., J. Exp. Med., 142 (1975) 151.
9. Chiller, J.M. and Weigle, W.O., J. Exp. Med., 137 (1973) 740.
10. Chiller, J.M., Skidmore, G.J., Morrison, D.C. and Weigle, W.O., Proc. Nat. Acad. Sci., 70 (1973) 1229.
11. Claman, H.N., New Engl. J. Med., 287 (1972) 388.
12. Claman, H.N., J. Immunol., 91 (1963) 833.
13. Cohen, J.J., Fischbach, M. and Claman, H.N., J. Immunol., 105 (1970) 1146.
14. Cohen, J.J. and Claman, H.N., J. Exp. Med., 133 (1971) 1026.
15. Coutinho, A. and Müller, G., Nature New Biol., 245 (1973) 12.
16. Coutinho, A. and Müller, G., Scand. J. Immunol., 3 (1974) 133.
17. Desaymard, C. and Feldmann, M., Eur. J. Immunol., 5 (1975) 537.
18. Desaymard, C. and Feldmann, M., Cell. Immunol., 16 (1975) 106.
19. Desaymard, C. and Howard, J.G., Europ. J. Immunol., 5 (1975) 541.
20. Diener, E. and Feldmann, M., Transplant. Rev., 8 (1972) 76.
21. Diener, E. and Armstrong, W.D., Lancet, 1 (1967) 1281.
22. Diener, E. and Armstrong, W.D., J. Exp. Med., 129 (1969) 591.
23. Diener, E. and Feldmann, M., Cell. Immunol., 5 (1972) 131.
24. Diener, E. and Feldmann, M., J. Exp. Med., 132 (1970) 31.
25. Diener, E., O'Callaghan, F. and Kraft, N.J., Immunol., 107 (1971) 1775.
26. Diener, E. and Paetkan, V.H., Proc. Nat. Acad. Sci., 69 (1972) 2364.
27. Diener, E., Kraft, N., Lee, K.C. and Shiozawa, C., J. Exp. Med., (in press)
28. Diener, E., Progress in Immunol., II., 3 (1974) 217.
29. Diener, E. and Lee, K.-C., Immunol. Tol. (Eds. D.H. Katz and B. Benacerraf) Academic Press, New York, New York (1974)
30. Diener, E., Kraft, N., Lee, K.C. and Shiozawa, C., (submitted).
31. Diener, E., Shortman, K. and Russell, P., Nature, 225 (1970) 731.
32. Edelman, G.M., Yahara, I. and Wang, J.L., Proc. Nat. Acad. Sci., 70 (1973) 1442.
33. Feldmann, M., Contemp. Topics in Mol. Immunol., 3 (1974) 57.
34. Feldmann, M., J. Exp. Med., 135 (1972) 735.
35. Feldmann, M. and Basten, A., J. Exp. Med., 134 (1971) 103.
36. Feldmann, M., J. Exp. Med., 136 (1972) 532.
37. Fujiwara, M. and Cinader, B., Cell. Immunol., 12 (1974) 194.
38. Gershon, R.K. and Kondo, K., Immunol., 21 (1971) 903.
39. Golub, E.S. and Weigle, W.O., J. Immunol., 98 (1967) 1241.
40. Hamaoka, T. and Katz, D.H., J. Immunol., 111 (1973) 1554.
41. Hellström, I. and Hellström, K.E., Transplant. Proc., 3 (1971) 721.
42. Kagnoff, M.F., Billings, P. and Cohn, M., J. Exp. Med., 139 (1974) 407.
43. Lee, K.-C., Shiozawa, C., Shaw, A. and Diener, E., Europ. J. Immunol., (in press)

44. Lee, K.-C., Shiozawa, C., Shaw, A. and Diener, E., Europ. J. Immunol., (in press).
45. Lee, K.-C., Europ. J. Immunol., (in press).
46. Lee, K.-C., Langman, R.E., Paetkan, V.H. and Diener, E., Eur. J. Immunol., 3 (1973) 306.
47. Levine, M.A. and Claman, H.N., Science, 167 (1970) 1515.
48. Louis, J.A., Chiller, J.M. and Weigle, W.O., J. Exp. Med., 138 (1973) 1481.
49. Lukic, M.L. and Leskowitz, S., Nature, 252 (1974) 605.
50. Mitchell, G.F., Contemp. Topics in Immunobiol., 3 (1974) 97.
51. Mosier, D.E., Johnson, B.M., Paul, W.E. and McMaster, P.R.B., J. Exp. Med., 139 (1974) 1354.
52. Ornellas, E.O., Sanfilippo, F. and Scott, D.N., Europ. J. Immunol., 4 (1974) 587.
53. Schrader, J.W., J. Exp. Med., 138 (1973) 1466.
54. Scott, D.W. and Diener, E., (submitted)
55. Shortman, K., Diener, E., Russell, P. and Armstrong, W.D., J. Exp. Med., 131 (1971) 461.
56. Unanue, E.R., Adv. Immunol., 15 (1972) 95.

RES BLOCKADE: EFFECTS ON IMMUNITY AND TOLERANCE

H. FRIEDMAN[1] and T. Y. SABET[2]

Albert Einstein Medical Center[1], Philadelphia,
Pennsylvania (USA) and University of Illinois College
of Dental Medicine[2], Chicago, Illinois (USA)

The cells of the reticuloendothelial system (RES) are involved
in adoptive immunity toward microbial pathogens as well as "endogenous" parasites represented by malignant cells. Macrophages are
considered the major cell type in the RES which "digests" and/or
processes antigens, especially particulate ones (7,8,17,19). The
importance of such "antigen-processing" cells for specific antibody formation is now widely accepted. Non-specific as well as
specific activities of macrophages are thought to be important in
many types of immune responses (1,2,4,9,17). In this regard, experimental models designed to study macrophage function have
involved the "inactivation" of macrophage function by chemical,
physical, and even immunologic methods. For example, removal of
macrophages from lymphoid cell suspensions by adsorption procedures
or treatment with anti-macrophage serum often markedly reduces
immunologic reactivity of the remaining cell populations (7,17-19).

Some of the earlier studies concerning the function of macrophages in immunity involved "blockade" of the RE system by particulate non-antigenic substances such as colloidal carbon. In some
cases suppression of immune responses occurred, whereas in other
instances no suppression or actual enhancement occurred. Such
differences, however, have been attributed by some to differences
in strain and species of animals used, differences in the source,
dose or route used for the "blockade" agent, difference in the
time, dose or nature of antigen used, as well as differences in the
methods utilized to detect the immune response (7,8,9,17). Nevertheless, there has been a continuing series of reports indicating
that under appropriate conditions "RES blockade" can indeed induce
suppression of immune responses to a large variety of antigens,

including sheep erythrocytes, bacterial vaccines or extracts, skin
allografts, transplantable tumors, etc.

The mechanism involved in immunologic impairment in RE blockaded individuals is not clear. There is very little evidence that
substances such as colloidal carbon affect immune responses by
impairing phagocytosis or uptake of antigen by macrophages. Indeed
some studies have shown that antigen uptake into lymphoid tissue is
greater in carbon treated as compared to control animals. In this
regard, analyses of the mechanisms involved in RE blockade affecting immune responses are complicated by the fact that macrophages
which may be involved in the uptake of both antigen and a colloidal
RES blockade substance are not those cells which are directly
involved in antibody synthesis or cell mediated immune responses.
Thus an RE blockading substance may influence immune responses only
indirectly, i.e., they may affect humoral factors which, in turn,
influence cells involved in immune responsiveness.

Regardless of the mechanism of how RES "blockade" affects
immune responses, it now seems reasonably clear that the manner in
which antigen is presented to lymphoid cells helps determine
whether and how the appropriate immunocytes are activated (1,2,4,9,
10,18). It seems likely that immunologic tolerance, i.e., a situation characterized by the failure of a specific immune response in
an antigen pre-treated individual, may also be related to antigen-
processing. Some investigators believe that if an antigen "bypasses"
macrophages and directly interacts with lymphoid cells, tolerance
rather than immunity ensues (7,8,17,19). Thus non-specific
"blockade" of macrophages with inert substances such as colloidal
carbon particles should facilitate tolerance induction to an antigen which otherwise would be expected to stimulate an immune response. Indeed, studies performed in this and other laboratories
during the last few years have shown that carbon pre-treated mice
evince a marked suppression of both primary and secondary immune
responsiveness in vivo, as well as in vitro to a specific antigen
such as sheep erythrocytes (2,9,10,11,12,13-16,18). Under appropriate
conditions "immune tolerance" was the ultimate outcome of pre-treat-
ing animals with carbon and sheep erythrocytes by the same route.

GENERAL METHODS AND EXPERIMENTAL PROCEDURES

For these studies inbred Balb/c mice obtained from Jackson
Memorial Laboratories, Bar Harbor, Maine, or Hunting Farms,
Conshohocken, Pennsylvania, were utilized. Sheep erythrocytes
obtained commercially in Alsever's solution were used as the
immunogen. Antibody responses to the erythrocytes were determined
by the standard hemolytic plaque assay in agar gel essentially
as described initially by Jerne et al. Direct 19S IgM plaque

forming cells (PFCs) were determined by the agar plating technique
after a single incubation at 37 C. Indirect low efficiency 7S IgG
PFCs were determined by the facilitation method utilizing rabbit
anti-mouse gamma globulin in a two-step procedure.

Colloidal carbon (Pelikan) was utilized to induce RE blockade
(10-14). Mice were injected with one mg carbon either intraperitoneally (i.p.) or intravenously (i.v.). The effect of carbon
treatment on clearance rates was determined by subsequent injection
of 1.0 mg carbon i.v., followed by the assessment of carbon in
the blood of the animals over a period of 60 minutes. Alternatively,
mice were injected with ^{51}Cr labeled SRBC and the clearance rate of
the radioactivity from the blood of the mice and/or the uptake of
radioactivity into the liver and spleen of the animals was determined at various times thereafter.

For tissue culture experiments the in vitro immunization procedures described initially by Dutton and Mishell and by Marbrook
were utilized. Suspensions of five or ten million viable splenocytes were cultured in medium in vitro and immunized with 0.1 ml
of a 0.1% suspension of SRBC. The number of PFCs per culture was
determined at various times thereafter by the direct and indirect
hemolytic plaque assays. In some experiments splenocytes from mice
pre-immunized with SRBC, with or without carbon treatment, were
cultured in vitro and the response to SRBC determined in a similar
manner.

The effects of carbon treatment on both primary and secondary
("memory") immune responses were determined by injecting groups of
five to ten mice each with SRBC plus carbon at various times relative to the day of immunization. The development of 19S and 7S
PFCs was determined at various times thereafter, both in vivo and
in vitro. In all cases the immune responses of splenocytes from
carbon treated animals or from cultures treated with carbon in
vitro were compared to immune responses of control splenocytes.

EXPERIMENTAL RESULTS AND CONCLUSIONS

Administration of Pelikan carbon to mice i.p. simultaneously
or prior to challenge immunization with 0.2 ml of a 1.0% suspension
of sheep erythrocytes markedly affected the immune response. Mice
given carbon one to three days prior to immunization showed markedly depressed PFC responses as compared to control mice given the
same dose of RBCs alone (Table I). The concentration of carbon
markedly affected the magnitude of the antibody response, with
the greatest level of depression occurring after injection of 5-10
mg carbon. However, significant depression also occurred with 1
or 2 mg carbon but not with less. The depressed numbers of hemo-

lytic PFCs in carbon treated animals was not due to a shift in the
peak day of the immune response since kinetic studies revealed a
suppression of antibody formation throughout a one to two week time
period after challenge immunization. For example, not only were
the numbers of PFCs depressed on day four after immunization,
generally considered the day of the peak response, but also during
the first two-three days after immunization and as late as 10 to 15
days afterwards (Table I). The depressed PFC responses were not due
to an accelerated appearance of low efficiency 7S antibody. Analyses of the number of direct and indirect PFCs in the spleens of
mice five, seven, ten and 15 days after immunization with the
SRBCs shown a depression of both forms of PFCs in carbon treated
animals as compared to the control animals.

TABLE I

Effect of Time of Injection of Pelikan Carbon on the Primary
Antibody Response of Mice to Sheep Erythrocytes

Day Pelikan Carbon Injected*	PFC Response per Spleen on Day**				
	+2	+4	+6	+10	+15
None (controls)	1200	68,500	21,310	15,100	3100
-10	985	71,200	19,500	12,500	450
- 5	760	59,350	17,300	11,100	1950
- 3	385	10,850	4,310	2,700	1100
- 1	315	4,700	2,100	1,350	950
0	650	31,500	11,200	5,600	1500
+ 1	896	48,500	20,500	9,100	3400
+ 2	750	62,100	22,800	14,900	5100

* Groups of Balb/c mice injected i.p. with 10.0 mg carbon on day
indicated relative to day of i.p. immunization with 0.2 ml 1% SRBC
** Average IgM PFC response for 3-6 mice per group on day indicated
after challenge immunization.

    The day of treatment of mice with carbon was a major factor
for altered immune responsiveness. Less suppression occurred when
mice were given carbon after immunization. Maximum suppression
developed in mice given carbon 1-3 days before the RBCs. Mice
given carbon 5 to 10 days or longer before immunization also showed
little alteration of the expected PFC response (Table I).

    The route and dose of SRBC used to immunize the mice affected
the immune response of carbon treated animals. Mice given carbon
i.p. and challenged with SRBCs by the i.v. route showed much less

alteration of their expected immune response as compared to mice
given SRBC by the i.p. routes. Furthermore, when mice were given
both carbon and RBCs by the i.v. route the alteration of immune
responsiveness was much less marked than when carbon was given i.p.
Increasing the dose of antigen from a 1% to a 10 or 20% suspension
resulted in much less suppression of PFC formation by the carbon
treated mice. This increase in antigen dose resulted in only a
slight to moderate increase in PFC formation in control mice.
A 50% suspension of RBCs markedly increased PFC formation in carbon
treated mice but not control mice.

TABLE II

Effect of Carbon Treatment on the Secondary PFC Response of
Mice to Sheep Erythrocytes

Day Pelikan Carbon Injected*	PFC Response per Spleen**					
	Day +3		Day +5		Day +8	
	IgM	IgG	IgM	IgG	IgM	IgG
None (controls)	48,750	82,100	41,300	98,700	21,000	55,000
−5	40,100	63,200	35,100	65,700	18,500	45,300
−3	12,600	9,300	15,300	10,500	6,340	5,200
−2	6,100	8,700	9,300	7,300	5,100	2,500
−1	4,200	6,100	4,700	3,500	4,000	2,700
0	8,300	12,700	11,700	18,500	10,600	14,500
+2	48,700	59,700	46,000	65,300	23,700	47,500

* Groups of mice injected eight weeks earlier with SRBCs i.p.
and then injected i.p. with 10.0 mg carbon on indicated day before
secondary i.p. challenge immunization with SRBC.
** Average PFC response per spleen for 3–5 mice per group on
indicated day after challenge immunization; PFCs determined by
direct assay for IgM and indirect assay for IgG hemolysin
secreting immunocytes.

Carbon treatment also suppressed the secondary immune response.
When Pelikan carbon was given one to two days prior to a secondary
injection of SRBC into mice pre-treated with the same RBC dose four
to eight weeks earlier, a marked suppression of antibody formation
occurred (Table II). Untreated mice given a second injection of
SRBC showed a markedly enhanced PFC response, especially 7S PFCs.
The total number of PFCs in the spleens of boosted mice was usually
several fold higher than that in mice given a single injection of
RBCs. Furthermore, the number of PFCs increased at a much more
rapid rate than after primary immunization and peak numbers often

occurred on the third to fourth day. At this time nearly twice as
many PFCs were detected by the facilitation assay with anti-globulin
serum, indicating that IgG antibody forming cells predominated. By
the fifth to eighth day after secondary immunization over 60-80%
of the PFCs were of the 7S class. When carbon was injected i.p.
one to two days before the booster immunization many fewer PFCs
(both 7S and 19S) developed. The depression was not as marked as
that occurring during the primary response in carbon pre-treated
mice. The dose of carbon and its route of administration also
markedly affected the secondary immune response; a larger dose of
carbon given i.p. resulted in greater immune suppression. However,
similar to the primary response, a higher dose of SRBC abrogated
the depression of the immune response. Injection of RBCs i.v.,
rather than i.p., also overcame partially the immune suppression
induced by the carbon.

TABLE III

Effect of Pre-treatment of Mice with Pelikan Carbon and Sheep
Erythrocytes on Subsequent Hemolytic Antibody Response to SRBC

| Treatment* | SRBC | PFC response** | |
		IgM	IgG
None	+	57,500	124,350
Carbon i.p.	−	43,700	1,000
	+	33,100	2,500
Carbon i.v.	−	46,100	1,000
	+	2,160	1,000

* Groups of mice injected as indicated with 10.0 mg carbon and
either unimmunized or given i.v. or i.p. injection of SRBC.
** Average splenic PFC response of 3-5 mice per group after challenge immunization with SRBC four weeks after initial injection
of carbon and/or SRBC; PFCs determined by direct (IgM) and indirect
(IgG) procedures.

Carbon treatment of mice simultaneously and one or two days
before a primary injection of SRBC interfered with development of
a normal secondary 7S IgG antibody response (Table III). For
these experiments, groups of animals given carbon during the time
of the primary immune response were challenged with SRBC two, four,
or six weeks later. Unlike control mice given SRBC only, these
animals showed mainly a 19S IgM type immune response, with very
little 7S antibody. This indicated that "immunologic memory"
characterized by appearance of IgG PFCs was inhibited. Furthermore,

the immunokinetics of the 19S IgM PFC response of these mice was
very similar to that occurring in animals given a single injection
of only SRBC.

TABLE IV

Effect of Carbon Treatment In Vivo or In Vitro on Responsiveness of Spleen Cells to In Vitro Immunization with Sheep
Erythrocytes

Spleen cell source*	Carbon in vitro**	PFC responses per culture***
Normal	–	5760 + 576
	+	431 + 71
Carbon pre-treated		
-5 days	–	7810 + 1410
-3 days	–	2130 + 438
-1 day	–	1100 + 276

* Spleens pooled from 5-10 mice per group and 5 - 10 x 10^6 cells
cultured in vitro with 2 x 10^6 SRBC.
** Cultures treated with 1.0 mg carbon on day of culture initiation.
*** Average PFC response of 3-5 cultures five days after in vitro
immunization.

A marked depression of both 7S and 19S PFC formation occurred
in mice given i.v. injections of carbon before an injection of
SRBC by the same route. This was most evident when mice were challenged two to four weeks after treatment. For example, mice injected with carbon one and two days before immunization and one day
thereafter, developed very few PFCs and also failed to develop significant numbers of PFCs when challenged with SRBC two or four
weeks later. Development of both 7S and 19S PFCs were suppressed
in these animals. However, by six to ten weeks after simultaneous
pre-treatment with carbon and SRBC, the animals recovered their
immune responsiveness to the sheep RBCs and developed the expected
numbers of PFCs. However, the characteristic of the immune responses of these animals indicated that they had not developed "immunologic memory," since most PFCs were of the 19S IgM class. This
indicated that abrogation of a primary type immune response to
sheep RBCs, as shown by depressed numbers of 19S and 7S PFCs
following injection of carbon shortly before a first immunization
with RBCs, preferentially prevented a normal secondary 7S response
to SRBCs at a later time. This type of unresponsiveness was specific

for SRBCs. In control experiments mice pre-treated with carbon
and sheep erythrocytes showed a normal response to chicken erythrocytes or E. coli LPS. Animals challenged with these unrelated
antigens, regardless of whether they were treated earlier with
carbon and SRBC, responded with similar numbers of PFCs to these
antigens.

DISCUSSION AND CONCLUSIONS

It does not seem necessary to discuss in detail the now
generally accepted concepts that macrophages are an important
component of the immune response mechanism. Both fixed and
wandering macrophages have been implicated in a wide variety of
immune responses, including those involving humoral antibody formation to microorganisms, soluble antigens, etc., as well as cell-
mediated responses to a wide variety of immunogens (7,8,9,17,19).
Macrophages serve both to regulate the nature and magnitude of the
immune response, perhaps by some humoral mechanism (soluble factors)
as well as by direct cell to cell interactions; they also may
"process" and/or present antigen to appropriate immunocompetent
lymphoid cells. Numerous studies have shown that antigen accumu-
lates in macrophages in distinct anatomical locations in the
organized lymphoid tissue. There is some controversy as to the
nature or type of macrophages which may actually be involved in
"presenting" antigen to immunocompetent lymphoid cells or their
precursors. Nevertheless, it now seems generally accepted that
the antigen handling function of macrophages is involved in the
initial stages of immune stimulation (1,2,4,9,18). However, many
investigators believe that the mere "uptake" of an antigen by
macrophages does not mean that these cells are directly involved
in antibody formation or cellular immunity. Some macrophages are
thought to take up antigen to remove it from the body. In this
regard, macrophages and other phagocytic cells may act as
"scavengers" and nonspecifically ingest inanimate particles as
well as particulate antigens. For example, the uptake of glass
or plastic beads, colloidal substances of all types, etc.,
certainly does not seem to be associated with a specific immune
function.

"Overloading" of macrophages by non-antigenic substances has
been considered for many years a possible mechanism explaining
depressed immune responses in certain situations (9-12). For
example, some forms of impaired immunity noted in a variety of
induced or natural pathologic conditions may be due not so much to
the absence or dysfunction of macrophages but rather to their "pre-
commitment" to large quantities of non-antigenic particulate
matter, including cellular debris, dead or inactive bacteria and
viruses, etc. Therefore, it seemed logical over the last few

decades to attempt to "modulate" antibody formation and other immune responses by manipulation of the antigen-processing mechanism,
represented by the macrophage system. Thus, over the years, a
variety of reports have appeared indicating that pre-treatment of
mice with substances such as colloidal carbon, thorotrast, saccharated iron, proferrin, silica, etc., interferes with antibody responses.
Most of these studies dealt with humoral immunity assessed by
serologic titrations. Studies concerning cell-mediated immunity
have also indicated that under appropriate conditions, pre-
treatment of animals with inanimate particulate susbstances
which "blockade" the RE system and macrophage function may
abrogate the expected cellular immune response, including allograft rejection.

Studies during the last decade in this laboratory have been
concerned with evaluating mechanisms of RES blockade in regards to
immune function and to utilize the RE blockade model as a tool
to dissect some of the pathways involved in humoral immune responses to SRBCs (3,6,10-16). As indicated before, injection of mice
with Pelikan carbon shortly before either primary or secondary
immunization with sheep erythrocytes markedly interfered with the
expected immune response at the cellular level as judged by hemolytic plaque assays. Both 19S and 7S Ig secreting immunocytes
were affected by carbon blockade. It appeared unlikely that the
carbon directly interfered with antibody secretion by precommitted
immunocytes, since injection of carbon on the day of the expected
peak of the antibody response failed to influence antibody formation. Similarly, addition of carbon to splenocyte cultures
containing already immunized lymphoid cells did not affect the
number of PFCs subsequently detected in vitro.

Injection of carbon particles may have so altered macrophage function that these cells were no longer capable of either
processing and/or presenting antigen to the appropriate lymphoid
cells, which normally would be stimulated into productive antibody
formation. This explanation seemed plausible since maximum
depression of immune responsiveness occurred when carbon was
introduced into an animal one to three days prior to challenge
immunization. A one day lag occurs before maximum accumulation
of carbon particles in phagocytic cells in the organized lymphoid
tissue; this time coincided with the time necessary for maximum
immunosuppression. Furthermore, greater suppression occurred when
carbon was administered i.p. in mice challenged with SRBC by the
same route. It seemed likely that the carbon preferentially
"blockaded" macrophages accumulating in the peritoneum, as well as
the spleen and other lymphoid organs draining the peritoneal
cavity. A "competition" between carbon particles and RBCs for
macrophages necessary for appropriate antigen processing could
account for the observed immunosuppression (5,9,10,11). Challenge

of mice with SRBC by the i.v. route resulted in more antibody
formation in carbon treated mice, probably because the antigen
localized directly into the spleen rather than in the "blockaded"
macrophages in the peritoneal cavity. Studies using radiolabeled
RBCs would be valuable to determine precisely the fate of the
erythrocyte in mice given RBCs by one route and carbon particles
by another route.

The in vitro experiments showed that intimate contact with
carbon did not affect the viability of the cultured spleen cells.
Microscopic examinations revealed that macrophages appeared to be
"engorged" with carbon particles; this was also evident in intact
mice where numerous spleen cells seemed to be filled with carbon
particles (16). However, it should be noted that in in vivo
experiments carbon treatment may enhance rather than depress
macrophage function. This could obviously result in enhanced
"processing" or degradation of subsequently administered sheep
erythrocytes, resulting in less antigen stimulating appropriate
lymphoid cells (5,9). In vivo experiments are difficult to design
which can support this view since assessment of accumulation of
radiolabeled erythrocytes in the spleen or other lymphoid organs
in carbon vs. normal control animals fails to show the actual
accumulation of immunogenic SRBC determinants. However, earlier
observations of increased uptake of labeled SRBCs in the liver and
spleen of carbon treated mice does suggest that antigen may be
processed in a more rapid manner; therefore, antigen may not reach
the appropriate antigen reactive B-cell in either an adequate
amount or form. Increased numbers of macrophages, even from
normal individuals, have been found to prevent a normal antibody
response to SRBC in vitro, suggesting that macrophages do have a
suppressive function when present in large numbers.

It should also be noted that carbon particles, similar to
other colloidal particles, may interfere with or inactivate humoral
factors, such as "recognition factor" or other serum opsonins
necessary for appropriate processing of antigen such as SRBC. A
large body of evidence indicates that colloids may adsorb such
humoral factors, which seem important in the early phases of antigen recognition and uptake.

The observation that 19S and 7S hemolytic antibody responses
following secondary challenge immunization of mice treated with
carbon and erythrocytes at an earlier time suggested that RES
blockade may lead to important alterations which may be unrelated to
a direct effect of the carbon on immunocytes. For example, pre-
treatment of mice with carbon particles ten days or longer
before primary immunization with sheep RBCs had no detectable
effect on the expected immune response. On the other hand, pre-
treatment of mice with carbon and SRBC affected development of

immunologic memory, as shown by failure of the animals to respond
with the expected numbers of 7S IgG PFCs upon subsequent challenge
immunization.   Thus interference with a primary immune response
may prevent development of "memory" lymphoid cells capable of recognizing or interacting with SRBC upon subsequent challenge immunization.   Such animals apparently responded to sheep erythrocytes
as a "new antigen" upon secondary immunization.   The partial or
complete abrogation of the expected immune response to a second
injection of SRBCs after carbon and antigen pre-treatment several
weeks earlier is reminiscent of "classic" immunologic unresponsiveness induced by injecting antigen into either neonatal animals or
animals which are irradiated or treated with immunosuppressive
drugs.   However, unlike the effects of radiation or chemicals,
pre-treatment with carbon particles did not deplete the animals of
lymphoid cells or cause physical destruction of cells.   Nevertheless, animals treated with carbon in such a manner failed to
develop typical 7S and 19S type immune responses upon challenge.
These animals could respond normally to other challenge antigens,
indicating that the unresponsiveness to SRBCs was specific and
mimicked "immunologic tolerance."

The mechanisms involved in such RES blockade effects on
antibody formation are unclear.   Nevertheless, it seems plausible
that antigen injected together with carbon or shortly thereafter
may so alter RE cell function that the mice do not develop appropriately stimulated lymphoid cells necessary for "recognizing"
antigen; thus the animals appear to be "unprimed."   The initial
injection of antigen and carbon could also result in enhanced
degradation of the SRBCs into small molecular weight soluble
determinants which function as a classic "tolerogen."   It is also
possible that the carbon and SRBC injected into "blockaded" mice
may induce a low level of "blocking factors," perhaps low titers
of hemolysins, which prevent or depress a subsequent immune response
to sheep red cells.   Regardless of the mechanisms involved, suppression of immunologic memory and induction of immunologic
tolerance in mice pre-treated with carbon and SRBC seems worthy
of further examination.   For example, specific involvement of
distinct cell classes, such as T and B lymphocytes, in these events,
as well as functional capability of macrophages to SRBC and other
antigens in unresponsive mice should be examined in detail.   In
addition, the role of "natural" blockade of the RES by "excessive"
amounts of other natural substances, including cellular "debris"
in certain immunopathologic conditions, should also be examined in
regards to immune unresponsiveness.

SUMMARY

Reticuloendothelial cell blockade has been studied for decades in regards to physiological and immunological effects. "Overloading" of RE cells with inert colloidal particles, such as carbon or other particulate substances, has often been used to analyze the role of phagocytic activities in antibody formation, often with contrasting results. In the present studies the effects of colloidal carbon treatment of mice on immunologic responsiveness to sheep erythrocytes was investigated. Pre-treatment of mice with carbon shortly before either primary or secondary immunization with SRBCs markedly suppressed the expected antibody response, as shown by depressed numbers of hemolytic antibody plaque forming cells. Carbon treatment did not affect antibody forming cells per se as shown by lack of an effect on plaque forming cells when carbon was given after SRBCs, either in vivo or in vitro. Carbon injection before primary immunization prevented development of "immunologic memory," as shown by an altered secondary immune response. Mice given carbon and SRBC several weeks before secondary immunization with RBCs developed a primary type antibody response characterized by appearance of 19S antibody with little or no 7S hemolysins, characteristic of a secondary response. Furthermore, by appropriate treatment of mice with carbon and SRBC, immunologic unresponsiveness to SRBCs could also be induced, as evident by absence of both 19S and 7S antibody formation after subsequent challenge immunization with sheep erythrocytes. The mechanisms involved in RE "blockade" induced aberrations of normal immune responses may be related to effects on macrophages or soluble humoral factors, or both. It is unlikely that carbon treatment affects immunocytes directly. Further studies concerning the nature and mechanism of RE blockade on cellular and humoral components of the immune response mechanisms seem warranted and should provide more insight concerning the role of macrophages in antibody formation.

ACKNOWLEDGMENTS

The capable technical assistance of Mrs. Leony Mills, Miss Marsha Israel, Mr. Leonard Silverman, Mr. Jerry Rosenzweig, and Mrs. Carol Newlin during various portions of these studies are acknowledged.

REFERENCES

1. Argyris, B.F., J. Exp. Med., 128 (1968) 459.
2. Cruchard, A., Lab. Invest., 19 (1968) 15.
3. Friedman, H. and Sabet, T.Y., Immunol., 18 (1970) 883.
4. Gallily, R. and Feldman, M., Immunol., 12 (1967) 197.
5. Gottlieb, A.A. and Waldman, S., In: Macrophages and Cellular Immunity (Ed. A. Laskin and H. Lechevalier) CRC Press, Cleveland, Ohio (1972) 13.

6. Melnick, H.D. and Friedman, H., Proc. Soc. Exp. Biol. Med.,
    133 (1970) 432.
7. Nelson, D.S., Macrophages and Immunity, North Holland Publish-
    ing Co., Amsterdam, Holland (1969).
8. Pearsall, N.N. and Weiss, R.S., The Macrophage, Lea and
    Fabiger, Philadelphia, Pennsylvania (1970).
9. Perkins, E. and Makinoden, O., J. Immunol., 99 (1965) 765.
10. Pross, H.F. and Eidinger, D., Adv. Immunol., 18 (1974) 133.
11. Sabet, T., Newlin, C. and Friedman, H., Proc. Soc. Exp. Biol.
    Med., 128 (1968) 274.
12. Sabet, T., Newlin, C. and Friedman, H., Immunology, 16 (1969)
    433.
13. Sabet, T. and Friedman, H., Proc. Soc. Exp. Biol. Med., 131
    (1969) 1317.
14. Sabet, T. and Friedman, H., Immunology, 17 (1969) 535.
15. Sabet, T. Y. and Friedman, H., Immunology, 19 (1970) 843.
16. Sabet, T. Y., Young, I. and Friedman, H., Adv. Exp. Med., 29
    (1973) 391.
17. Schwartz, R.S., Ryder, R.J.W. and Gottlieb, A.A., Progr.
    Allergy, 14 (1970) 81.
18. Unanue, E.R. and Askonas, J., J.R.E.S., 4 (1967) 440.
19. Van Furth, R. (Ed.) Mononuclear Phagocytes, F.A. Davis Co.,
    Philadelphia, Pennsylvania (1970) 1.

MODULATION OF THE EXPRESSION OF A SECONDARY IgG ANTIHAPTEN

RESPONSE IN VITRO

A. ROMAN* and M. B. RITTENBERG

University of Oregon Health Sciences Center
Portland, Oregon (USA)

Despite several observations which indicate that lymphocytes originally producing IgM may switch to the production of IgG (9,6) some investigators still term the switch hypothetical (10). Thus the significance of these observations to antigen driven immune processes remains unclear. It is possible to view the single cell M→G switch as one which occurs during ontogeny (6), and presumably is not antigen and to view subsequent shifts in antibody class during immunization as reflections of antigenic influences on the expression of precommitted percursor cells. This raises the question of how antigen influences expression of Ig classes; antibody and interacting cells or cell products have all been implicated in the regulatory process, but again their precise role remains unelucidated (10, 7, 18, 8).

In vitro culture techniques have provided much insight into the initiation of the immune response, but as pointed out recently IgG responses have been difficult to induce in vitro (7). Even when IgG responses are obtained in vitro (1), their level compared to IgM is lower than would be predicted from in vivo results. As Mitchell has indicated (7) caution must be used to interpret results of in vitro secondary responses if the IgG level obtained is minimal. Thus it appeared that if in vitro techniques were to be applied to the study of the M→G relationship, conditions which reproducibly establish in vivo-like G/M ratios would be required.

*Present address: Indiana University School of Medicine, India-
  napolis, Indiana (USA).

We have attempted to circumvent this difficulty by beginning the secondary immune response in vivo and then boosting it by antigen challenge in vitro. While this strategy has been used by others (11,16), it has not previously led to secondary IgG/IgM ratios which approximate those seen in vivo. However, we have found that appropriately administered antigen can be used to achieve in vivo-like IgG/IgM ratios in vitro and that this approach appears to provide a suitable model for the study of many of the questions associated with modulation of the secondary immune response.

MATERIALS AND METHODS

Mice

Adult female Balb/c mice were obtained from Simonsen Laboratories, Gilroy, California and were caged in groups of 6 with free access to water and food.

Antigens

Trinitrophenylated-keyhole limpet hemocyanin (TNP-KLH) was prepared as described previously (12). Mole ratios of TNP-KLH ranged from $TNP_{1176}KLH$ to $TNP_{965}KLH$. Particulate TNP-KLH was prepared by coating TNP-KLH onto bentonite (T-K-B) as described previously (14).

Immunization

Mice were primed at 2-3 months of age with 3 injections of T-K-B as previously described (1).

Cell Culture

Spleen cells from at least 3 mice were cultured using microplates (5) which reduced all volumes to 0.1 of those used previously (1). Culture medium was supplemented with $5 \times 10^{-5}M$ 2-mercaptoethanol (5). The dose of antigen and the mode of administration whether in vivo, in vitro or both is indicated in the results.

Plaque Assays

Anti-TNP plaque-forming-cells (PFC) were detected using TNP-haptenated sheep red blood cells (TNP-SRBC) as prepared previously (14). Cells from 8 replicate microcultures were pooled and plated

for PFC by the technique of Cunningham and Szenberg (2). Cultures
were tested for both IgM and IgG. Cells producing IgG anti-TNP
antibody were detected by adding goat anti-mouse IgG antiserum (3).
Cells producing IgM anti-TNP antibody were detected by direct plating. The anti-IgG serum inhibited some but not all IgM PFC; therefore, the indirect plaque assay does not represent the net IgG PFC.
When the ratio of indirect to direct plaques is sufficiently large,
the total facilitated plaques can be considered to represent total
IgG PFC. This was verified in several separate experiments not
shown using the 2 anti-mouse μ chain antisera kindly provided by
Drs. M. Feldmann and A. Malley. When appropriately diluted anti-μ
was incorporated into the plaque assay medium to suppress IgM PFC
(11); the indirect PFC were unaffected while the number of direct
PFC was suppressed > 99% indicating that the indirect plaques were
IgG. We subsequently refer to indirect PFC as IgG and direct PFC
as IgM with the understanding that when the ratio G/M is low (<3),
the number of IgG PFC may be artificially high because of incomplete
suppression of IgM.

Plaque assays performed on spleen cells directly upon removal
from the animal are called in vivo responses regardless of the in
vivo antigen regimen.

RESULTS

Effect of Secondary Antigen Challenge in Vivo versus in Vitro

Animals primed to T-K-B were either given a secondary challenge with TNP-KLH in vivo or spleen cells from animals of the same
group were placed in culture on the same day and challenged in vitro.
Table I shows the effect of these 2 routes of administration of antigen on the ratio of IgM and IgG anti-TNP PFC. The in vitro route
results in an IgG/IgM ratio of approximately 1. In contrast in vivo
administration of the antigen resulted in a high G/M ratio. Similar
results were obtained regardless of whether TNP-KLH of T-K-B were
used to challenge in vivo.

The low G/M ratio in vitro is not due to an inappropriate antigen dose. As shown in Table II the antigen dose which stimulates
the maximum number of IgM plaques also stimulates the maximum number of IgG plaques.

The data in Tables I and II were obtained on day 5 of culture;
however other experiments not shown yielded similar results in the
comparative G/M when assays were carried out on other days of the
response.

TABLE I

A Comparison of the Class of Anti–TNP PFC Produced when
Secondary Antigenic Challenge is Given in Vivo or in Vitro

Experiment #	PFC Class	PFC/10^6 Cells	
		In Vitro[1]	In Vivo[2]
1	IgM	4246	1016
	IgG	4539	8580
	G/M	1	8
2	IgM	621	205
	IgG	664	3017
	G/M	1	15
3	IgM	1650	200
	IgG	1750	2000
	G/M	1	10

1 – Optimum dose in vitro = 0.02 µg TNP-KLH. Assay on day 5 of
culture.
2 – In vivo dose = 100 µg TKB. Assay on day 5 after injection.

TABLE II

Secondary Dose Response in Vitro

µg TNP-KLH in Vitro	Anti-TNP PFC/10^6 Cultured Cells Assayed[1]		
	IgM	IgG	G/M
2.0	480	525	1
0.2	883	1180	1
0.02	1472	1611	1
0.002	1175	1485	1
0.0002	1600	900	1

1 – Assay on day 5 of culture

TABLE III

Effect of in Vivo Boost on in Vitro Response*

In Vitro Culture Conditions	PFC Class	Anti-TNP PFC/10^6 Cultured Cells Assayed						
		Days Post Boost						
		0**	1	2	3	4	5	6
No Antigen	IgM	109	469	372	153	167	127	134
	IgG	64	800	1285	3687	2060	661	322
	G/M	1	2	3	24	12	5	2
0.2 µg TNP-KLH	IgM***	158	0	478	333	364	262	719
	IgG	641	0	681	0	5822	4797	4303
	G/m	4	–	1	–	16	18	6

*Animals were boosted with 100 µg TKB. On consecutive days their spleens were placed in culture with or without further antigen. All cultures were assayed 5 days after initiation of culture.
**Spleens taken from animals of the same group not receiving a boost.
***In order to determine the influence of antigen the values obtained in the absence of antigen were subtracted from those obtained in the presence of antigen.

The Effect of in Vivo Challenge Followed by in Vitro Culture

Since there was a marked difference in the G/M ratio depending on whether the secondary antigenic challenge was administered
in vivo or in vitro, experiments were performed to initiate the
secondary response in vivo and then examine it in vitro. Primed
animals were boosted with 100 ug TKB and their spleens removed on
consecutive days and cultured with or without added antigen. After 5 days in culture cells were assayed for anti-TNP PFC. The
results of such an experiment are shown in Table III. In the absence of an in vivo boost (day 0 post boost) the G/M ratio in vitro
in the presence of 0.2 µg TNP-KLH is 4. In marked contrast if
spleens are removed 4 days post boost and placed in culture with
0.2 µg TNP-KLH the G/M ratio is 16. In this experiment the number
of IgM PFC increased by a factor of 2 while the number of IgG PFC
increased by a factor of 9. In 3 similar experiments the IgM response remained relatively constant while the IgG response increased dramatically. The average IgG/IgM ratio for all 4 experiments in cells from unboosted animals was 5. In these same experiments the average IgG/IgM ratio in cultures from animals boosted
4 days earlier and receiving further antigen in vitro was 20.
Since the average IgG/IgM ratio in the boosted animals assayed in
vivo was 21, the procedure of boosting prior to culture resulted
in an in vitro system which mimicked the in vivo response.

Effect of Varying the Dose of the in Vivo Boost
on the in Vitro Response

In Table III there is a 3-fold stimulation of PFC by the addition of antigen in vitro above that seen in its absence. Since
the degree of stimulation was variable, the effect of the dose of
antigen given in vivo on the number of PFC developed in vivo and in
vitro was examined. Primed animals were boosted with 0,10,100 or
1000 µg of TNP-KLH in vivo. Spleens were removed 4 days later and
assayed for anti-TNP PFC. Table IV shows that while the dose of
antigen in vivo had a variable effect on the number of IgM PFC, the
number of IgG PFC increased with increasing dose of antigen.

A portion of each of the spleen cell pools assayed for the in
vivo response shown in Table IV was cultured with or without added
TNP-KLH. The plaque assay was performed on days 4,5 and 6 to determine the kinetics of appearance of PFC in vitro under different
boosting conditions. The results are shown in Table V. The number of IgM and IgG anti-TNP PFC for a given in vivo boost dose remained constant throughout the culture period in the absence of
further antigen (Table V,A). In contrast, the addition of antigen
in vitro resulted in a marked increase in the number of IgG PFC

between days 4 and 5 of culture (Table V, B). Although the number
of IgM PFC in some instances was also increased the effect was less
pronounced.

Two other relationships are shown in Table V. First, the
greatest antigenic stimulation (Ag/0) in vitro was seen when the
lowest dose or no antigen is given in vivo and this stimulation is
mainly IgM since the G/M is approximately 1. In contrast, however,
the greatest G/M ratio is achieved in vitro when the largest dose
of antigen was given in vivo. This trend has been found repeatedly
although in some groups of mice the maximum G/M in vitro was
achieved after a 100 µg in vivo boost. The results are not due to
an inappropriate combination of in vivo and in vitro antigen doses
as in all cases a 5 $\log_{10}$ range of TNP-KLH (beginning with 2.0 µg)
was tested in vitro to determine the optimal dose.

TABLE IV

In Vivo Dose Response Curve

Experiment	PFC Class	PFC/10^6 Spleen Cells*			
		Antigen Dose in Vivo			
		0	10	100	1000
1	IgM	14	25	39	850
	IgG	10	780	2200	8000
	G/M	1	31	56	9
2	IgM	15	15	15	35
	IgG	5	200	380	1505
	G/M	1	13	25	43

*Pooled spleen cells from 3 mice were assayed at each point

TABLE V

Kinetics of Appearance <u>in Vitro</u> of Anti-TNP PFC

A. No Antigen Added <u>in Vitro</u>				
µg TNP-KLH <u>in Vivo</u>	PFC Class	PFC/10^6 Cultured Cells Assayed Days in Culture		
		3	4	5
0	IgM	13	18	26
	IgG	3	21	7
10	IgM	16	25	42
	IgG	100	72	54
100	IgM	28	35	48
	IgG	181	161	167
1000	IgM	26	15	106
	IgG	642	656	586

B. 0.2 µg TNP-KLH Added <u>in Vitro</u>				
µg TNP-KLH <u>in Vivo</u>	PFC Class	Days in Culture		
		3	4	5
0	IgM	0	96	287
	IgG	7	23	277
	G/M	–	0	1
	Ag/0*	0	1	39
10	IgM	0	61	198
	IgG	0	2	394
	G/M	–	0	2
	Ag/0	0	0	7
100	IgM	39	213	315
	IgG	0	314	1483
	G/M	0	1	5
	Ag/0	0	2	9
1000	IgM	27	133	112
	IgG	0	324	1742
	G/M	0	2	15
	Ag/0	0	1	3

Ag/0 = <u>Number of PFC in the presence of antigen added in vitro</u>
          Number of PFC in the absence of added antigen

DISCUSSION

The data in Table I reflect the difficulty of obtaining adequate IgG responses in vitro as discussed recently by Mitchell (7). The average G/M ratio obtained after in vivo secondary challenge was 11 whereas that obtained after in vitro challenge was only 1. The deficiency in immune memory demonstrable in vitro appears limited to the IgG response since IgM responses are equivalent to and may exceed those observed in vivo. As indicated in Table II the low in vitro G/M ratio was not affected by the antigen dose used for in vitro challenge; neither was it affected by the day on which the culture was assayed. There is also no obvious correlation with the absolute magnitude of the response since in about 50% of all of our experiments the in vivo response has been significantly greater than the in vitro response even though in all of the experiments the in vivo G/M ratio was much larger.

The IgG deficiency may be overcome by in vivo boosting prior to culture but there is a lag period of several days before the effect of the boost can be detected; days 4 and 5 consistently provide the boosted cells necessary to achieve a high G/M in vitro. The critical nature of the timing could explain why others who have used the technique of initiating secondary responses in vivo to study the secondary response in vitro have failed to note in vivo-like G/M ratios (11,16). The ability of the appropriate in vivo boost to overcome the IgG deficiency indicates that tissue culture per se is not inherently inadequate and that the deficiency lies within the spleen itself.

The time after in vivo boost which is most effective for establishing in vitro secondary responses coincides with the peak of the secondary in vivo response. While we have not excluded the possibility that antibody in some way positively mediates the development of IgG PFC in vitro extensive washing of cells (10 times) prior to culture failed to alter the results. Furthermore, in several instances antibody has been shown to have a negative effect on the development of IgG PFC and IgG synthesis (18,8).

The most obvious explanation for the ability of the boost to repair the splenic deficiency is that boosting restores or recruits cells necessary for IgG expression. It has been suggested that there is compartmentalization of memory cells and that T helper cells leave the spleen sooner than B cells (15) leaving a deficiency which would be reflected in the inability of the primed spleen to generate an adequate secondary IgG response in vitro. This would be in keeping with the observation that IgG responses are frequently more dependent on T cell help than are IgM responses (7). Recently Romano and Thorbecke (16) reported that the IgM and IgG secondary responses to SRBC were equally thymus dependent.

Also we have reported previously that mice primed to TNP-KLH according to the regimen used here showed a vigorous secondary IgM response in vitro and that the latter was thymus dependent (13). Thus if we are to accept a T cell deficiency as explanation for the failure of the primed but unboosted spleen to provide in vitro IgG responses, we must also consider the possibility that there are either quantitative or qualitative differences in the help required to initiate secondary IgM and IgG responses. Suppressor T cells or accessory cell involvement (19,4) would similarly demand a differential susceptibility in expression of the 2 classes of anti-hapten memory.

Compartmentalization of B cell memory is an alternative explanation which must also be considered since primed B cells do recirculate in the rat (17). If the latter explanation proves correct, it would appear to be the mouse IgG memory cell which recirculates and localizes outside of the spleen thus giving rise to the splenic IgG deficiency detected by in vitro culture and which is repaired by in vivo boosting prior to culture.

ACKNOWLEDGEMENTS

We thank K. Pratt for excellent technical assistance and M. Baltz for valuable discussion. This work was supported by grants from the Medical Research Foundation of Oregon, USPHS Grant AM 13173 and USPHS Grant HD 00165.

REFERENCES

1. Bullock, W.W. and Rittenberg, M.B., J. Exp. Med.,132 (1970) 926.
2. Cunningham, A. and Szenberg, A., Immunology,14 (1968) 599.
3. Dresser, D.W. and Wortis, H.H., Nature, 208 (1965) 859.
4. Gershon, R.K., Contemporary Topics in Immunobiology (Ed. M.D. Cooper and N.L. Warner) Plenum Publishing Co., New York, 3 (1974) 1.
5. Kappler, J.W., J. Immunol., 112 (1974) 1271.
6. Lawton, A.R., III and Cooper, M.D., Contemporary Topics in Immunobiology (Ed. M.D. Cooper and N.L. Warner) Plenum Publishing Co., New York, 3 (1974) 193.
7. Mitchell, G.F., Contemporary Topics in Immunobiology (Ed. M.D. Cooper and N.L. Warner) Plenum Publishing Co., New York, 3 (1974) 97.
8. Möller, G., J. Exp. Med., 127 (1968) 291.
9. Nisonoff, A., Fudenberg, H.H., Wilson, S.K., Hopper, J.E. and Wang, A.C., Fed. Proc., 31 (1972) 206.
10. Nossal, G.J.V. and Shrader, J.W., Transplant. Rev., 23 (1975) 138.

11. Pierce, C.W., Solliday, S.M. and Asofsky, R., J. Exp. Med.,
    135 (1972) 698.
12. Rittenberg, M.B. and Amkraut, A.A., J. Immunol., 97 (1966) 421.
13. Rittenberg, M.B. and Bullock, W.W., Immunochemistry, 9 (1972)
    491.
14. Rittenberg, M.D. and Pratt, K.L., Proc. Soc. Exp. Biol. Med.,
    132 (1969) 575.
15. Romano, T.J., Mond, J.J. and Thorbecke, G.J., Eur. J. Immunol.,
    5 (1975) 211.
16. Romano, T.J. and Thorbecke, G.J., Cell. Immunol., 17 (1975) 240.
17. Strober, S., Transplant. Rev., 24 (1975) 84.
18. Wigzell, H., J. Exp. Med., 124 (1966) 953.

EXTERNAL CONTROL OF A SHIFT IN ANTIBODY SYNTHESIS

S. HINCHMAN and J. R. BATTISTO

.   Cleveland Clinic Foundation
Cleveland, Ohio (USA)

Two types of differentiation of B cells involved in humoral immunologic responses were described: that which develops naturally
with time (3,8) and that which is antigenically driven (17,24).
With time, cells become immunoglobulin bearers in a definitive order: IgM cells appear first, followed by IgG and then IgA (3). In
this case, antibodies of each class would be expected to emanate
from the respective cell. On the other hand, when the immunological
system is directed by antigen to differentiate on a short term basis,
the result may be stepwise synthesis of IgM and IgG (12,18). Many
observations attest to the fact that T cells are required to cooperate with B cells for IgG production (5,6,9,14,20,21). The switch
from M to G could conceivably occur within a single cell or by stimulation of separate M and G producing cells. Of course, both mechanisms of antibody production may co-exist.

This study deals with that differentiation which is induced by
antigen, with the way in which the shift in antibody synthesis is
brought about and with the nature of the cells involved.

MATERIALS AND METHODS

Animals and Immunization

Six to eight week old CBA female mice (Jackson Labs, Bar Harbor, Maine) were used in all experiments. Animals were routinely
injected intraperitoneally with 1.0 ml of a 0.025% suspension of
sheep red blood cells (SRBC) washed in physiological saline and
their splenic cells were assayed for plaque forming cells (PFC).

Hemolytic Plaque Assay

Antibody response to SRBC was determined 4.5 days after immunization using the technique of Jerne and Nordin (7) as modified by Cunningham and Szenberg (4). For reasons that are unclear, the shift attributable to RPMI-1640 was observed only when the liquid monolayer plaquing technique was used but not with the agarose plating method. IgM producing cells were identified as direct plaque forming cells (DPFC) by development with guinea pig serum alone as a source of complement. Total number of PFC, i.e., both IgM and IgG producers, was measured by developing plaques with guinea pig complement (GPC) and normal mouse serum according to the method of Silver and Winn (22). They have reported that C_1 of GPC is unable to fix to mouse IgG rendering GPC incapable of detecting IgG producing cells. On the other hand, C_1 of mouse complement is able to fix to murine IgG permitting the remainder of the components of GPC to carry out the lytic process and thereby identifying both IgM and IgG producers. For the statistical data reported here, we used mouse complement and GPC to develop IgG plaques, but identity of IgG was additionally confirmed by using heavy chain specific anti-IgG (Cappel Laboratories, Downingtown, Pennsylvania). The value for indirect plaque forming cells (IPFC) or IgG producers, was achieved by subtracting DPFC from the total count. The suspending media for plaquing were Earle's BSS, RMPI-1640, Minimal Essential Medium with Earle's Salts (MEM) and Earle's BSS with the addition of various components of RPMI in approximately the same concentrations as those found in RPMI-1640 (all media were from Grand Island Biologicals, Grand Island, New York).

Anti-θ-Serum Preparation

Antibody to θ-antigen was prepared by the method of Reif and Allen (19). Each AKR mouse was injected intraperitoneally with 2 x 10^7 $C3H/H_eJ$ thymus cells on 3 occasions separated by 3 week intervals and bled 10 days after the last injection. Pooled sera were assayed on thymus cells in vitro in the presence of GPC by trypan blue exclusion. The serum dilution used in experiments was that which gave maximum kill of thymus cells. As a test upon the specificity of the anti-θ serum it was ascertained that incubating immune spleen cells for 1 hr in the anti-serum and GPC did not adversely affect the number of spleen cells producing antibody to SRBC in the plaque assay system.

Elimination of θ-Bearing Cells

T cells were removed prior to incubation in the final suspending medium in which cells were to be plaqued. This was achieved

by incubating 20 x 10^6 spleen cell suspended in Earle's with anti-
θ (final dilution, 1:5) and agarose-absorbed GPC (final dilution,
1:11) for 30 min at 37 C in a 5% CO_2 atmosphere. Controls were
similarly treated with normal mouse serum and GPC. After incubation,
cells were washed twice in Earle's and resuspended at the appropriate number in either RPMI or Earle's where they were incubated for
1.5 hr prior to plaquing. Between 30 to 34% of the spleen cells
were killed by this procedure. Preparation of agarose-absorbed GPC
was as described by Cohen and Schlesinger (2).

RESULTS

Initially a difference in the number of cells making antibodies of the M and G varieties was seen when spleen cells from immunized mice were suspended in 2 different media. At 4.5 days following an intraperitoneal injection of SRBC, half of a spleen cell
suspension was placed in RPMI-1640 medium while the remainder was
put into Earle's BSS. Following incubation at room temperature,
the cells were plaqued and the results seen in Table I were observed.

TABLE I

A Shift from IgM to IgG Synthesis can be Made to Occur by
the Medium in which Immunized Spleen Cells are Suspended

| Suspending Medium | DPFC* | | IPFC* | | Total |
	per 10^6	%Total + S.E.	per 10^6	%Total + S.E.	per 10^6
Earle's BSS	265	51 ± 4	254	49 ± 4	519
RPMI-1640	61	12 ± 2	447	88 ± 2	508

*Average of 20 trials.

Whereas in Earle's BSS the DPFC accounted for 51% of the total, in RPMI they comprised only 12% of the total. Furthermore,
in each trial the total number of PFC remained the same regardless
of the medium in which the cells were suspended. The difference
between the total PFC and the DPFC is accounted for by IPFC which
were seen to be about 40% more numerous in RPMI than in Earle's.

Results of a paired T test applied to DPFC showed that a significant difference exists between the number of DPFC in the 2 media (i.e., p < 0.001). The same test applied to the total PFC indicates no significant difference between the number in the 2 media (i.e., p = NS). We have noted that despite rather wide differences in the immune responses between individuals, the ratio of DPFC to IPFC remains rather constant for the medium in which the cells were suspended. For instance, in Earle's the DPFC to IPFC ratio was about 1:1 whereas in RPMI the ratio was 1:9. Therefore, in all of the work to be reported we expressed this ratio as a percent of the total PFC plus or minus the standard error (S.E.)

To investigate the thought that anti-complement effects might be exerted by ingredients of RPMI, complement titrations using SRBC and a constant amount of antibody to SRBC were conducted in both RPMI and Earle's. This test (the data of which are not shown) confirmed that at the concentrations of complement used in the plaque assay, there was no anti-complementary effect exerted by RPMI. Thus, an excess amount of complement is known to be present with the cell suspensions and complement cannot be the factor limiting the number of direct plaques developed in RPMI.

To determine whether the shift brought about by RPMI was constant or could be made to revert to baseline, immune cells were first incubated in RPMI for 1 hr and thereafter washed copiously in Earle's and resuspended in the latter for plaquing. As may be seen from the data of Table II, the shift away from IgM production persisted despite the fact the cells experienced RPMI for only a brief time interval.

In order to identify the agents in RPMI responsible for the immunoglobulin shift, we first tried a medium that possessed some, but not all of the ingredients of RPMI, namely MEM. It caused cells to produce much the same ratio of M to G as does RPMI (Table III). A comparison of the ingredients of MEM to RPMI suggested that those items common to both media and absent in Earle's could be the responsible agents. Inclusion of 2 vitamins, thiamine and nicotinamide, into Earle's in amounts comparable to those found in MEM and RPMI resulted in a medium that caused cells suspended in it to shift antibody synthesis from M to G (Table III).

To further enumerate the items capable of causing this switch, individual substances present in RPMI but absent in Earle's were incorporated into Earle's and the resulting solutions tested on immunized spleen cells. As may be seen in Fig. 1, biotin as well as para-amino benzoic acid were as effective as thiamine and nicotinamide to cause the immunoglobulin shift. In contrast, pantothenate, riboflavin, pyridoxal and inositol were less effective. Thus, certain of the B vitamins, but not all, are capable of triggering the change in synthesis from M to G.

TABLE II

The IgM to IgG Shift Observed in RPMI-1640 is Irreversible

Cells in Medium	Incuba- tion (hr)	DPFC*		IPFC*		
		per 10^6	%Total $\pm$ S.E.	per 10^6	%Total $\pm$ S.E.	p**
Earle's	3	131	39 $\pm$ 7	237	61 $\pm$ 7	---
RPMI	3	16	5 $\pm$ 2	360	95 $\pm$ 2	<0.02
RPMI	1	55	12 $\pm$ 3	346	88 $\pm$ 3	<0.05
then						
Earle's	2	27	8 $\pm$ 3	416	92 $\pm$ 3	<0.02

*Average of 5 trials
**p was calculated for DPFC using the value in Earle's versus the values in each of the other media
NOTE: In each trial, total PFC remained the same in each medium

TABLE III

Comparison of Shift-Inducing Properties of Different Media

Suspending Medium	DPFC*		IPFC*		
	per 10^6	%Total $\pm$ S.E.	per 10^6	%Total $\pm$ S.E.	p**
Earle's	361	62 $\pm$ 3	225	38 $\pm$ 3	---
RPMI	80	14 $\pm$ 5	443	86 $\pm$ 5	<0.001
MEM	109	20 $\pm$ 4	408	80 $\pm$ 4	<0.001
Earle's plus Thiamine and Nicotinamide	168	25 $\pm$ 8	502	75 $\pm$ 8	<0.001

*Average of 5 trials
**p was calculated for DPFC using the value in Earle's versus the values in each of the other media.
NOTE: In each trial, total PFC remained the same in each medium

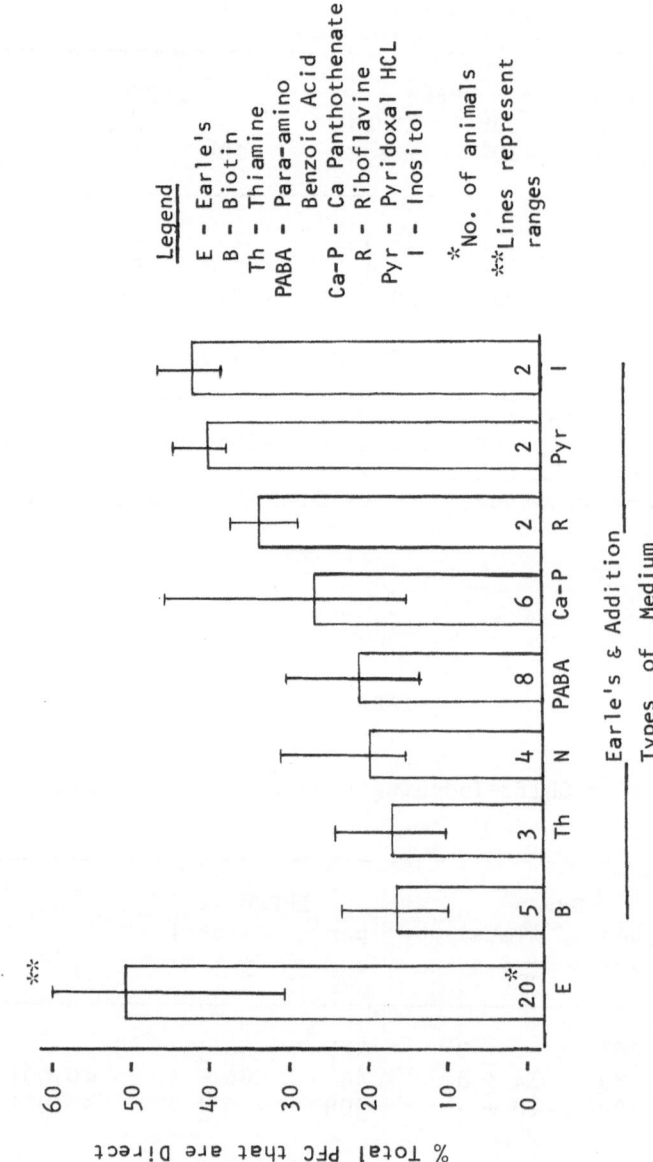

Fig. 1.   Identification of substances in RPMI triggering the IgM to IgG shift.

Attention was now directed to determining which cells were being affected by RPMI. Although the ultimate effect is experienced by B cells, it was conceivable that other cells could initially be involved. We thought to eliminate θ-bearing cells by treating the cell pool in Earle's medium with anti-θ antibodies and complement. As control, other portions of cells were treated with normal mouse serum and complement or Earle's and complement alone. Thereafter, the cells in each group were washed, divided, and half put into Earle's with the remainder placed in RPMI for still another incubation prior to plaquing.

At the outset it should be noted (Table IV), that although the extended incubations in mouse serum plus complement and additional manipulations of the cells required for this experiment had reduced the overall plaquing efficiency, the proportion of cells making each type of antibody was not altered (54% DPFC, 46% IPFC). Furthermore, the shift in synthesis attributable to RPMI was still observable (4% DPFC, 96% IPFC). Finally, it should be noted (Table IV) that eliminating θ-bearing cells from the cell pool in no way altered the switch in synthesis brought about by RPMI. Thus, T cells appear not to be required at this point in time to initiate the shift.

TABLE IV

The Shift Induced by RPMI does not Require Presence
of T Cells at This Point in Antibody Synthesis

Treatment of Cells in Earle's	Suspending Medium After Treatment	DPFC*		IPFC*		
		per 10^6	%Total $\pm$ S.E.	per 10^6	%Total $\pm$ S.E.	p**
Anti-θ + C'	Earle's	97	42 $\pm$ 14	138	58 $\pm$ 14	< .05
	RPMI	20	9 $\pm$ 6	199	91 $\pm$ 6	
NMS + C'	Earle's	98	54 $\pm$ 15	84	46 $\pm$ 15	< .02
	RPMI	10	4 $\pm$ 2	250	96 $\pm$ 2	

*Average of 5 trials
**p was calculated for DPFC using the values in Earle's versus the values in RPMI for each category of treatment
NOTE: In each trial, total PFC remained the same for each medium

DISCUSSION

The conversion from IgM to IgG antibody synthesis during a
response to an antigenic stimulus has been well documented (1,12,
17,18,20,23,24). The number of cells making IgM is maximal at 4
days whereas that producing IgG is most plentiful at 6 days (12,18).
Our data suggest that the number of cells in these 2 categories may
be altered at 4.5 days by exposing the antibody-producing cells to
various media. Thus, a model has become available by which this
shift in antibody synthesis may be readily studied. We call attention to the thought that the shift reported here is in response to
stimuli during an immune response to an antigen rather than the result of developmental differentiation. The latter area has been extensively studied (3,8).

The observations that spleen cells from mice injected intraperitoneally 4.5 days earlier with SRBC would produce M and G antibodies in a ratio of about 1 to 9 when suspended in RPMI medium
rather than the customary ratio of 1 to 1 when suspended in Earle's
medium were made using the liquid monolayer plaquing technique.
Parallel studies using the agarose plating method were not fruitful
for reasons that are presently obscure.

Our initial, trivial explanation for the diminution in the
number of cells producing IgM in RPMI was that the medium was exerting an anti-complementary effect. However, that possibility was
dismissed in several ways. First, cells incubated in RPMI, then
washed and resuspended in Earle's still retained their bias away
from M synthesis. For this to be explained on an anti-complementary
basis would necessitate that the cells would have to carry over into Earle's the hypothesized anti-complementary components from RPMI
despite thorough washing. We considered this unlikely. Nevertheless, complement titrations using SRBC and a constant amount of antibody to SRBC were conducted in both RPMI and Earle's. These tests
revealed sufficient amounts of complement were present in both media
so as to eliminate complement as a limiting factor. Furthermore,
in some plaque assays we doubled the amount of GPC present with cells
in RPMI without altering the biased ratio.

The fact that those spleen cells first exposed to RPMI and
then suspended in Earle's retained their bias to G production indicates that the shift could not be readily reversed. This should not
be interpreted to mean that reversal is not possible, simply that a
modicum of stability exists at this stage of cellular differentiation that may relax with time.

The constituents of RPMI medium that were found to be responsible for initiating the shift in immunoglobulin synthesis were some,
but not all, of the B vitamins. Thus, biotin, thiamine, para-amino-

benzoic acid and nicotinamide effected the shift, but pantothenate, riboflavin, pyridoxal and inositol did not. Slightly better results were achieved when some vitamins were used together than when they were utilized separately. Whether these accomplished the task by entering the cells to help activate enzymatic reactions or simply interacting with cell membranes to provide external signals has not been investigated. We were more immediately concerned with which cells represented the targets of the medium constituents. To this end, θ-bearing cells were removed from the spleen cell suspensions prior to exposing the remainder to RPMI. When the shift was seen to persist, the T cell was eliminated as being involved at this late stage in immunization. That is not to say that IgG antibody production is independent of the T cell, for such dependency early in immunization has been repeatedly demonstrated in a variety of systems (5,9,14,21). Even though the T cell presumably is operative in the same manner in this model, it apparently is not necessary at the point in time when IgM synthesis ceases and IgG is begun. Although the macrophage has not been rigidly eliminated as a target, those medium constituents operational in the M to G shift appear to be affecting B cells directly.

Several models for B cell activation have been described, some suggesting that B cells differentiate into IgM-, IgG- and IgA-producing cells during the course of development (3,8), others indicating that 1 cell may have the capability to produce more than 1 type of immunoglobulin (11,13). The idea that 1 cell may be synthesizing more than 1 class of antibody was first suggested by Nossal (11) who reported the identification of cells simultaneously producing M and G antibodies to SRBC. Double producers composed approximately 4% of the total population of cells (13). This thought was subsequently supported by the identification of IgM and IgG antibodies of identical idiotypes (15,16) as well as by evidence for 1 gene controlling production of the variable heavy chain region of both IgM and IgG (10,25). The data presented here also suggest that 1 cell may be synthesizing both classes of antibody since the total numbers of antibody producing cells in both types of media remained identical. If one were to assume existence of B cells of 2 types, those producing only IgM and those producing only IgG, this would make stringent requirements upon our system. Precisely the same number of cells that were initially producing IgM and then turned off would have to be recruited from another population programmed to synthesize IgG (top Fig. 2). This possibility seems remote thus lending credibility to the thought that those cells already synthesizing IgM are stimulated sequentially to produce IgG. Although these pluri-potential cells may only be producing 1 class of antibody at any one point in time, they retain the capability of producing a different class when triggered by the appropriate stimulus (bottom Fig. 2).

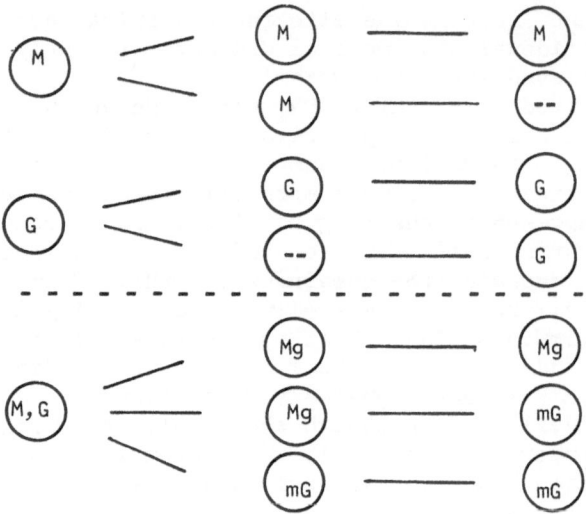

Fig. 2. Possible mechanisms for shift in class of antibody.

        In summary, in studying humoral antibody responses of murine
splenic cells, we have found that the number of cells producing
antibodies of the M and G classes can be changed by suspending cells
in different media prior to plaquing. When Earle's BSS was the suspending medium for splenic cells from CBA/J mice that had been injected intraperitoneally with SRBC 5 days earlier, approximately
equal numbers of PFC producing M and G SRBC-specific antibodies were
obtained. In contrast, another portion of the cell pool placed in
RPMI medium produced the M and G antibodies in a ratio of 1 to 9.
In both media the total number of PFC were equal. Attempts to reverse the shift in antibody synthesis were not productive. For example, cells first incubated in RPMI, then washed copiously and replaced in Earle's still demonstrated the bias toward G synthesis.

        To identify the responsible agents, components of RPMI absent
from Earle's were added singly to Earle's prior to introducing cells.
Certain B vitamins, but not others, triggered the change in antibody
production. Eliminating T cells from the pool with anti-θ-serum
and complement indicated that, at this late time, T cells were not
essential for the effect which presumably is directed to the synthesizing cell. Thus, certain substances classifiable as common
cellular constituents can, when incubated with immunized B cells in
vitro, control quantity and quality of antibodies made.

        These data suggest that a single cell synthesizing IgM anti-
body directed against a particular antigenic specificity also may
produce, during the course of an immune response, IgG antibody of

the same specificity. Thus, the initial shift in antibody synthesis from M to G may represent further differentiation of those cells already involved in M antibody synthesis rather than cessation of the IgM synthesis by 1 population of cells and initiation of IgG by another.

ACKNOWLEDGEMENT

This research was supported by NIH Grant No. AI-12468.

REFERENCES

1. Bauer, D.C. and Stavitsky, A.B., Proc. Nat. Acad. Sci., 47 (1961) 1667.
2. Cohen, A. and Schlesinger, M., Transplantation, 10 (1970) 130.
3. Cooper, M.D., Lawton, A.R. and Kincaid, P.W., Clin. Exp. Immunol., 11 (1972) 143.
4. Cunningham, A.J. and Szenberg, A., Immunology, 14 (1967) 599.
5. Davis, J.M. and Paul, W.E., J. Immunol., 113 (1974) 1438.
6. Grumet, F.C., J. Exp. Med., 135 (1972) 110.
7. Jerne, N.K. and Nordin, A.A., Science, 140 (1963) 405.
8. Lawton, A.R., Asofsky, R., Hylton, M.B. and Cooper, M.D., J. Exp. Med., 135 (1972) 277.
9. Mitchell, G.F., Grunet, F.C. and McDevitt, H.O., J. Exp. Med., 135 (1972) 126.
10. Nisonoff, A., Fudenberg, H.H., Wilson, S.K. and Hopper, J.E., Fed. Proc., 31 (1972) 206.
11. Nossal, G.J.V., Szenberg, A., Ada, G.L. and Austin, C.M., J. Exp. Med., 119 (1964) 485.
12. Nossal, G.J.V., Warner, N.L. and Lewis, H., Cell. Immunol., 2 (1971) 13.
13. Nossal, G.J.V., Warner, N.L. and Lewis, H., Cell. Immunol., 2 (1971) 41.
14. Ordal, J.C. and Grumet, F.C., J. Exp. Med., 136 (1972) 1195.
15. Oudin, J. and Cazenave, P.A., Proc. Nat. Acad. Sci., 68 (1971) 2616.
16. Oudin, J. and Michel, M., J. Exp. Med., 130 (1969) 169.
17. Pierce, C.W., Soliday, S.M. and Asofsky, R., J. Exp. Med., 135 (1972) 75.
18. Plotz, P.H., Talal, N. and Asofsky, R., J. Immunol., 100 (1968) 744.
19. Reif, A.E. and Allen, J.M., Cancer Res., 26 (1966) 123.
20. Romano, T.J. and Thorbecke, G.J., J. Immunol., 115 (1975) 332.
21. Schimpl. A. and Wecker, E., J. Exp. Med., 137 (1972) 547.
22. Silver, D.M. and Winn, H.J., J. Immunol., 111 (1973) 128.
23. Sterzl, J., Cold Spring Harbor Symposia Quantitative Biology, 32 (1967) 493.

24. Vitetta, E.S., Grundke-Igbal, I., Holmes, K.V. and Uhr, J.W.,
    J. Exp. Med., 139 (1974) 862.
25. Wang, A.C., Wilson, S.K., Hopper, J.E., Fudenberg, H.H. and
    Nisonoff, A., Proc. Nat. Acad. Sci., 66 (1970) 337.

EFFECT OF ANTIGEN ON LOCALIZATION OF IMMUNOLOGICALLY SPECIFIC B CELLS

N. M. PONZIO, J. M. CHAPMAN and G. J. THORBECKE

New York University School of Medicine
New York, New York (USA)

Injection of antigen, either intravenously (i.v.), intraperitoneally (i.p.) or locally, can enhance temporarily the influx of lymphoid cells into antigen-containing sites (2,3,4,6,7,10,11,16,17, 22,27,29,30,35). This has been examined early, usually only 24 to 48 hr, rarely as late as 72 hr after antigen (35). In many cases, a nonspecific increase of T cells (4,16) and/or macrophages (22) is involved, but a specific retention of T cells (3,11,30) and/or B cells (6,7,10,27,29,35) has also been shown. Homing of cells into allografts or sites of delayed reactions is mainly nonspecific as has been reviewed by McCluskey (22).

Antigen specific cells may get detained in sites of antigen deposits on the basis of 2 different mechanisms: a) the cells become stimulated into blast cell transformation after local contact with antigen and therefore stay out of the recirculating pool temporarily, and b) the cells are held in place through surface receptors reacting with antigen. Cells, such as helper T cells and those cells responding with blast transformation in MLR or GVH reactions have not been shown to be absorbed out on fibroblast monolayers containing specific antigen (21) and should therefore only undergo specific accumulation during blast cell responses (first mechanism only), whereas cytotoxic T cell precursors (8) as well as B cells can be absorbed out on antigen and could presumably be detained on the basis of both mechanisms. In the absence of helper T cells, memory B cells would be expected to bind to antigen (such as to hapten on the wrong carrier) without responding and be detained purely on the basis of the second mechanism.

The known localization of B cells in follicles of lymphoid tissue, particularly in the corona of follicles, appears to be the same site where the only detectable long term antigen retention occurs (1). This suggests that antigen might cause specific retention of B cells in such areas without necessarily inducing their activation into antibody production. The antigen present in these sites is frequently in the form of antigen-antibody complexes, which localize in follicles more effectively than antigen alone (12, 18). Studies directed at specific retention of B cells by antigen in the follicles may take advantage of the fact that, within a short time after antigen-antibody complex injection, and somewhat later after antigen alone, the bulk of the antigen is found in follicles.

Previous observations on rabbits suggest that increased follicular accumulation of cells occurs at the site of antigen localization 2 weeks after its injection (5). Specific homing of functional memory B cells to such lymph nodes was never shown after transfer. However, an influence due to a local antigen injection on the development and maintenance of immunological memory in the draining lymph node as compared to the contralateral side has been clearly demonstrated in sheep (28) and in rabbits (9,13,31,34). In rabbits such differences in the level of immunological memory between the 2 lymph nodes are maintained for periods of more than 1 year after a single antigen injection (34). These memory differences are thought to be due to memory B cell retention by local antigen (9,31,32,34). The present studies in mice are aimed at demonstrating homing of memory B cells to sites of antigen localization in lymph nodes, using functional criteria to detect local presence of memory cells at varying intervals after their i.v. injection.

MATERIALS AND METHODS

Male LAF$_1$ mice were purchased from Jackson Laboratories (Bar Harbor, Maine) and used at 8-12 weeks of age. Donors of trinitrophenylated bovine IgG (TNP-BGG)-primed and hemocyanin (KLH)-primed cells were given a single i.p. injection of 500 µg of either antigen in complete Freund's adjuvant (CFA) and spleens were excised 4-6 weeks later, teased into cell suspensions, washed and injected into groups of 6 syngeneic recipients, which had been prepared in 1 of 2 ways:

Protocol I (Fig. 1). 100 µg TNP-KLH were injected into the left front footpad and 100 µg para-azobenzoate-KLH (PABA-KLH) into the right front footpad. Both injections included 10 µg E. coli endotoxin (Difco Laboratories, Detroit, Michigan). A second similar injection was given after a 2 week interval and 1 week later

these recipients were given 920R ^{137}Cs γ-irradiation from a Gammator
M (Radiation Machinery Corp., Parsippany, New Jersey), and 4 x 10^7
TNP-BGG-immune or normal spleen cells were injected i.v. Two to
5 days later, 4 x 10^7 KLH-primed spleen cells $\pm$ 50 μg TNP-KLH were
injected i.v.

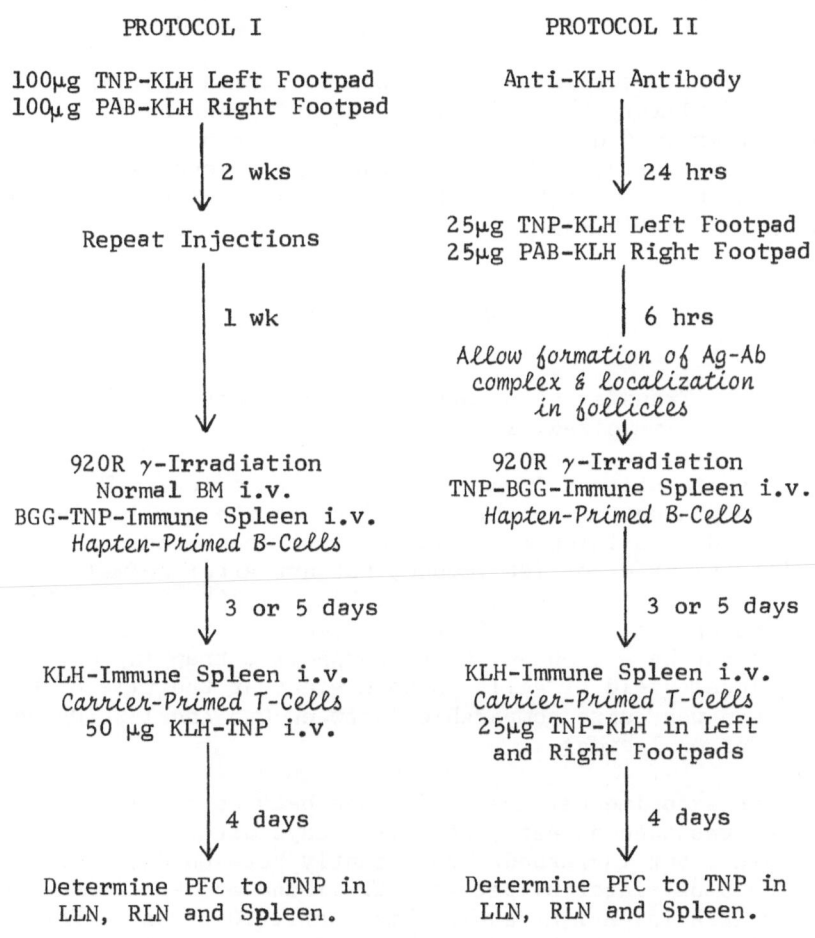

PROTOCOL I

100μg TNP-KLH Left Footpad
100μg PAB-KLH Right Footpad

↓ 2 wks

Repeat Injections

↓ 1 wk

920R γ-Irradiation
Normal BM i.v.
BGG-TNP-Immune Spleen i.v.
Hapten-Primed B-Cells

↓ 3 or 5 days

KLH-Immune Spleen i.v.
Carrier-Primed T-Cells
50 μg KLH-TNP i.v.

↓ 4 days

Determine PFC to TNP in
LLN, RLN and Spleen.

PROTOCOL II

Anti-KLH Antibody

↓ 24 hrs

25μg TNP-KLH Left Footpad
25μg PAB-KLH Right Footpad

↓ 6 hrs

*Allow formation of Ag-Ab
complex & localization
in follicles*

920R γ-Irradiation
TNP-BGG-Immune Spleen i.v.
Hapten-Primed B-Cells

↓ 3 or 5 days

KLH-Immune Spleen i.v.
Carrier-Primed T-Cells
25μg TNP-KLH in Left
and Right Footpads

↓ 4 days

Determine PFC to TNP in
LLN, RLN and Spleen.

Fig. 1. Experimental protocols I and II.

Protocol II (Fig. 1). Recipient mice were passively immunized with anti-KLH sera derived from LAF$_1$ mice immunized with
KLH in CFA. Each mouse received the equivalent of 0.06 ml un-
diluted antiserum which showed a 2-mercaptoethanol resistant log$_2$

titer of 12 as determined by passive hemagglutination using KLH-
(Mann Laboratories, Orangeburg, New York) sensitized sheep erythrocytes (SE) (14). Twenty-four hr later, these mice were given 25 μg
TNP-KLH in the left front footpad and 25 μg PABA-KLH in the right
front footpad. After 6 hr (to allow formation of antigen-antibody
complexes in the draining lymph nodes) these recipients were γ-ir-
radiated (920R) and received 4 x 10^7 TNP-BGG-immune or normal spleen
cells i.v. KLH-primed spleen cells ± TNP-KLH were injected 3 or 5
days thereafter, but while the cells were given i.v., the TNP-KLH
challenge (25 μg) was given locally in each footpad.

Single cell suspensions were obtained from a pool of the 12
lymph nodes (axillary + brachial) on the left side of mice in an
experimental group 4 days after transfer of KLH-primed spleen cells
± TNP-KLH. Similarly, right lymph nodes and spleens were separately pooled and teased. Antibody forming cells to TNP were enumerated using TNP-SE (19,26).

RESULTS

Transfer of Memory Cells into Locally
Immunized, Irradiated Recipients

The results of experiments performed according to Protocol I
(Fig. 1) are presented in Table I. Although the challenging dose
of antigen had been injected i.v., the left lymph nodes challenged
5 days after transfer of TNP-immune, but not after normal spleen
cell transfer, had a higher number of PFC per 10^6 plated cells
than did the spleen. The difference between left and right lymph
nodes could not have been due to a nonspecific trapping effect,
since the local yield of cells on both sides at the time of PFC
determination was quite comparable (between 0.5 and 1 x 10^6 cells/
2 lymph nodes on 1 side).

In both experiments, the difference between the left and right
lymph nodes was seen as early as 2 or 3 days after TNP-immune spleen
cell transfer, but increased significantly between day 2 or 3 and
day 5. The 5 day interval between TNP-immune and KLH-immune spleen
cells + TNP-KLH injection resulted in nearly 10 times higher PFC
in the left nodes in both experiments. In experiment number 2,
however, normal spleen cell transfer with or even without i.v. TNP-
KLH challenge resulted in unilateral PFC detection as well. While
this could be due to unilateral localization of TNP-specific normal B cells, it also seemed possible that γ-irradiation-resistant
antibody formation by host cells to TNP-BGG was persisting on the
left side.

TABLE I

Effect of Persisting Antigen in Axillary Lymph Nodes
After Footpad Injection of Antigen on the
Distribution of Intravenously Injected Memory Cells

Expt. No.	Day 0* Spleen Cells	Day of KLH-Immune** Spleen Cells	TNP-KLH I.V.	Total PFC/10^6 Cells**		
				LLN	RLN	Spleen
1	TNP-BGG Immune	2	+	17	0	56
		5	+	124	14	77
	Normal	2	+	0	0	28
		5	+	0	0	155
2	TNP-BGG Immune	3	+	4	0	5
		5	+	41	0	20
		5	−	25	0	12
	Normal	5	+	28	2	28
		5	−	19	0	25

*LAF_1 mice were immunized in the front footpads with TNP-KLH (left side) and PABA-KLH (right side) on days -21 and -7 (Protocol I). On day 0 they received 800-920R whole body γ-irradiation, followed 2 hr later by i.v. injection of 4×10^7 TNP-BGG immune or normal spleen cells.
**PFC to TNP-SE determined on day 4 after i.v. injection of 4×10^7 KLH-immune spleen cells ± TNP-KLH (50 µg) as indicated.

Since the spleens of these irradiated, repopulated mice contained approximately 10^8 cells, the largest part of the immune response in these animals was occurring in the spleen. In Protocol II, therefore, the challenging TNP-KLH injections were given in the left and right front footpads.

The experiments in Table II were performed to examine further the possibility of persistent antibody formation in the host lymph nodes. Some locally immunized mice, according to the same antigen dose schedule as used before (Protocol I), were irradiated but received no cell transfers. The lymph nodes and spleens of both non-irradiated and irradiated immunized mice were examined for PFC content on day 14 or 16 after the second antigen injection. The lymph nodes of the nonirradiated animals at this late date after immunization contained very low PFC, 5 to 10 per 10^6 cells, and 15 to 50

PFC per combined axillary and brachial lymph nodes. Two days after irradiation no PFC were detected per 50,000 plated lymph node
cells and the combined lymph nodes on 1 side contained 250,000 cells
per mouse. These data demonstrate clearly that the previously detected PFC on the left side of mice receiving TNP-immune cells
(Table I) could not have been caused by locally persisting antibody forming cells of the host only.

On the other hand, the data on spleen PFC (Table II) with and
without irradiation show a striking radiation-resistance of the
persisting antibody-forming cells in these animals. It was therefore considered desirable to use recipients for the experiments
which had not been actively preimmunized.

TABLE II

Persistence of Antibody-forming Cells
to TNP after Footpad Immunization*

920R (Day 0)	Day	Total PFC/10^6 Cells			Total PFC/spleen
		LLN	RLN	Spleen	
–	0	10	0	15	1725
–	2	5	0	15	801
+	2	0	0	672	1646

*Immunization in front footpads with 100 µg antigen (TNP-KLH on
left side and PABA-KLH on right side) + 10 µg E. coli endotoxin
followed 2 weeks later by 100 ug of antigen alone. Last injection
was 2 weeks before the day of γ-irradiation.

Transfer of Memory Cells into Passively Immunized,
Antigen-Antibody Complex Containing Hosts

In order to perform experiments in which host cell responses
would be absent while recipients would still contain depots of
antigen in their lymph nodes, we decided to use passive immunization to the carrier protein prior to injection of the hapten-car-
rier conjugate (Protocol II). In this way, antigen could become
localized, presumably in follicles (12,18) of recipients' draining
lymph nodes, without a response on the part of the recipient itself, which was irradiated 6 hr after antigen injection.

TABLE III

Effect of Antigen-Antibody Complexes in Axillary Lymph Nodes
on the Distribution of Intravenously Injected TNP-Memory Cells

Expt. No.	Day 0* Spleen Cells	TNP-KLH in footpads	Total PFC/10^6 Cells**			
			Day 3 Interval*		Day 5 Interval*	
			LLN	RLN	LLN	RLN
3	TNP-BGG Immune	+	51	49	1553	475
		−	0	0	52	2
	Normal	+	38	36	436	158
		−	0	0	15	0
	None	+	0	0	0	0
4	TNP-BGG Immune	+	211	98	1078	122
		−			258	0
	Normal	+	24	11	339	17
	TNP-BGG*** Immune	+			438	53

*Antibody and antigen injected LAF$_1$ mice (Protocol II) were ir-
radiated (920R) on day 0, received 4 x 10^7 spleen cells as indicated, and 25 μg TNP-KLH in each footpad on day 3 or day 5 depending on the interval as stated in table. Regardless of whether
TNP-KLH was given or not, all mice received 4 x 10^7 KLH-primed
spleen cells i.v. on this day.
**Determined 4 days after i.v. injection of KLH-immune spleen cells
+ TNP-KLH.
***This group received normal instead of KLH-primed spleen cells.

Table III shows the results of 2 such experiments. In the
first place, it should be noted that in the absence of transfer
of spleen cells on the day of irradiation no PFC developed in lymph
nodes on either side although carrier primed cells were transferred
i.v. and TNP-KLH was injected in each footpad. Thus, at either the
3 or the 5 day interval, a host cell component of the responses
listed in Table III could be excluded. In addition, the local TNP-
KLH injections indeed brought out differences between lymph nodes
more effectively and caused far fewer PFC to appear in the spleen
(not shown in tables) than did the i.v. challenge.

Although there were slight differences between left and right lymph nodes on day 3 (especially in experiment 4), these differences were much more marked upon challenge on day 5 after TNP-immune cell transfer (primarily indirect PFC) and could also be detected--with that interval--after normal spleen cell transfer (majority of direct PFC). The antigen-antibody complexes were apparently not very effective in challenging the transferred TNP-immune cells, since responses were much higher with local TNP-KLH challenge than without.

In experiment 4, 1 group of mice received normal instead of KLH-primed spleen cells. The response on both sides in these mice was approximately 50% of that in the mice receiving KLH-primed cells, suggesting that normal spleen cells were less able to provide helper function for the TNP-immune B cells in the local response to TNP-KLH.

DISCUSSION

The results in this paper demonstrate a unilateral homing of adoptive memory B cells to antigen, presumably primarily on the basis of binding to antigen. The lymph nodes draining the left side, which had contained TNP prior to challenge with TNP-KLH, gave a higher response to the hapten. This meant that at the time of challenge more cells responsive to TNP were present in those lymph nodes, since conditions for both specific and nonspecific influx of cells from the circulation were similar for left and right lymph nodes from the moment of challenge onwards. Both sides must have responded to KLH, since both TNP-KLH and cells sensitized to KLH had been injected.

In the first 2 experiments recipients were able to respond prior to irradiation and cell transfers and thus part of the TNP-response on the left side might have been due to recipient cells. The marked resistance to irradiation of such antibody-forming cells also contributed to this possibility. The recipient component of the higher response on the left side is merely another expression of local memory such as has previously been shown in mice and sheep (28) and in rabbits (13,34) and which is probably due to accumulation of long-lived B memory cells (23,24) in sites of antigen localization (31). In the second series of experiments, however, the recipients of the antibody and antigen were irradiated before they could give a response. The higher response on 1 side must therefore have been due to a greater influx and/or retention of specific B cells during the time interval between transfer of hapten-sensitized cells and challenge with antigen and carrier-sensitized cells.

Previous studies in rabbits have shown unilateral homing of labeled cells to antigen under experimental conditions similar to the ones used here (5). Preliminary results in mice aimed at demonstrating such unilateral homing and the intranodal localization of labeled immune B cells isolated by adherence on nylon wool columns (15) have not shown a significant difference between left and right draining lymph nodes of irradiated recipients on day 1 or 2 after B cell transfer. Since the functional (PFC response) difference on day 5 after B cells is much higher than on day 3, a longer time interval might also be needed in these homing experiments.

The objection could be raised that irradiation might have caused a lack of retention of antigen in follicles (25,36). However, it should be pointed out that the recipients in the present study were irradiated only after the antigen or antigen-antibody complexes had localized. Besides, Williams (36) has shown that a small amount of passively transferred antibody completely restores antigen fixation in follicles of irradiated animals. In addition, injection of spleen cells after irradiation, as had been done in the present studies, also counteracts the loss of antigen retention in follicles (25). Results from studies not included here indicate that, even without spleen cell injection, the retention of 125I-KLH in draining nodes after irradiation of mice injected twice with KLH in the footpads is quite significant.

Thus, while the intranodal localization of the cells retained by specific antigen needs further study, the phenomenon of passively transferred unilateral B cell memory seems established. The development over a period of several days of this difference between left and right lymph nodes suggests that recirculating memory B cells (33) are being progressively selected by antigen in the lymph node, rather than that this difference is due to a specific exit of cells from the circulation towards the antigen, which presumably should have taken less time.

ACKNOWLEDGEMENTS

Several of the hapten-protein conjugates (20) were kindly donated by Dr. A. Nisonoff (Rosenstiel Medical Sciences Research Center, Brandeis University, Waltham, Massachusetts). This work was support by USPHS Grant No. AI-3076.

REFERENCES

1. Ada, G.L. and Williams, J.M., Immunology, 10 (1966) 417.
2. Asherton, G.L., Allwood, G.J. and Mayhew, B., Immunology, 25 (1973) 485.

3. Atkins, R.C. and Ford, W.L., Cell. Immunol., 141 (1975) 664.
4. Bhan, A.K., Reinisch, C.L., Levey, R.H., McCluskey, R.T. and
    Schlossman, S.F., J. Exp. Med., 141 (1975) 1210.
5. Durkin, H.G. and Thorbecke, G.J., Nature New Biol., 238 (1972)
    53.
6. Emeson, E.E. and Thursh, D.R., J. Immunol., 113 (1974) 1575.
7. Ford, W.L., Clin. Exp. Immunol., 12 (1972) 243.
8. Golstein, P., Svedmyr, E.A.J. and Wigzell, H., J. Exp. Med.,
    134 (1971) 1385.
9. Greene, E.J., Tew, J.G. and Stavitsky, A.B., Cell. Immunol.,
    18 (1975) 476.
10. Griscelli, C., Vassalli, P. and McCluskey, R.T., J. Exp. Med.,
    130 (1969) 1427.
11. Hay, J.B., Cayhill, R.N.P. and Trnka, Z., Cell. Immunol., 10
    (1974) 145.
12. Herd, Z.L. and Ada, G.L., Aust. J. Exp. Biol. Med. Sci., 47
    (1969) 63.
13. Jacobson, E.B. and Thorbecke, G.J., J. Exp. Med., 130 (1969)
    287.
14. Johnson, H.M., Brenner, K. and Hall, H.E., J. Immunol., 97
    (1966) 791.
15. Julius, M.H., Simpson, E. and Herzenberg, L., Europ. J. Immunol.,
    3 (1973) 645.
16. Koster, F.T., McGregor, D.D. and Mackaness, G.B., J. Exp. Med.,
    133 (1971) 400.
17. Lance, E.M. and Frost, P., Progress in Immunology (Ed. L. Brent
    and J. Holboro), North Holland Publishing Co., Amsterdam, 2
    (1974) 157.
18. Lang, P.G. and Ada, G.L., Immunology, 13 (1967) 523.
19. Lerman, S.P., Romano, T.J., Mond, J.J., Heidelberger, M. and
    Thorbecke, G.J., Cell. Immunol., 15 (1975) 321.
20. Little, J.R. and Eisen, H.N., Methods in Immunology and Im-
    munochemistry (Ed. C.A. Williams and M.W. Chase), Academic
    Press, New York, New York (1967) 128.
21. Mage, M.G. and McHugh, L.L., J. Immunol., 111 (1973) 652.
22. McCluskey, R.T. and Leber, P.D., Mechanism of Cell-Mediated
    Immunity (Ed. R.T. McCluskey and S. Cohen), John Wiley and
    Sons, Inc., New York, New York (1974) 1.
23. Miller, J.J., J. Immunol., 92 (1964) 673.
24. Miller, J.J., Cell. Immunol., 8 (1973) 413.
25. Nettesheim, P. and Hammons, A.S., J. Immunol., 107 (1971) 518.
26. Rittenberg, M.B. and Pratt, K.L., Proc. Soc. Exp. Biol. Med.,
    132 (1969) 575.
27. Rowley, D.A., Gowans, G.L., Atkins, R.C., Ford, W.L. and Smith,
    W.E., J. Exp. Med., 136 (1972) 499.
28. Smith, J.B., Cunningham, A.J., Lafferty, K.J. and Morris, B.,
    Aust. J. Exp. Biol. Med. Sci., 48 (1970) 57.
29. Sprent, J. and Miller, J.F.A.P., J. Exp. Med., 139 (1974) 1.
30. Sprent, J., Miller, J.F.A.P. and Mitchell, G.F., Cell. Immunol.,

    2 (1971) 171.
31. Stavitsky, A.B. and Folds, J.D., J. Immunol., 108 (1972) 152.
32. Stecher, V.J. and Thorbecke, G.J., J. Exp. Med., 125 (1967) 33.
33. Strober, S. and Dilley, J., J. Exp. Med., 138 (1973) 1331.
34. Thorbecke, G.J. and Bell, M.K., J. Immunol., 111 (1973) 1043.
35. Thursh, D.R. and Emeson, E.E., J. Exp. Med., 135 (1972) 754.
36. Williams, G.M., Immunology, 11 (1966) 475.

LOCALIZATION OF IMMUNOGLOBULINS IN GERMINAL CENTERS OF HUMAN TONSILS

M. KOJIMA and R. TSUNODA

Fukushima Medical College
Fukushima (Japan)

Although localization of immunoglobulins in tissues and cells has been determined by the immunofluorecent technique, determination with this technique is limited in accuracy and previous studies have been made only at the light microscopic level.

With the recently established enzyme-labelled antibody technique and its application to electron microscopy, the present authors have observed the localization of immunoglobulins in tonsils, the intracellular production process of these immunoglobulins, the differentiation and cytokinetics of immunoglobulin-forming cells as well as the mechanisms of antibody production.

MATERIALS AND METHODS

Materials

Hypertrophied palatine tonsils were surgically removed from a consecutive series of 60 patients ranging from 4 yr to 14 yr.

Preparation of Tissues

Tonsils were fixed in the periodate-lysin-2% paraformaldehyde solution (4) at 4 C for 4 hr, and washed overnight in 0.05 M sodium phosphate buffer (PB), pH 7.4, containing 7% sucrose. For light microscopy, tissues were embedded in polyethylene glycol, according to the method of Mazurkiewicz (5). For electron microscopy,

other tissue specimens were put in 0.05 M PB containing 15% sucrose
for 4 hr and in 0.05 M PB containing 25% sucrose and 10% glycerol
for 2 hr then embedded in Ames O.C.T. and quickly frozen at 70 C.
Thin sections cut 4 μ in thickness for light microscopy and frozen
sections 6-8 μ in thickness for electron microscopy were placed to
dry on albumin-coated microscope glass slides.

Identification of Immunoglobulins

The slides were washed in cold PBS for 10 min with 3 changes.
Then, the sections were reacted with the specific labelled-antibody
for 2 hr in a moist chamber at room temperature. After washing 3
times with cold PBS for 30 min, they were fixed with 2.5% glutaraldehyde in PBS for 30 min. The sections were placed in Karnovsky's
solution (1) without hydrogen peroxide and then for 5 min in complete
Karnovsky's solution at room temperature. After washing with 3
changes of cold PBS, the tissue specimens were post-fixed in osmium
tetroxide, dehydrated in a graded series of ethanol up to 100%. A
gelatin capsule filled with Epon was inverted and the Epon was polymerized on the tissue specimens. By heating the blocks at 90 C, the
capsules were removed from the slides by hand. Two preparations were
used as the controls, 1 of which was obtained by immersion of the
sections in the Karnovsky's solution without Ag-Ab reaction, and the
other was obtained by reaction with normal rabbit labelled IgG.

Preparation of Peroxidase Labelled Antibody

The IgG fraction was purified from each anti-human immunoglob-
ulin (-γ, -α, and -μ) serum (Behringwerke) by precipitation in half-
saturated ammonium sulfate and chromatography on DEAE-Sephadex. The
peroxidase-aldehyde was formed from horseradish peroxidase Type VI
(Sigma) by the method of Nakane (6). This was added to the IgG to
conjugate, then NaBH4 was used to stabilize. After dialyzing in
PBS, the mixture was fractionated on Sephadex G-200 to remove unreacted peroxidase.

RESULTS

Light Microscopy

Immunoglobulin-containing cells were found in the epithelium
of crypts, subepithelial layer, perifollicular areas and germinal
centers of the palatine tonsils. Localization of immunoglobulins
was clearly demonstrated in the germinal centers showing a lacy
pattern. Ratio of whole IgG, IgA and IgM containing cells was
approximately 5:3:1.

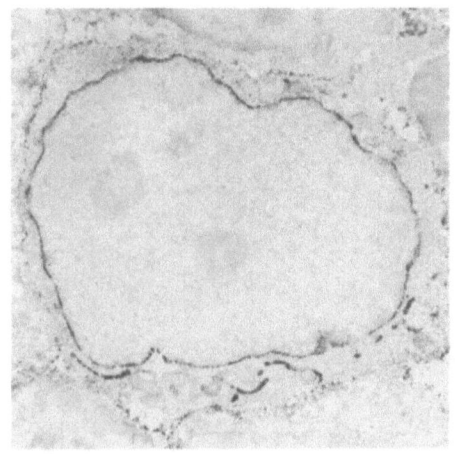

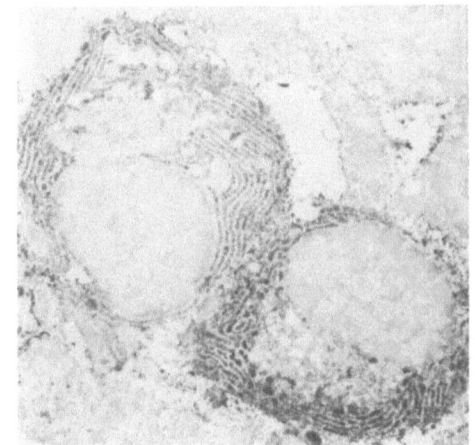

Fig. 1. An immature IgG containing cell outside the germinal center. IgG is present
in the perinuclear cistern (PC)
and a few short, thread-like
rough endoplasmic reticula
(rER). (4,000 x)

Fig. 2. Mature IgG containing
cells outside the germinal center.
IgG is contained in abundant lamellar well-developed rER. (3,500 x)

Figures 1-14 have been reduced 15% for reproduction.

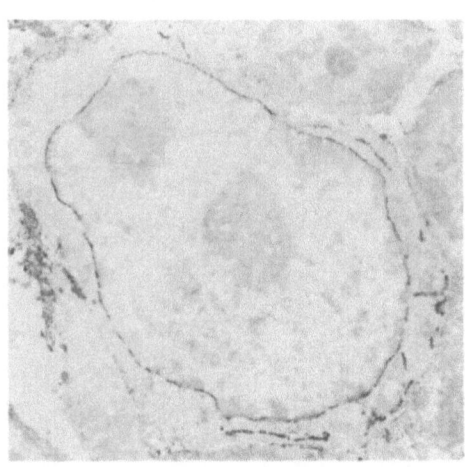

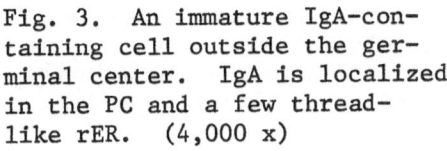

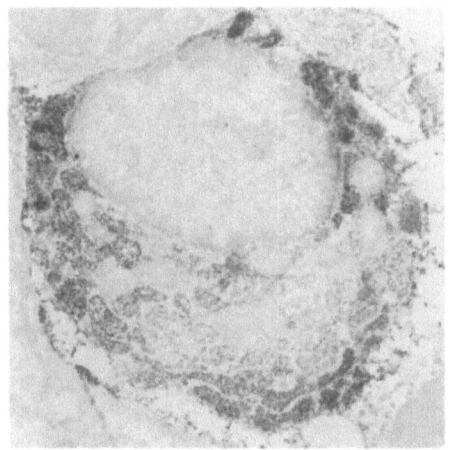

Fig. 3. An immature IgA-containing cell outside the germinal center. IgA is localized
in the PC and a few thread-
like rER. (4,000 x)

Fig. 4. A mature IgA-containing
cell outside the germinal center.
Distended rER is filled with IgA.
(6,300 x)

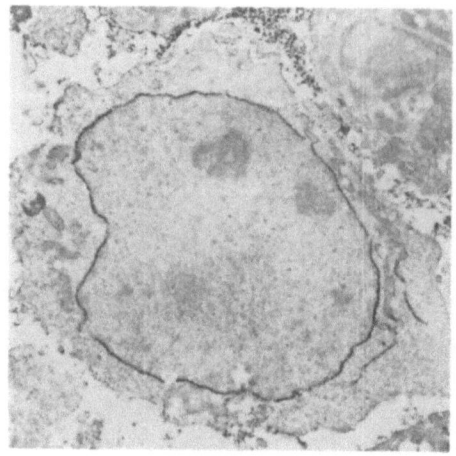

Fig. 5.   An immature IgM-containing cell outside the germinal center.   IgM is present
in the PC and a few thread-like
rER.   (4,000 x)

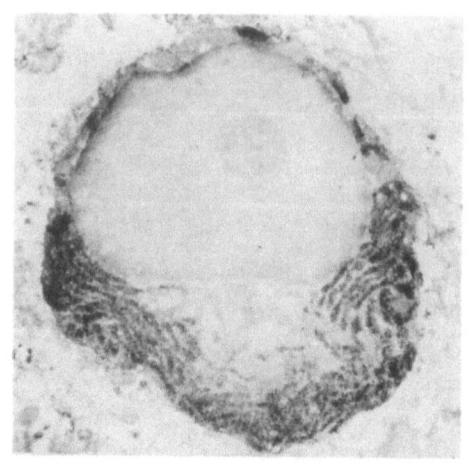

Fig. 6.   A mature IgM-containing
cell outside the germinal center.
The well-developed rER is filled
with IgM but IgM is absent in the
Golgi apparatus.   (6,300 x)

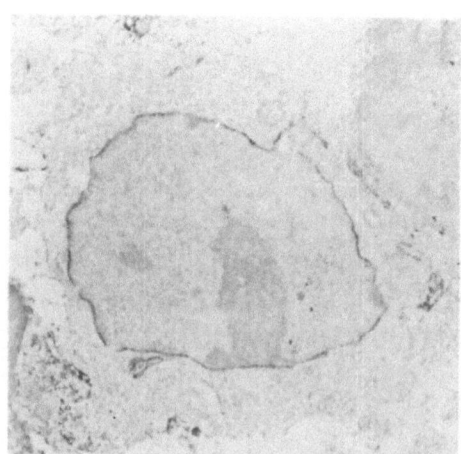

Fig. 7.   An immature IgM-containing cell in the germinal
center.   IgM is localized in
the PC and a few thread-like
rER.   (3,500 x)

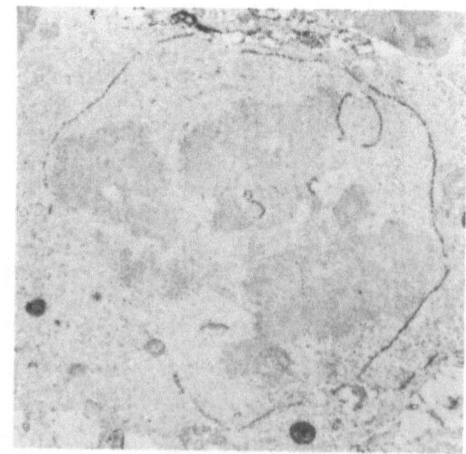

Fig. 8.   An IgM-containing cell
undergoing mitosis in the germinal
center.   IgM is contained in thread-
like rER.   (3,500 x)

The IgG-containing cells were chiefly distributed outside the germinal centers and were inclined to form colonies particularly in the subepithelial layer. In proportion to the increase in the number of these cells outside the germinal centers, IgG-containing cells tended to increase in number and the lacy pattern appeared more prominent within the germinal centers. IgA-containing cells were found more frequently in the epithelium of tonsillar crypts, but extremely rare in the germinal centers, unaccompanied by any lacy patterns. IgM-containing cells were seen scattering throughout the tonsillar tissues, and prominent localization of the lacy pattern was found in the germinal centers in all cases examined.

Electron Microscopy

Regardless of the classes of immunoglobulins, the immunoglobulin-containing cells revealed cell morphology of varying stages of maturation and there were no definite morphological differences among them. Outside the germinal centers, almost all the immunoglobulin-containing cells were mature in ultrastructure and immature cells were rare, while within the centers most of them were immature cells.

Ultrastructure of the Immunoglobulin-containing Cells Outside Germinal Centers and Intracytoplasmic Localization of the Immunoglobulins

Ultrastructure of the most mature cells showed cell characteristics identical with those previously described as plasmocytes. The immunoglobulins were found to be localized in the majority of abundantly developed rough endoplasmic reticulum but there were some plasmocytes in which the immunoglobulins were localized in a small part of the endoplasmic reticula. Perinuclear localization of the immunoglobulins was extremely rare. Immature cells containing the immunoglobulins were large and round, measuring 15 μ to 18 μ in diameter, and had a large round clear nucleus with slight indentations, containing 1 or 2 nucleoli. In the cytoplasm, the Golgi apparatus was indistinct, and short and thread-like rough endoplasmic reticulum was randomly distributed. Immunoglobulins were localized in both the perinuclear area and all the rough endoplasmic reticula. Besides the mature and immature immunoglobulin-containing cells mentioned above, there were intermediary cells of various stages of maturation, in which perinuclear localization of the immunoglobulins was partial or absent. The immunoglobulins were also observed in the rough endoplasmic reticulum of cells exhibiting mitosis.

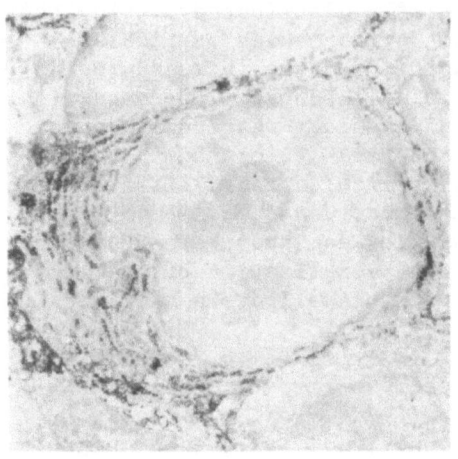

Fig. 9.   A mature IgM-containing cell in the germinal
center.   IgM is present in
the well-developed rER and
intracellular spaces.
(5,000 x)

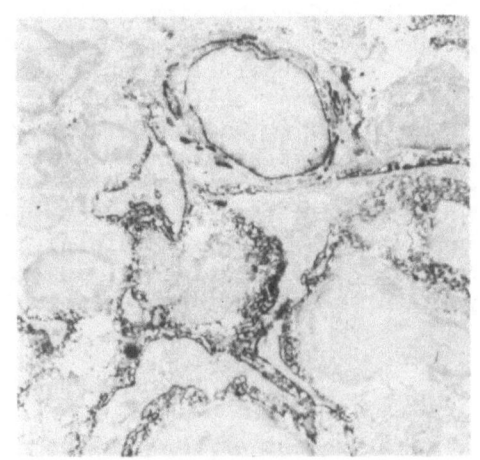

Fig. 10.   An intermediate-matured
IgM-containing cell in the germinal center, and the cytoplasmic
processes of dendritic reticulum
cells in which IgM is localized
on the cell membrane.   (1,800 x)

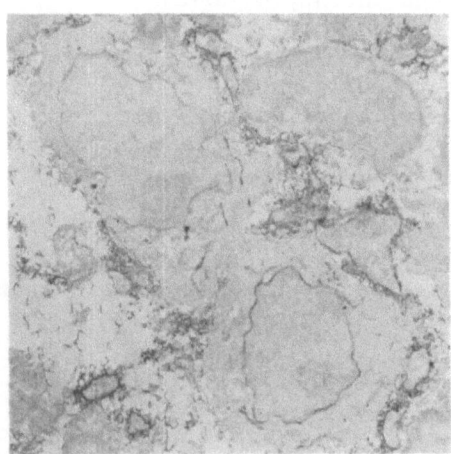

Fig. 11.   Immature IgA-containing cells in the germinal
center.   IgA is localized in
the PC and a few thread-like
rER and in the intercellular
spaces.   (2,000 x)

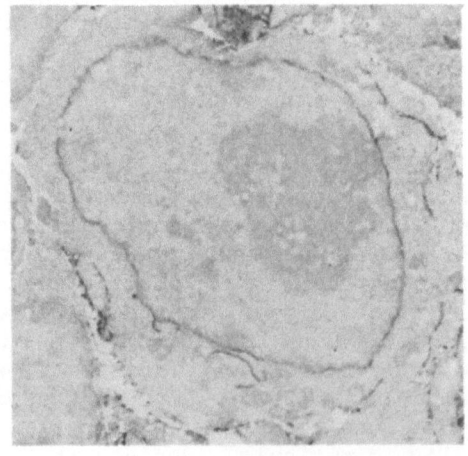

Fig. 12.   An immature IgG-containing cell in the germinal center.
IgG is present in the PC and a
few thread-like rER.   (3,500 x)

Ultrastructure of Immunoglobulin-containing Cells
Within the Germinal Centers and
Sites of Immunoglobulin Localization

Although a quantitative analysis with the aid of the electron
microscope was impossible, the presence of cells containing each
class of immunoglobulin was confirmed in the germinal centers, but
none of the germinal centers containing only 1 series of immunoglobulin-containing cells was observed. Among the mature immunoglobulin-containing cells, besides typical plasmocytes, there was
another type of cells having a markedly indented nucleus. Intracytoplasmic localization of the immunoglobulins was confirmed mostly in the rough endoplasmic reticulum but extremely rare in the
perinuclear area. Immature immunoglobulin-containing cells were
large, measuring 15 μ to 18 μ in diameter, which had an oval nucleus
with small or large indentations forming occasional intranuclear
pockets of the cytoplasm. The nucleus was on the whole clear and
usually contained 2 rod-shaped giant nucleoli. In the relatively
abundant cytoplasm, numerous mitochondria were distributed but the
Golgi apparatus was unremarkable. A small number of comma-like or
short thread-like rough endoplasmic reticula were present, some of
which showed unilateral vesicular dilatation, loop-like appearance
or ramification. The immunoglobulins were found to be localized
in the perinuclear regions and in all of the rough endoplasmic reticula. In medium-sized lymphocyte-like cells which contained
smaller nucleoli and in which the rough endoplasmic reticulum appeared more poorly developed than in the larger cells, localization
of the immunoglobulin was found with less frequency. Besides, varying numbers of immunoglobulin-containing cells possessing cytological
features intermediate between the mature and immature cells were seen
showing a variegated cell morphology. These cells, although their
nucleus was young, contained in their cytoplasm increased numbers
of thicker rough endoplasmic reticula which partly showed a lamellar
arrangement. Localization of immunoglobulins was mostly demonstrated
in the perinuclear cistern and rough endoplasmic reticulum, but was
sometimes absent or partial in the former. Moreover, the immunoglobulins were also observed in the rough endoplasmic reticulum of cells
in mitosis or phagocytized by macrophages.

Immunoglobulins in the Intercellular Spaces
of Germinal Centers

The present electron microscopic study has revealed that such
localization of immunoglobulin seen by light microscopy as the lacy
pattern coincides with that in the intercellular spaces among its
constituent cells of the germinal centers. Though localization of
IgM was confirmed in the spaces by light microscopy, the present
study clearly indicated that IgG and IgA were also consistently

intermixed in varying degrees. These immunoglobulins were disposed
to markedly deposit on the cell membrane of reticulum cells with
multidendritic cytoplasmic processes existing in the germinal centers and mantel zone. Generally speaking, we have found an inverse
relationship between the intensity of immunoglobulin localization
in the intercellular spaces and the number of the immature, immunoglobulin-containing cells within the centers.

DISCUSSION

The present ultrastructural study with the enzyme labelled
antibody technique has clearly shown that various kinds of cells
containing each class of immunoglobulins (IgG, IgA, IgM) within the
tonsillar tissue are almost consistent in cytological features with
each other and show varying degrees of maturation from immature to
mature cells in each series.

However, it is uncertain whether all these series of immunoglobulin-producing cells originated from a common precursor cell.
From the fact that the majority of the immunoglobulin-containing
cells were immature in the germinal centers despite that such immature cells were rare outside the centers, it appears that the
germinal centers are a major source of the immunoglobulin-producing
cells.

As for cytokinetics of such immature immunoglobulin-containing
cells in the germinal centers during their maturation process, it
may be certain that there are some of the immunoglobulin-containing
cells destined to die within the centers because they were occasionally phagocytized by macrophages. Since an ultrastructural transi-
tion between the immunoglobulin-containing germinal center cells
and mature immunoglobulin-containing cells was traced in the present
study, it seems that their maturing processes are present in the
germinal centers. Furthermore, the finding of the present study
that young immunoglobulin-containing cells were present in passage
ways formed by fine fibers and cytoplasmic processes of reticulum
cells into the mantel zone, suggests, together with the results of
previous experiments of tonsils with ^{3}H-thymidine (2), the possibility that the germinal center cells are able to migrate and to
move outside the germinal centers. In this connection, it is ques-
tionable whether all of the mature immunoglobulin-containing cells
found outside germinal centers originated in the germinal centers,
because a small number of immature immunoglobulin-containing cells
are found in tissues other than the centers, and because morphological transitions between the immature cells and mature immunoglobulin-containing cells were confirmed.

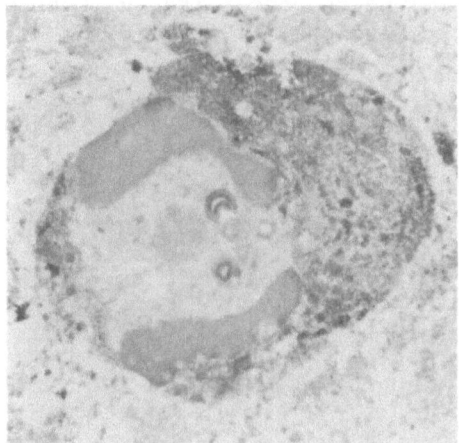 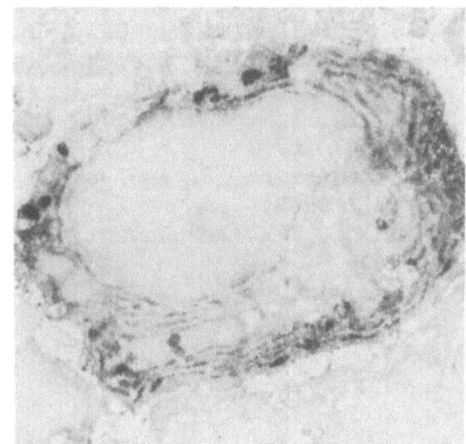

Fig. 13. A mature IgG-containing cell in the germinal
center. Well-developed rER
and some part of PC is filled with IgG. (6,300 x)

Fig. 14. A degenerated mature
IgG-containing cell is found to
be phagocytized by the tingible
body macrophage in the germinal
center. (8,000 x)

     The lacy intercellular deposition of immunoglobulin specific
for the germinal centers was confirmed in this study to preferably
occur on the cytoplasmic membrane of the reticulum cells. In addition to this finding, from the results of previous experimental
studies on antigen-trapping mechanism of the germinal centers (7)
and from the intimate relationship of development of the centers
with antigen-antibody complex (3), it may be strongly assumed that
the lacy pattern of immunoglobulins consists not only of single antibody molecules but also of antigen-antibody complexes.

     In summary, each series of immunoglobulin- (IgG, IgA, IgM)con-
taining cells were found in tissues of the human tonsils and some
differences of distribution among these cells were confirmed. Each
series of these cells revealed cell morphology of various stages of
maturation from immature cells. It seems that the germinal center
is a collection of the immature, immunoglobulin-containing cells
and is a major source of each kind of these cells in the tonsils.

REFERENCES

1. Graham, R.C. and Karnovsky, M.Y., J. Histochem. Cytochem.,
    14 (1966) 291.

2. Koburg, E., Germinal Centers in Immune Responses (Ed. H. Kottier) Springer Publishing Co., New York, New York, (1967) 176.
3. Laissue, J., Cottier, H., Hess, M.W. and Stoner, R.D., J. Immunol., 107 (1971) 822.
4. McLean, I.W. and Nakane, P.K., J. Histochem. Cytochem., 22 (1974) 1077.
5. Mczurkiewicz, J. and Nakane, P.K., J. Histochem, Cytochem. 20 (1972) 969.
6. Nakane, P.K. and Kawaoi, A., J. Histochem. Cytochem., 22 (1974) 1084.
7. Nossal, G.J.V., J. Exp. Med., 127 (1968) 277.

β_2-MICROGLOBULIN OF LYMPHOCYTES

T. P. CONWAY and M. D. POULIK

William Beaumont Hospital and Wayne State School
of Medicine, Detroit, Michigan (USA)

Recent developments in the chemistry and function of membranes have been of particular interest to immunologists because of the primary events of the immune response (antigen recognition) as well as other immune phenomenon e.g. complement binding, allergic reactions, phagocytosis, etc. occur at the cell surface. The largest constituent of most cell membranes is protein and undoubtedly the composition, conformation and organization of this protein determines membrane function. A wide variety of proteins of the lymphocyte membrane has been described, usually in functional e.g. histocompatible or immune response (Ia) antigens, Fc receptor, complement receptor, rather than in physicochemical terms. Probably the most thoroughly characterized of the membrane-associated proteins are the immunoglobulins (Ig) and β_2-microglobulin (β_2m).

As a reflection on the state of the art of membrane research, it is interesting to note that β_2m, a ubiquitous membrane protein, was initially isolated from human urine obtained from patients with renal tubular dysfunction (3). Human β_2m is a globular protein with molecular weight of 11,600 daltons and pKI of 5.4 to 5.6; it is devoid of carbohydrate and free sulfhydryl groups (3). The amino acid sequence of human β_2m as well as the partial sequence of rabbit, dog and rat β_2m indicate a single 100 amino acid peptide with a 57 amino acid disulfide loop (9,10,33). There is extensive sequence homology between human, dog and rabbit β_2m (8) and some immunologic cross reactivity between human, chimpanzee and mouse β_2m (21,37). The considerable homology between β_2m and the constant domains of IgG, in particular the 28% homology with the C_H3 domain, indicates a close structural and evolutionary relationship between β_2m and Ig (26,38). Despite extensive sequence homology between human β_2m and the Ig constant domains there is little if any antigenic cross reactivity (3). The similarity in structure between β_2m and the effector domains of Ig apparently does extend to complement fixation. Aggregated β_2m can fix complement (17) and free β_2m is more effective than C_H3 and Fab fragments in binding complement (25).

β_2m is found in normal serum (1-2 mg/L), cerebral spinal fluid (1.7 mg/L) and urine (0.11 mg/24 hr vol) and in greater amounts in colostrum, saliva, amniotic fluid, fetal and cord blood (3,12,18). In pathologic conditions β_2m levels vary, with high concentrations of circulating β_2m in serum of patients with renal disease and lym-phoproliferative disorders (4,18). Typical of low molecular weight proteins, β_2m is rapidly filtered through the glomerulus and ca-tabolized in proximal convoluted tubules (5).

Fluorochrome or radio-labeled anti-β_2m indicate β_2m to be on the membranes of lymphocytes (11,28,32) as well as fibroblasts, (23,24,30). With only 1 major exception (Daudi, an IgM-producing Burkitt lymphoma cell line) all nucleated cells appear to have membrane bound β_2m (24). Current estimates of the number of β_2m molecules on the lymphocyte surface range from 3×10^5 to 6×10^7 (8,11,24). In addition to the membrane associated molecules, cells also appear to secrete β_2m into the medium. This secreted β_2m appears identical by radioimmunoassay, immunoelectrophoresis and gel chromatography with β_2m isolated from urine (23,30). The amount secreted varies, with nonmalignant lymphocytes synthesizing 20 to 65 ng/ml, 5×10^5 cells, 65 hr (24), and some lymphoblastoid and carcinoma cell lines secreting 200 to 2850 ng/ml 5×10^5 cell lines, 65 hr (24). There is no apparent correlation between amount of Ig and the amount of β_2m secreted into the medium (16,24). β_2m could be released from cells by direct secretion into the medium (exo-cytosis) and/or membrane shedding. Membrane shedding is probably not the sole mechanism since comparison of the amount of protein on the cell surface and released into the medium for some carcinoma cell lines would require complete regeneration of the membrane every 24 min (24). Actually β_2m rather than membrane is probably shed from the cell surface. Thoracic duct lymphocyte membrane proteins radio-labeled with ^{125}I by the lactoperoxidase method re-lease β_2m into the medium with membrane bound T 1/2 of 4-5 hr (6).

Even more interesting than the finding of β_2m on the lympho-cyte membrane was the discovery of the similarity (amino acid com-position, pKI, antigenic identity) of β_2m and the common antigenic determinant of the HL-A antigens (22,27).

The antigenic specificities (SD antigens) of the human (HL-A) and mouse (H-2) histocompatability locus are expressed on the cell surface in a complex composed of 2 types of polypeptide chains (7,

34). The large subunit (MW 44,000) contains carbohydrate, is partially imbedded in the membrane and carries the alloantigenic determinants. The small subunit (MW 11,500) is noncovalently bound to the large subunit, is probably not under genetic control of the histocompatibility locus (14) and has been identified as β_2m (15, 22,27,29).

Probably the most convincing demonstration of the in situ association of HL-A antigens and β_2m was the co-capping experiments of Poulik et al. (29) and Solheim and Thorsby (39). Capping is a manifestation of in plane mobility of membrane proteins and is initiated by the reaction of multivalent antibody with cell surface antigen to form a patch on the membrane. This is followed by the energy dependent migration of these patches to a cap at one of the cell's poles, internalization of the aggregated protein, and within 3 to 8 hr regeneration of the membrane protein (36). In these experiments (29), lymphocytes were treated with cyclohexamide to prevent regeneration. In the first phase the cells were reacted with rabbit anti-β_2m and incubated at 37 C for 2 hr, the caps were then detected by the addition of rhodamine conjugated goat anti-rabbit IgG. More than 50% of the cells exhibited capping. In the co-capping experiment the cells were incubated with human anti-HL-A antisera and the caps visualized with fluorescein labeled anti-human IgG and the rhodamine anti-β_2m system, both fluorochromes overlapped i.e. co-capped. As a control the absence of co-capping of membrane Ig and HL-A antigens was demonstrated. The close association between β_2m and HL-A peptide was also demonstrated by surface labeling thoracic duct lymphocytes with ^{125}I followed by immune precipitation of the NP-40 solubilized membrane proteins (15). Examination of the immunoprecipitates with electrophoresis on SDS polyacrylamide gels revealed that anti-β_2m precipitates 2 radio-labeled peptides. A 44,000 MW peptide with HL-A antigenic activity and a 11,000 MW peptide presumably β_2m. The estimated ratio of chains was 1 to 1. The association between HL-A and β_2m appears to be noncovalent as β_2m can be separated from HL-A by low pH without reduction and alkylation (27). Thus β_2m appears to be a peripheral membrane protein rather than an integral protein (36).

The converse question i.e. are β_2m and HL-A peptides always associated on the cell membrane, also warrants inspection. One of the cell lines which does not synthesize β_2m (Daudi) has been reported to snythesize HL-A antigens (24); however, this report has not been verified (35). Although HL-A peptides have been isolated which do not contain β_2m, the isolation conditions may be such as to induce dissociation between these noncovalently associated peptides. Thus, most of the evidence indicates that β_2m and HL-A are associated in the membrane and they form a 1 to 1 molar complex (7,15,31). Given that β_2m and HL-A antigens are associated, little is known as to the function of this structural relationship.

β_2m may be required for the surface expression of HL-A antigens (6) or it may alter the availability of some HL-A antigenic determinants (40). Current best estimates of the relative amounts of β_2m and HL-A peptide molecules in the membrane indicate an excess of β_2m over HL-A e.g. 6 to 1 as reported by Nilsson et al. (24). With the exception of the TL antigen of mouse spleen cells (41) it is not known with which other membrane proteins β_2m is associated. An intriguing possibility is that β_2m is associated with membrane proteins which function in the immune response. Anti-β_2m is capable of inhibiting the stimulation of sensitized cells by PPD and PHA (1,24) and stimulating blast formation in B lymphocytes (24). Anti-β_2m blocks responder lymphocyte reactivity against allogenic cells in the mixed lymphocyte reaction (1,20). Anti-β_2m probably does not kill the responder cell as no complement is present and Fab anti-β_2m blocks the reaction. While anti-β_2m may inhibit cell-cell contact either directly or through redistribution of cell surface proteins it may also be reacting with an antigen receptor-β_2m complex.

Anti-β_2m does not inhibit the effector cell in cell mediated lympholysis (19), or T lymphocyte rosette formation with sheep red blood cells (20).

In conclusion, structural evidence indicates that β_2m is related to the precursor or ancestral gene from which Ig may have evolved. The facts that β_2m is associated with the major histocompatibility antigens, and that the immune response genes also appear to be closely associated with the histocompatible region on the genome (2) provide interesting basis for speculation on the function of β_2m. Gally and Edelman (13) postulated, on the basis of sequence data and the genetics of the immunoglobulins, the existence of 4 loci coding for Kappa, Lambda, heavy and immune response genes. Of these gene products the membrane structures responsible for antigenic recognition and stimulation or suppression of the immune response are most poorly understood. One of the requirements for the antigenic recognition subunit would be that it approach the same range of specificity to antigens as exhibited by the Ig. Since the specificity of the Ig is reflected by its amino acid sequence in the variable regions, it would be expected that the Ir gene products also exhibit similar variation. Thus far, this degree of variation has not been observed for β_2m, diminishing its role in antigen recognition. However, the possibility still exists that it may be associated with the recognition unit and it may function in stabilization of the binding site.

ACKNOWLEDGEMENTS

This work was supported in part by grants from USPHS Grant

No. AI-11335, the Children's Leukemia Foundation of Michigan and
the William Beaumont Hospital Research Institute.

REFERENCES

1. Bach, M.L., Huang, S.W., Hong, R. and Poulik, M.D., Science,
   182 (1973) 1350.
2. Benaceraff, B. and McDevitt, H., Science, 175 (1972) 273.
3. Berggård, I. and Bearn, A.G., J. Biol. Chem., 243 (1968) 4095.
4. Bernier, G.M. and Conrad, M.E., Am. J. Physiol., 217 (1969)
   1359.
5. Bernier, G.M. and Conrad, M.E., Nature, 218 (1968) 598.
6. Cresswell, P., Springer, T., Strominger, J.L., Turner, M.J.,
   Grey, H.M. and Kubo, R.T., Proc. N.A.S.,71 (1974) 2123.
7. Cresswell, P., Turner, M.J. and Strominger, J.L., Proc. N.A.S.,
   70 (1973) 1803.
8. Cunningham, B.A. and Berggård, I., Transplant, Rev., 21 (1974)
   3.
9. Cunningham, B.A. and Berggård, I., Science, 187 (1975) 1079.
10. Cunningham, B.A., Wang, J.L., Berggård, I. and Peterson, P.,
    Biochemistry, 12 (1973) 4811.
11. Evrin, R.E. and Pertoft, H., J. Immunol., 111 (1973) 1147.
12. Evrin, P.E., Peterson, P.A., Wide, L. and Berggård, I., Scand.
    J. Clin. Lab. Invest., 28 (1971) 439.
13. Gally, J.A. and Edelman, G.M., Ann. Rev. Genetics,6 (1972) 1.
14. Good fellow, P.N., Jones, E.A., van Heyningen, V., Solomon, E.,
    Bobrow, M., Miggiano, V. and Bodmer, W.E., Nature, 254 (1975)
    267.
15. Grey, H.M., Kubo, R.T., Colon, S.M., Poulik, M.D., Cresswell,
    P., Springer, T., Turner, M. and Strominger, J.L., J. Exp. Med.,
    138 (1973) 1608.
16. Hütteroth, T.H., Cleve, H., Litwin, S.D. and Poulik, M.D., J.
    Exp. Med., 137 (1973) 838.
17. Isenman, D.E., Painter, R.H. and Dorington, K.J., Proc. N.A.S.,
    72 (1975) 548.
18. Kithier, K., Gejka, J., Belamaric, J., Al-Sarraf, M., Peterson,
    W.D., Jr., Vaitkevicius, V.K. and Poulik, M.D., Clin. Chem.
    Acta, 52 (1974) 293.
19. Lightbody, J.J., Urbani, L. and Poulik, M.D., Nature, 250 (1974)
    227.
20. Lindblom, B., Östberg, L. and Peterson, P.A., Tissue Antigens,
    4 (1974) 186.
21. Moller, E. and Persson, U., Scan. J. Immunol., 3 (1974) 5.
22. Nakamuro, K., Tanigaki, N. and Pressman, D., Proc. N.A.S.,70
    (1973) 2863.
23. Nilsson, K., Evrin, P.E., Berggard, I. and Ponten, J., Nature,
    244 (1973) 44.
24. Nilsson, K., Evrin, P.E. and Welsh, K.I., Transplant. Rev.,

    21   (1974) 53.
25. Painter, R.H., Yasmeen, D., Assimeh, S.N. and Poulik, M.D.,
    Immunol. Commun., 3 (1974) 19.
26. Peterson, P.A., Cunningham, B.A., Berggård, I. and Edelman, G.
    M., Proc. N.A.S., 69 (1972) 1967.
27. Peterson, P.A., Rask, L. and Lindblom, J.B., Proc. N.A.S., 71
    (1974) 35.
28. Poulik, M.D. and Motwani, N., Clin. Res., 70 (1972) 795.
29. Poulik, M.D., Bernoco, M., Bernoco, D. and Ceppellini, R.,
    Science, 182 (1973) 1352.
30. Poulik, M.D. and Bloom, A.D., J. Immunol., 110 (1973) 1430.
31. Poulik, M.D., Ferrone, S., Pellegrino, M.A., Sevier, D.E., Oh,
    S.K. and Reisfeld, R.A., Transplant. Rev., 21 (1974) 106.
32. Poulik, M.D. and Motwani, N., Clin, Res., 70 (1972) 795.
33. Poulik, M.D., Shinnick, E.S., Smithies, O., Fed. Proc., 34
    (1975) 945.
34. Rask, L., Lindbloom, J.B. and Peterson, P.A., Nature, 249 (1974)
    833.
35. Reisfeld, R.A., Sevier, E.D., Pellegrino, M.A., Ferrone, S. and
    Poulik, M.D., Immunogenetics, 2 (1974) 183.
36. Singer, S.J., Adv. Immunol., 19 (1974) 1.
37. Smithies, O. and Poulik, M.D., Proc. N.A.S., 69 (1972) 2914.
38. Smithies, O. and Poulik, M.D., Science, 175 (1972) 187.
39. Solheim, B.G. and Thorsby, E., Tissue Antigens, 4 (1974) 83.
40. Tanigaki, N. and Pressman, D., Transplant, Rev., 21 (1974) 15.
41. Vitetta, E.S., Uhr, J.W. and Boyse, E.A., J. Immunol., 114 (1975)
    252.

EFFECT OF RESERPINE ON SERUM HEMOLYSIN RESPONSE IN MICE

E. G. BLIZNAKOV

New England Institute
Ridgefield, Connecticut (USA)

Descriptions of the use of extracts from rauwolfia plants may be traced back to ancient Hindu ayurvedic writings. These extracts were used in "primitive" Hindu medicine for a variety of central nervous derangements, both psychic and motor, and for treatment of a truly astonishing list of other ailments including hypertension, dysentery, diarrhea, cholera, insect bites and fevers with a variety of etiologies (18).

Today, the plant has been stripped of some of its legendary power but it has lost none of its interest. In modern clinical medicine reserpine is used as a hypotensive, tranquilizing and sedative agent and is an ingredient in numerous pharmaceutical formulations taken by an estimated 4 million Americans.

The clinical use of rauwolfia alkaloids including reserpine is under review by various U. S. Governmental Health and Regulatory Agencies as a result of published clinical reports indicating a possible association between prolonged use of reserpine for lowering blood pressure in hypertensive patients and an increased risk of breast cancer in women over the age of 50 (2,11,25). Retrospective studies also indicated further association of reserpine with malignancies of the brain, uterus, pancreas, skin and kidney. Thus, the centuries-old argumentation on the pharmacological effects of reserpine is entering a new phase.

The present study is part of our attempt to evaluate the effect of reserpine, the most commonly used representative of the group of rauwolfia alkaloids, on various parameters of the reticuloendothelial system.

MATERIALS AND METHODS

Animals

Male CF-1 mice weighing 20 g were obtained from a single commercial breeder (Carworth Farms, Inc., New York, New York). At least
20 mice in each experimental group were housed in an air-conditioned
(22 ± 1 C) room with uniform humidity. For some experiments requiring elevated ambient temperature, mice were kept in an air-conditioned walk-in Environmental Room (Model 1270 H, Hotpack Corp.,
Philadelphia, Pennsylvania) 96 hr before beginning the experiments
at uniform humidity and a temperature of 33 ± 0.5 C. They were allowed to rest for 1 week before experimental use. Food and water
were freely available and consumed despite the state of sedation and
reduction of spontaneous activity resulting from the reserpine administration. For the acute toxicity studies, the mortality was
recorded daily for 28 days.

Antigen

Fresh, sterile sheep red blood cells (SRBC) were centrifuged
and washed 3 times with sterile nonpyrogenic 0.9% sodium chloride
solution (saline). Primary immunization of each mouse at day 0 was
given intravenously using the tail vein with a standard saline suspension of $3 - 5 \times 10^7$ SRBC/0.2 ml.

Hemolytic Antibody

At suitable intervals after the administration of the SRBC,
blood was collected from each mouse with a capillary tube by a
retroorbital venous plexus puncture. Equal volumes of blood from
all animals in a group were pooled and the serum was separated and
stored at -40 C until hemolysin titers could be determined. Titration of all serum samples was carried out in a single day using
the 50% end point method (15). The best fitting regression line
between probit percent hemolysis and the log of the serum dilution
(8-10 serum dilutions) was determined by computer analysis. The
experimental points shown on the figures represent the determined
values, with standard deviations indicated.

Drugs

The following drugs were used: reserpine (Sandril, Eli Lilly
and Co., Indianapolis, Indiana); pentazocine lactate (Talwin, Winthrop Laboratories, New York, New York); imipramine hydrochloride

(Tofranil, Geigy Pharmaceuticals, Ardsley, New York); methysergide
(courtesy of Sandoz Pharmaceuticals, Hanover, New Jersey) and DL-
p-chlorophenylalanine, methyl ester, hydrochloride (Sigma chemical
Co., St. Louis, Missouri).

Reserpine was further diluted, using a formulation kindly
disclosed by the Lilly Research Laboratories (Indianapolis, Indiana) containing polyethylene glycol 300, citric acid, ascorbic acid,
monothioglycerol and benzyl alcohol. The dose required was administered intramuscularly in a volume of 0.1 ml per mouse. All other
drugs were diluted further in saline and 0.2 ml was injected intraperitoneally.

The time schedule of drug injections and dose levels for each
mouse in the study on modification of reserpine antibody depression
were as follows: reserpine was administered 24 hr before SRBC at a
dose of 40 μg/mouse (2 mg/kg); pentazocine and imipramine were given
30 min before and 4 and 24 hr after reserpine at dosages of 1.7 mg/
mouse (85 mg/kg) and 200 ug/mouse (19 mg/kg), respectively; methysergide at 50 min before and 10 min and 24 hr after reserpine at a
dose of 100 μg/mouse (5 mg/kg) and p-chlorophenylalanine at 4, 3 and
2 days before reserpine at a dose of 3 mg/mouse (150 mg/kg). The
animals in the control groups received the same volume of the diluent
alone.

General

All glassware was heated for 5 hr at 170 C. Nonpyrogenic sterile saline, distilled water (Travenol Laboratories, Deefield, Illnois)
as well as syringes, needles and pipets were used throughout.

RESULTS

Intramuscular administration of reserpine 1 day before and
4 days after injection of the antigen (SRBC) resulted in a marked
and prolonged depression of the hemolytic primary humoral immune
response. It was accompanied by a delayed appearance of the antibody titer peak. The dose-response relationship of this depression
is presented in Fig. 1.

The suppression of the hemolytic antibody titer as a function
of the time of reserpine administration is presented in Fig. 2.
The results in this group of experiments were obtained by the intramuscular administration of a single dose of 40 μg/mouse (2 mg/
kg) of reserpine.

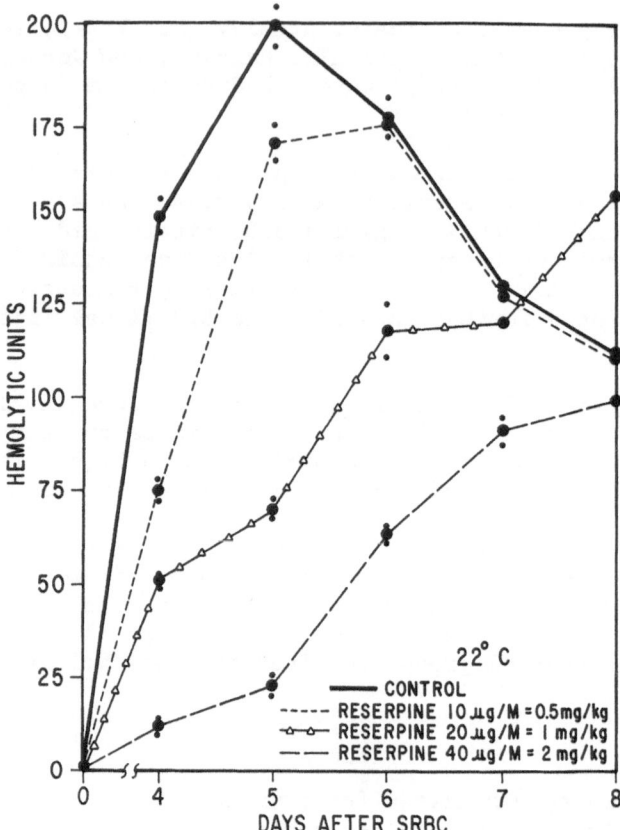

Fig. 1. Depression of the hemolytic primary humoral immune response and postponement of the antibody titer peak in male CF-1 mice after intramuscular administration of reserpine 1 day before and 4 days after injection of antigen (SRBC).

The animals in the 2 groups of experiments described above were maintained at 22 C. However, when the animals were maintained at the elevated ambient temperature (33 C) for the duration of the experiment, the profound hemolytic antibody depression resulting from the administration of reserpine was prevented. These results are shown in Fig. 3, and were obtained in parallel to those shown in Fig. 2, except for the difference in temperature.

Results in regard to our attempt to modify the antibody depression using 4 drugs known to affect some of the pharmacological manifestations of reserpine are presented in Fig. 4 demonstrating that the depression produced by reserpine (40 µg/mouse, 2 mg/kg) could not be modified by the administration of pentazocine, imipramine, methysergide and p-chlorophenylalanine.

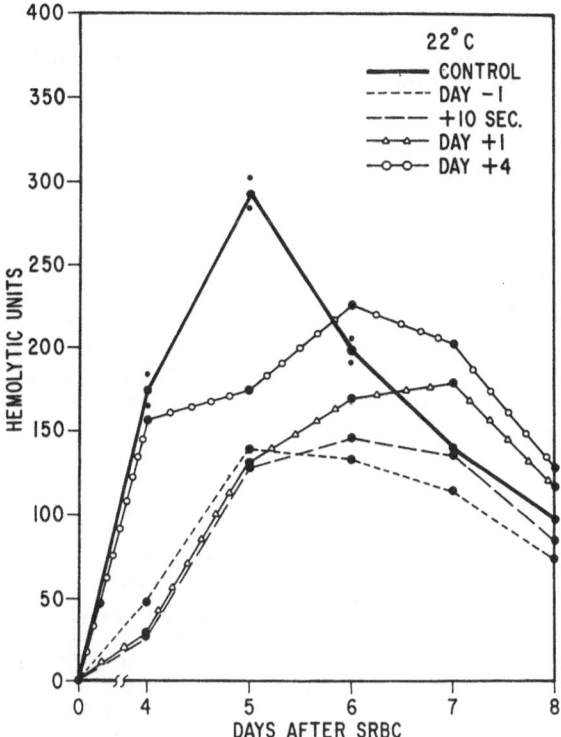

Fig. 2. Hemolytic antibody titer in male CF-1 mice maintained at
22 C as a function of the time of reserpine administration as a
single intramuscular dose of 40 µg/mouse=2 mg/kg. (See Fig. 3
for comparison.)

Fig. 5 and Fig. 6 illustrate the modifications of the acute
reserpine toxicity at elevated ambient temperature (33 C). Two
different single doses of reserpine were administered intramuscularly: 320 µg/mouse (16 mg/kg) and 640 µg/mouse (32 mg/kg) as
shown in Fig. 5 and Fig. 6, respectively. The mortality rate at
both dose levels of reserpine was first accelerated but subsequently decelerated, resulting in lower mortality at the end of the observation period (28 days).

DISCUSSION

It is well documented that reserpine interferes with the stor-

age of serotonin (5-hydroxytryptamine) (21,24) and of catecholamines
(3,4). Reserpine probably produces its antihypertensive effects
through depletion of tissue stores óf catecholamines (epinephrine
and norepinephrine) from peripheral sites (13). Its sedative and
tranquilizing properties are believed to be dependent on depletion
of serotonin from the brain.

Few published studies have been focused on the effect of reserpine on the immune response in general. Draškoci and Janković
(6) speculated that this effect might be mediated through depletion
of serotonin. In contrast, Devoino, et al. (5) in a more recent
study based on the widely accepted hypothesis originated by Shore,
et al. (23) attributed the effect of reserpine not to depletion of
serotonin but rather to an increased level of free serotonin resulting from impaired binding.

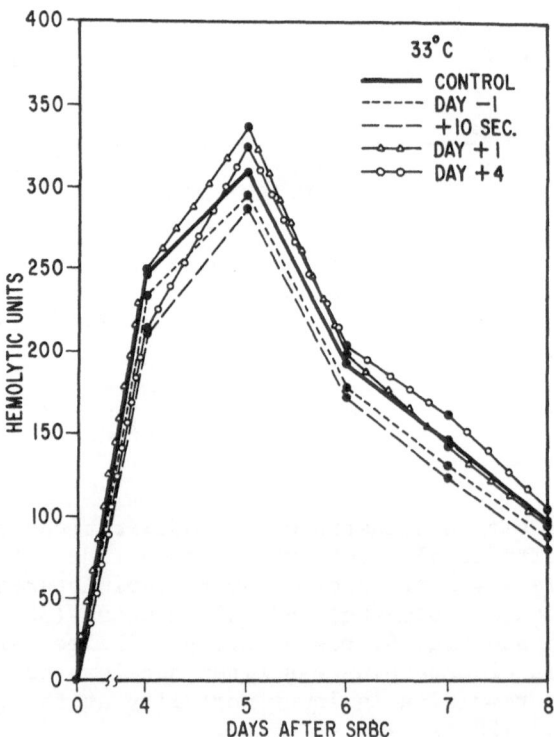

Fig. 3. Hemolytic antibody titer in male CF-1 mice maintained at
33 C as a function of the time of reserpine administration as a
single intramuscular dose of 40 μg/mouse=2 mg/kg. (See Fig. 2 for
comparison.)

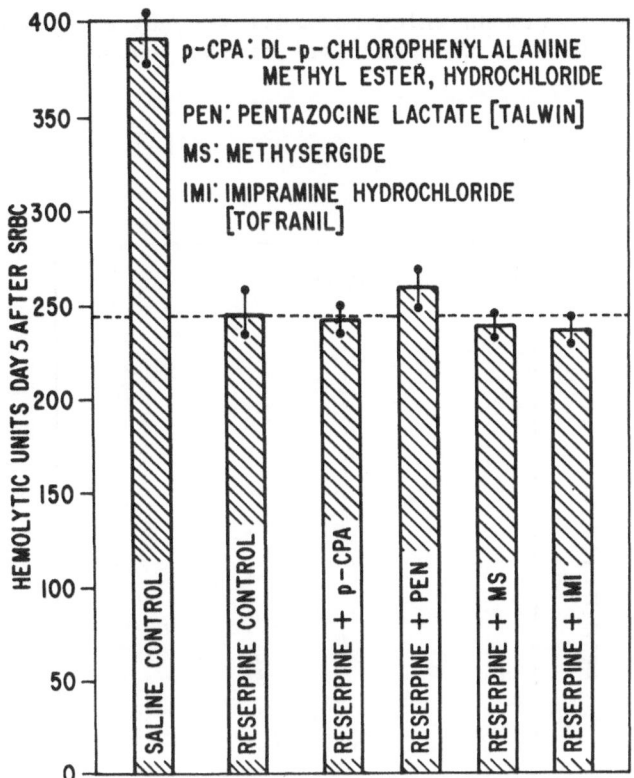

Fig. 4. Failure to modify the suppressive effect of reserpine on the hemolytic antibody titer in male CF-1 mice by the use of 4 drugs.

It has been reported that methysergide, a serotonin antagonist, completely blocks the clinical effects of reserpine or of serotonin (8). In contrast, p-chlorophenylalanine is a potent specific inhibitor of brain serotonin biosynthesis, presumably by inhibiting tryptophan hydroxylase (16).

The results of our study (Fig. 4) indicated that the administration of either methysergide or p-chlorophenylalanine did not potentiate or abrogate the depression of the antibody response produced by reserpine, thus not supporting the possibility of a major involvement of serotonin in the depression of antibody response by reserpine.

Reserpine produces an inhibition of the hypothalamus-pituitary-

adrenocortical functions which is manifested by a profound hypothermia, impairment of the functions of the thymus and adrenal hypertrophy (10,12).

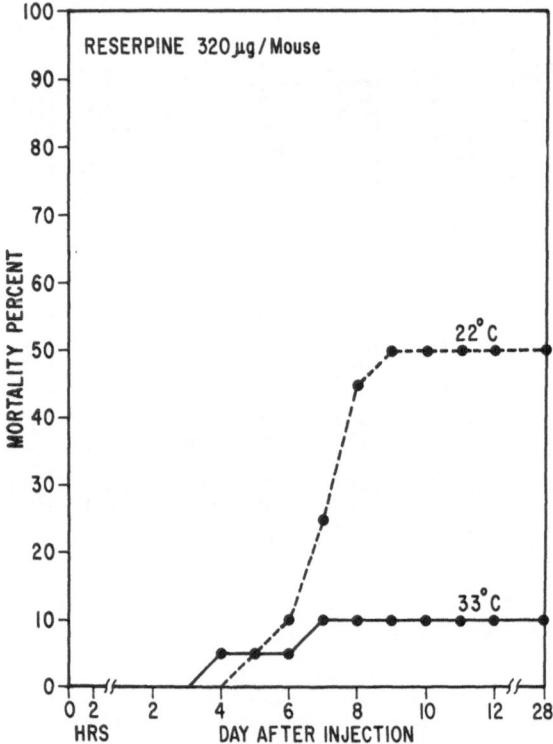

Fig. 5. Mortality rate of male CF-1 mice maintained at 22 C and
33 C following a single intramuscular reserpine dose of 320 μg/mouse=
16 mg/kg. (See Fig. 6 for comparison.)

        Fondy, et al. (9) reported an antileukemic effect of reserpine
in the L 1210 leukemia model. They concluded that hypothermia alone
was responsible for this effect.

        It would be an oversimplification to ascribe the suppression
of the antibody response as resulting directly from the hypothermic
effect of reserpine. Admittedly, we demonstrated (Fig. 3) that this

suppression could be abolished if the animals were maintained at 33 C instead of 22 C. In this connection, Plummer, et al. (22) showed that elevated ambient temperature compensated the hypothermic effect of reserpine.

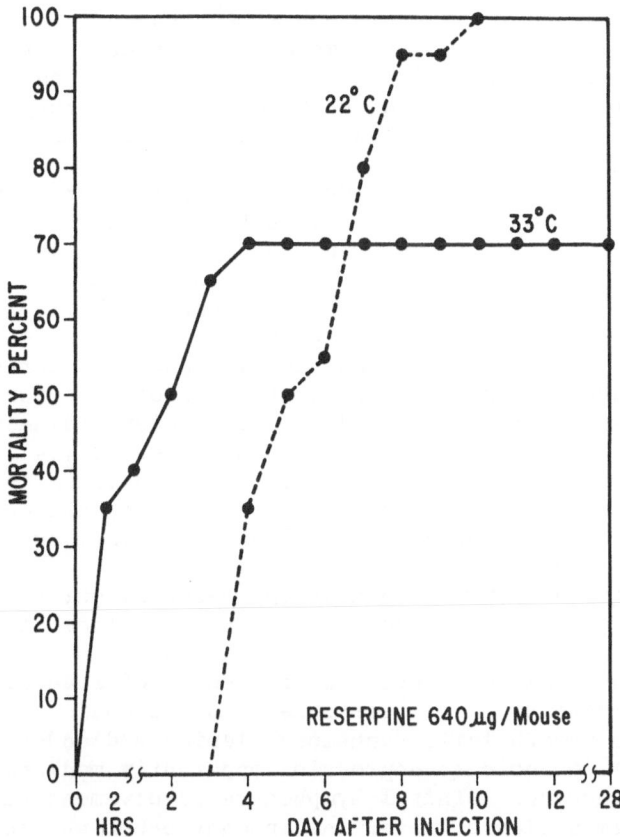

Fig. 6. Mortality rate of male CF-1 mice maintained at 22 C and 33 C following a single intramuscular reserpine dose of 640 μg/mouse-32 mg/kg. (See Fig. 5 for comparison.)

        In a detailed study in mice, Ahtee, et al. (1) showed that imipramine prevented reserpine-induced hypothermia while pentazocine potentiated it. The results of our study demonstrated (Fig. 4) that using this method for temperature response manipulation, administration of imipramine or pentazocine failed to potentiate or abrogate the depression of the antibody response caused by reserpine.

The most logical explanation for the suppression of the immune response to SRBC, a thymus dependent antigen (7,17,19,20), by reserpine is based on the functional impairment of the thymic T-lymphocyte pool which probably results from inhibition of some hypothalamic activities, directly or indirectly, via adrenal stimulation.

Altered ambient temperature has been shown to have pronounced effects on the distribution and metabolism of drugs, the duration and the intensity of the biological response elicited, and ultimately on their toxicity. It is now accepted that laboratory animals including mice in confinement adapt to heat or cold metabolically (26).

Johnson, et al. (14) studied the effect of drugs on the metabolic rate in animals and showed that the metabolic rate response to the administration of reserpine is biphasic, an initial increase and subsequent profound decrease. When the animals were treated with reserpine, but maintained at 30 C, the metabolic rate was increased. These findings offer probably the best logical interpretation at the present time of our results showing depression of the immune response by reserpine and prevention of this depression when the animals were maintained at 33 C. This elevated ambient temperature overcompensated, as shown by Johnson, et al. (14), the decreased metabolic rate produced by reserpine. The resulting increased metabolic rate accelerated the metabolic degradation and detoxification of reserpine and prevented the development of an immune depression.

This possibility was further suggested by our results (Fig. 5 and Fig. 6) showing a reduced mortality rate from high doses of reserpine when the animals were maintained at 33 C. Our study supports the non-specific occurrence of a chain of reactions resulting from the administration of reserpine. Such a chain of events commences with a hypothalamic functional lesion and ends with a profound humoral immunologic depression apparently mediated via the thymic T-lymphocytes. This T-lymphocyte involvement facilitates further incrimination of reserpine in many other physiological and pathological processes depending on intact T-cell function.

In this respect, we would like to propose that reserpine should be considered as a member of the immunosuppressive drugs with an effect probably mediated via the hypothalamus.

ACKNOWLEDGEMENTS

This work was derived from research programs supported in part by grants from the following institutions: Greenaway, Griffis, Heddens-Good, Landegger, Virginia and D. K. Ludwig, J. M. McDonald,

Roy R. and Marie S. Neuberger, Fannie E. Rippel, Rockledge, Scheider, Arnold Schwartz Fund for Health and Education Research, United Order of True Sisters, Gilbert Verney, Wahlstrom, Wallace Genetic, Raymond J. Wean and Whitehall.

The assistance and advice of Dr. S. J. Tao for the computer program and statistical analysis is gratefully acknowledged. We thank C. Bellus, L. Easton, N. Hastings, G. Katopodis, A. Santini and C. Torcellini for technical assistance.

REFERENCES

1. Ahtee, L., Jounela, A.J., Saarnivaara, L. and Simola, I., Pharmacology, 12 (1974) 39.
2. Armstrong, B., Stevens, N. and Doll, R., Lancet, ii (1974) 672.
3. Bertler, A., Carlsson, A. and Rosengren, E., Naturwissenschaften, 43 (1956) 521.
4. Carlsson, A., Resengren, E., Bertler, E. and Nilsson, J., Psychotropic Drugs, Vol. 6 (Ed. S. Garattini and V. Ghetti) Elsevier Publishing Co, Amsterdam (1957) 363.
5. Devoino, L.V. and Yeliseyeva, L.S., Europ. J. Pharmacol. 14 (1971) 71.
6. Draškoci, M. and Janković, B.D., Nature, 202 (1964) 408.
7. Feldman, M. and Basten, A., J. Exp. Med., 136 (1972) 49.
8. Fischer, E. and Heller, B., Nature, 216 (1967) 1221.
9. Fondy, T.P., Karker, K.L., Calcagnino, C. and Emlich, C.A., Cancer Chemother. Repts., 58 (1974) 317.
10. Gaunt, R., Renzi, A.A., Antonchak, N., Miller, G.J. and Gilman, M., Ann. N. Y. Acad. Sci., 59 (1954) 22.
11. Heinonen, O.P., Shapiro, S., Tuominen, L. and Turunen, M.I., Lancet, ii (1975) 675.
12. Hodges, J.R. and Vellucci, S.V., Brit. J. Pharmac., 53 (1975) 555.
13. Jarwik, M.E., Rauwolfia Alkaloids in the Pharmacological Basis of Therapeutics (Ed. L.S. Goodman and A. Gilman) Macmillan Co., New York, New York (1965) 178.
14. Johnson, G.E., Sellers, E.A. and Schoenbaum, E., Fed. Proc., 22 (1963) 745.
15. Kabat, E.A. and Mayer, M.M., Experimental Immunochemistry, 2nd Edition, Charles C. Thomas Publishing Co., Springfield, Illinois (1961) 1.
16. Koe, B.K. and Weissman, A., J. Pharm. Exp. Ther., 154 (1966) 499.
17. Miller, J.F.A.P. and Mitchell, G.F., J. Exp. Med., 128 (1968) 801.
18. Miner, R.W., Ann. N. Y. Acad. Sci., 59 (1954) 1.
19. Mitchell, G.F. and Miller, J.F.A.P., J. Exp. Med., 128 (1968) 821.

20. Nossal, G.J.V., Cunningham, A., Mitchell, G.F. and Miller, J. F.A.P., J. Exp. Med. 128 (1968) 839.
21. Pletscher, A., Shore, P.A. and Brodie, B.B., Science, 122 (1955) 374.
22. Plummer, A.J., Earl, A., Schneider, J.A., Trapold, J. and Barrett, W., Ann. N. Y. Acad. Sci., 59 (1954) 8.
23. Shore, P.A., Pletscher, A., Tomich, E.G., Carlsson, A., Kuntzman, R. and Brodie, B.B., Ann. N. Y. Acad. Sci., 66 (1957) 609.
24. Shore, P.A., Silver, S.L. and Brodie, B.B., Science, 122 (1955) 224.
25. Surveillance Boston Collaborative Drug Program, Lancet, ii (1974) 669.
26. Weihe, W.H., Ann. Rev. Pharmacol., 13 (1973) 409.

THE PREPARATION OF AN ANTISERUM AGAINST HUMAN MONOCYTES

A. STUART, G. YOUNG and P. GRANT

University of Edinburgh Medical School
Edinburgh (Scotland)

Antisera of varying specificity have been prepared against animal macrophages (1,2,3,5) but there appears to be no reference in contemporary literature to an antiserum against human cells. One reason is the difficulty in obtaining starting material.

Recently, a patient presented with leukemia in which the majority of the cells were monocytic, as judged by morphology, cytochemistry, cell receptor and cell culture studies. These cells were used to prepare an antiserum whose specificity is now briefly described.

MATERIALS AND METHODS

Leukemic monocytes were obtained from a patient with acute monocytic leukemia of the Schilling type by Ficoll-Hypaque density gradient centrifugation of a peripheral blood sample. The mononuclear cells had the characteristics shown in Table I. Normal peripheral blood mononuclear cells were incubated overnight in large glass vessels to provide adherent and non-adherent populations.

Immunization Schedule

Six outbred female guinea pigs were given an intraperitoneal injection of 80×10^6 leukemic monocytes followed by 2 further injections of 40×10^6 cells at weekly intervals by the same route. The animals were bled out 7 days after the final injections, the sera were pooled and heat inactivated at 56 C/30 min.

TABLE I

Characteristics of Cells Used in This Study

Cell Type	E	EAIgG	EAC	Surface membrane immunoglobulin	Neutral red ingestion
Leukemic monocytes	1%	47%	72%	2%	90%
Thymocytes	80%	ND*	ND*	5%	5%
B cells (CLL)	4%	64%	ND*	74%	2%

*ND = not done

Absorption

Human thymus tissue was obtained from children undergoing cardiac surgery. A thymocyte suspension was prepared which had the characteristics shown in Table I.

B cells were obtained by Ficoll-Hypaque density gradient centrifugation of a peripheral blood sample from a patient with chronic lymphatic leukemia (CLL), which had the characteristics shown in Table I.

Procedure

The antiserum was absorbed twice for 1 hr at 37 C with thymocytes and B cells using 1×10^9 cells per 1 ml antiserum. After each absorption the serum was centrifuged at 2,000 x g for 15 min.

The serum was also absorbed with group O-ve erythrocytes to which it had an agglutinating titer of 1:256.

After absorption, the antiserum was inactive against the cells with which it had been absorbed.

Indirect Immunofluorescence Test

The method used was a modification of that described by Nairn
(4). Air-dried cell smears were fixed in 90% ethanol for 1 min
and stored at 4 C. The smears were washed in phosphate buffered
saline (PBS) for 15 min before incubation with the test serum for
30 min at 37 C in a moist box. Normal guinea pig serum was used
as a control. The smears were rinsed and washed 3 times in PBS
for 15 min. A 1:20 dilution of FITC conjugated swine anti-guinea
pig IgG was then applied to the smear and incubated for 30 min at
37 C in a moist chamber. After rinsing and washing 3 more times
in PBS for 15 min the preparations were mounted in glycerol saline.

RESULTS

The antiserum before absorption reacted with a wide variety
of cells: leukemic monocytes, thymocytes, polymorphonuclear neutro-
philic leukocytes (PMNL), all peripheral blood lymphocytes and CLL
cells (Table II). After absorption with thymocytes and CLL cells,
the serum lost its activity against the latter but still fluoresced
strongly with the original leukemic monocytes and a small proportion,
approximately 12%, of peripheral blood mononuclear cells. Cells
whose size and nuclear morphology were acceptable as monocytic always showed bright fluorescence. PMNL also fluoresced but were
readily distinguished from mononuclear cells. Some "round cells"
gave a positive reaction and it was not possible to decide if these
were lymphocytes or monocytes. The number of fluorescing cells was
diminished by separation of the glass adherent population (Table
III).

TABLE II

Titer of Activity of Anti-Macrophage Serum
Against Human Cells Before Absorption*

Leukemic monocytes	Thymocytes	Neutrophil polymorphs (PMNL)	Platelets	Peripheral blood mononuclears	CLL cells
128	128	1024	256	1024	16

*Figures are the reciprocal of the highest dilution showing spe-
cific fluorescence.

TABLE III

Proportion of Fluorescent Cells Before and
After Separation of the Glass Adherent Population

Treatment	Percent fluorescent cells
None	26
Non-glass adherent	5
Glass adherent	85

DISCUSSION

The preparation of this serum has been entirely dependent on appropriate starting material and adequate absorption procedures. The leukemic cells used as a source of antigen were well differentiated in the sense of retention of surface receptors and the ability to segregate neutral red in a characteristic manner. Furthermore, there was minimal contamination by both B and T cells as judged by the low values recorded for surface immunofluorescence and sheep rosetting cells. Absorption with human thymocytes and B lymphocytes gave a serum which reacted only with a small number of peripheral blood mononuclear cells. Phase contrast microscopy showed that many of these had the morphology of monocytes but others were simply round cells without distinguishing features. Separation of blood mononuclear cells into glass adherent and glass non-adherent populations provided sound support for the view that the fluorescent cells were largely glass adherent. Although there is general agreement that monocytes and macrophages are glass adherent cells, we have frequently noted that some lymphocytes have the same property. Accordingly, we cannot at this time completely exclude the possibility that the antiserum may react with a small and minor population of lymphocytes. Nevertheless, the serum showed a remarkable specificity for human monocytes.

ACKNOWLEDGEMENTS

This work was supported by a grant from the Cancer Campaign.

The authors are indebted to Dr. A. Parker for his help in providing clinical blood samples. Mr. W. Bisset is thanked for his supply of fresh thymus glands.

REFERENCES

1. Despont, J.P. and Cruchaud, A., Nature, 223 (1970) 838.
2. Loewi, G., Temple, A., Nind, A.P.P. and Axelrad, M., Immunology, 16 (1969) 99.
3. Montfort, I. and Perez Tamayo, R., Proc. Soc. Exp. Biol. Med., 138 (1971) 204.
4. Nairn, R.C., Fluorescent Protein Tracing (Ed. R. C. Nairn) E. and S. Livingstone, London, England (1969) 61.
5. Schroit, A.J. and Gallily, R., Immunology, 26 (1974) 971.

LYMPHOCYTE SUBPOPULATIONS: ANALYSIS OF T-CELL ROSETTE CHARACTERS

D. T. Y. YU, C. YU and A. KACENA

UCLA School of Medicine
Los Angeles, California (USA)

An important step in the elucidation of the thymus-derived
(T) lymphocyte system was the discovery of specific cell surface
antigens. The most thoroughly studied of these are those in mice.
There, it is recognized that the θ antigen, among others, is specific to T-cells. Further, individual subpopulations of T-cells
carry different quantities of θ antigen on their cell surfaces.
This antigen is higher in immature thymus cells than in cells of
peripheral lymphoid organs (2). The T-cells of guinea pigs can
also be distinguished from the rest by their ability to form
rosettes with rabbit red blood cells (RRBC) (4). This paper deals
with the characteristics of RRBC rosettes in various lymphoid organs. It was found that the rosettes of thymus cells possessed
more RRBC than those of lymph node and spleen cells. The modulation of this parameter by various factors was the subject of
this investigation.

MATERIALS AND METHODS

Guinea pigs were of the Hartley strain and weighed 350-400 g
each. The method of assaying RRBC rosette forming cells was similar to that of Stadecker (4). A rosette was defined as a viable
lymphocyte surrounded by 3 or more RRBC's.

To count the number of RRBC per rosette, a drop of rosette
suspension was covered with a coverslip on glass slide and allowed to dry partially by standing at room temperature for about
30 min. The RRBC spread out as a monolayer around the lymphocytes
and could be counted easily with a 40x objective and 10x eyepieces

under light microscopy.   Assessment of 100 rosettes per sample was
made.

Lymphocytes were cultured at 4×10^6 cells/ml in plastic tubes
(Falcon 2059) in Minimum Essential Media (MEM) with 5% heat-inactivated fetal calf serum at 37 C.   The reagents sodium salts of
Dibutyryl Cyclic AMP (N^6,O^2-Dibutyryl Adenosine 3:5-cyclic monophosphoric acid) and Cyclic GMP (N^2,O^2-Dibutyryl Guanosine 3:5-
cyclic monophosphoric acid), theophylline and DL-Isoproterenol were
purchased from Sigma Co., Missouri.   Thymosin fraction V was kindly
supplied by Dr. A. Goldstein (University of Texas, Galveston),
Prostaglandin E_1 was obtained from Dr. J. Pike (Upjohn).   After incubation, lymphocytes were washed twice with Hank's Balanced Salt
Solution prior to making rosettes.

Results were expressed in percentages and standard errors.
Statistical comparison was made by means of Student's and paired
observation t tests.

RESULTS

Rosette proportions.   The average percentage of rosettes
formed with 18 thymus, 18 lymph node and 13 spleen samples were
$93.8 \pm 1.3\%$, $63.6 \pm 3.3\%$ and $40.6 \pm 2.6\%$, respectively.

Number of RRBC per rosette.   Rosettes were distinguished as
having 3-5, 6-8, 9-11, ≥ 12 RRBC per rosette.   A total of 15 thymus,
15 lymph node and 6 spleen samples were assessed.   Lymph node rosettes resembled those of spleen rosettes.   Thymus rosettes had
higher numbers of RRBC per rosette.   The percentage of thymus-
rosettes with more than 9 RRBC per rosette was significantly higher
than that of lymph nodes ($p < 0.001$) (Fig. 1) and spleens ($p < 0.01$)
(Fig. 2).   The most consistent difference was in the percentage of
the smallest rosettes, i.e. those with only 3-5 RRBC per rosette
(abbreviated subsequently as small rosettes).   These were $21.2 \pm$
1.9% in thymus, $55.7 \pm 2.9\%$ in lymph nodes and $44.8 \pm 4.5\%$ in
spleens ($p < 0.001$ for thymus vs lymph nodes or thymus vs spleens).
Whenever thymus rosettes were assessed together with those of lymph
nodes or spleens from the same animals, the percentages of small
rosettes in thymus were always less than that of the other organs.

For ease of comparison, the data presented in subsequent sections will be confined to those of small rosettes.

Effect of culture.   Thymus and lymph node cells were cultured
in MEM-5% FCS for 30 to 1440 min.   The percentages of small rosettes of the thymus progressively increased so that at 4 hr they
resembled that of lymph nodes (Fig. 3).   This parameter was not
affected in the lymph node cell cultures.

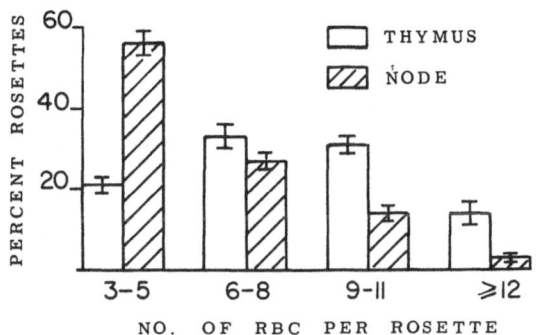

Fig. 1. Number of RRBC per rosette of thymus and lymph node cells.

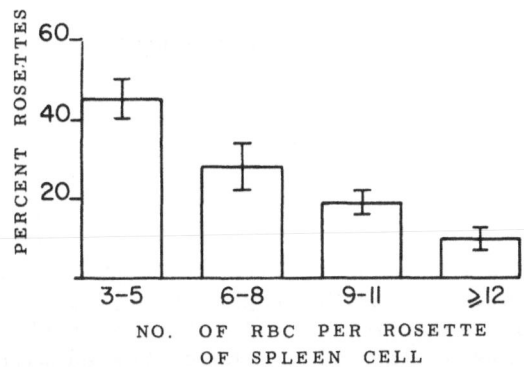

Fig. 2. Number of RBC per rosette of spleen cells.

Effect of cyclic nucleotides. Thymus cells were incubated
with or without Dibutyryl Cyclic AMP at $10^{-3}M$ for varying periods
of time. The normal increase in the precentages of small rosettes
with culture was suppressed by the reagent. Maximum effect was observed at 2-4 hr (Fig. 3). In all subsequent experiments comparing
effects of various reagents, cells were incubated for 4 hr. Results
described here and subsequently represented those of 3 separate
experiments.

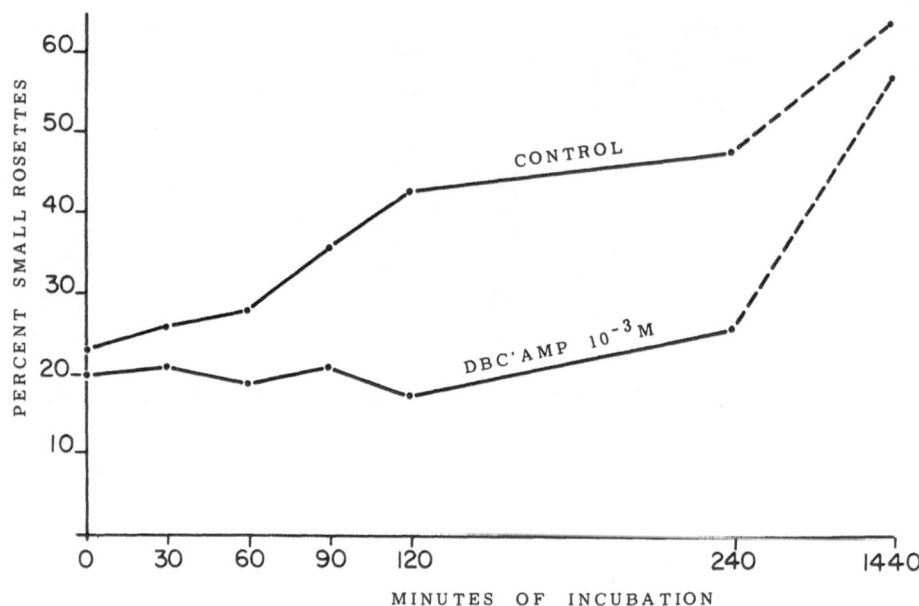

Fig. 3.   Incubation time.

Thymus cells were incubated with Dibutyryl Cyclic AMP, Theo-phylline, Isoproterenol, and Prostaglandin E_1 at 10^{-3}, 10^{-4}, 10^{-5}M. The increase in percentages of small rosettes which occurred on culture was suppressed (Table I).   Those treated with 10^{-3}M of the reagents were $54.7 \pm 7.0\%$, $61.2 \pm 10.7\%$, $60.2 \pm 0.2\%$ and $50.5 \pm 8.7\%$ of control samples ($p < 0.0025$, < 0.05, < 0.0025, and < 0.025, respectively).   Incubation with Dibutyryl Cyclic GMP at 10^{-3} to 10^{-9}M did not cause any significant change.

Culture of lymph node cells with or without the reagents did not elicit any changes.   The percentages of small rosettes in sam-ples treated with 10^{-3}M Dibutyryl Cyclic AMP, Theophylline, Isopro-terenol, Prostaglandin E_1 and GMP were 78.3 ± 15.2, 98.3 ± 10.1, 84.7 ± 8.3, 98.3 ± 10.1 and $97.0 \pm 4.0\%$ of the control samples ($p > 0.05$).

In all experiments described here, there was no change in cell viability and rosette proportions on culture with or without the reagents.

TABLE I

Effect of Reagents on Percent of Small Rosettes

Reagent		Thymus	Lymph Nodes
		Percent Small Rosettes in Thymus	
Dibutyryl Cyclic AMP	0	38	49
	10^{-3}M	18	28
	10^{-4}M	49	38
	10^{-5}M	35	39
Theophylline	0	38	49
	10^{-3}M	19	57
	10^{-4}M	35	52
	10^{-5}M	39	54
Isoproterenol	0	38	49
	10^{-3}M	17	35
	10^{-4}M	28	53
	10^{-5}M	26	52
Prostaglandin E_1	0	46	47
	10^{-3}M	31	38
	10^{-4}M	37	54
	10^{-5}M	41	72
Thymosin V	0	64	67
	50 μgm/ml	54	65
	100 "	50	65
	250 "	20	64
	500 "	31	70

Effect of Thymosin. Thymus and lymph node cells were cultured with thymosin fraction V at concentrations of 50, 100, 250, 500 μgm/ml for 4 hr at 37 C. The increase in the percentage of small rosettes of thymus cells was again suppressed. The small rosettes of the samples treated by 500 μgm/ml thymosin V was $59.8 \pm 6.2\%$ of untreated samples ($p < 0.001$). The lymph node cells were not affected. The small rosettes of the samples treated with 500 μgm/ml of thymosin V was $113.7 \pm 8.2\%$ of control samples ($p > 0.05$).

DISCUSSION

This study compared the surface characters of T-cells in the thymus on the one hand, and in lymph node and spleen on the other. All these cells shared the common ability to form rosettes with RRBC (4).

There was a very significant difference in the number of RRBC per rosette in the rosettes of the thymus cells and in those of peripheral lymphoid organs. More RRBC were present in the thymus rosettes. An equivalent phenomenon has been observed in mice. In mice, the θ antigen is T-cell specific and exists in highest quantities on the surface of thymocytes (2).

The significance of these findings can be enlightened by observations reported on mouse T-cells. In the mouse thymus, the majority of cells are functionally "immature", with a small proportion of apparently more mature ones. The latter resemble the peripheral cells in having less quantity of θ antigen. Because most of the peripheral T-cells are derived from emigration of thymocytes, it is believed that T-cells first undergo a partial maturation process inside the thymus, with partial loss of θ antigen and then migrate to the peripheral organs (2). The data here suggested that a similar phenomenon might exist in guinea pigs.

If one accepts this model of T lymphocyte development, the question that arises is: What is the factor(s) controlling the modulation of surface markers? In this paper it was attempted to answer this question using the 2 patterns of RRBC rosettes as a parameter of measurement. Lymphocytes with low percentages of small rosettes were assumed to have a thymus pattern. Those with large proportions of small rosettes were assumed to have a peripheral lymphoid cell type or mature cell pattern.

As thymocytes were cultured _in vitro_, the rosette characters changed from that of thymus pattern to peripheral lymphoid pattern. This strongly suggested that the _in vivo_ thymocytes are under the influence of a factor which prevents their change to a lymph node pattern. When cultured _in vitro_, they are partially released from this influence and undergo the change rapidly. To explore this possibility further, the thymocytes were incubated with Dibutyryl Cyclic AMP and other agents which could increase intracellular cyclic AMP activity. These consistently suppressed the change of thymocyte rosettes from a thymus pattern to peripheral pattern. These reagents did not have any effect on cell numbers, viability and rosette percentages. This indicated that the effect was not a result of cell death, cell proliferation or suppression of rosette formation. Cyclic AMP serves as a second messenger to hormonal stimulation in many tissues. When a hormone stimulates a cell, it first interacts with the hormonal receptors on the cell surface. The interaction causes an increase in intracellular cyclic AMP and secondarily leads to the activities programmed for the cell (3). The fact that the reagents tested here could suppress the changes in rosette characters indicated that this surface character might be under the influence of a naturally occurring mediator in vivo. The fact that thymosin V could also suppress the in vitro change of rosette character suggested that it might be the natural mediator.

In summary, the data indicated that 2 populations of thymocytes existed: immature and mature types, identifiable by their rosette characters. The immature type was capable of changing spontaneously to the mature type, but was partially suppressed in vivo by the high concentration of thymic hormone present in the intrathymic environment. The mature types of thymocytes emigrated to the peripheral organ, accounting for their high percentage of small rosettes. Alternately, cells of peripheral organs might have originated from the immature type in the thymus. Once emigrated and exposed to a lower concentration of thymic hormone, they changed into the mature pattern. The fact that lymph node rosettes were less affected by culture in vitro indicated that once cells have changed to the mature pattern they have less ground for further differentiation.

ACKNOWLEDGEMENTS

This work was supported by USPHS Grant No. GM 15759. David Yu is a Research Fellow of the Arthritis Foundation.

REFERENCES

1. Goldstein, A.L., Giuha, Z., Zatz, M.M., Hardy, M.A. and White, A., Proc. Nat. Acad. Sci., 69 (1972) 1800.
2. Greaves, M.F., Owen, J.J.T. and Raff, M.C., T and B Lymphocytes: Origins, Properties and Roles in Immune Responses, American Elsevier Publishing Company, New York, New York (1973) 113.
3. Robinson, G.A., Butcher, R.W. and Sutherland, E.W., Cyclic AMP, Academis Press, Inc., New York, New York (1971) 36.
4. Stadecker, M.J., Bishop, G. and Wortis, H., J. Immunol., III (1973) 1834.

The RES in

Immunopathology and Autoimmunity

IMMUNOPATHOLOGY OF THE NERVOUS SYSTEM

R. M. MORRELL

Baylor College of Medicine, Houston, Texas (USA)

The assignment for this presentation is construed as providing
a framework of fact and concept from which the current excitement
and directions of research in modern neuroimmunopathology can be
perceived. The presentation will be that of a general review with
great selectivity exercised in the choice of topics, bearing in
mind that comprehensiveness, although desirable, leads to genera-
lizations which may not take into account the tremendous subspeciali-
zation which has occurred in all branches of clinical and experimental pathology. Just as the failure of the instructive theory
of immunity came about from the analysis of immunity at the molecular level without simultaneously considering cell-cell interactions,
so immunopathology of the nervous system is now moving toward the
elaboration of molecular concepts, having begun and continuing with
the investigation of difficult systemic and intracellular interactions. Immunopathology of the nervous system includes not only
the intracranial contents and those of the spinal canal but also
the autonomic and peripheral nervous systems and the neuromuscular
apparatus.

Current research can best be set in the historical framework
of the topics considered important in 1969 when the Association for
Research in Nervous and Mental Disease (ARNMD) treated "immunological"
disorders of the nervous system." In his introductory comments,
Frank Dixon pointed out the need for precision and the avoidance
of terms like "hypersensitivity" in which the qualitative and
quantitative components were not adequately described. It was
possible then and is still possible, in a broad sense, to consider
immunopathology as the study of the results, at the systemic, tissue
and cellular level, of the interaction between antigen (Ag) and

antibody (Ab) molecules and most importantly their products, in a
complex matrix comprised of immunocompetent cells of a number of
types, including their subpopulations, and the cells of the organ,
tissue, or matrix in which the above-mentioned reactions take place.
Therefore, we may consider in an orderly fashion, the Ags and Abs
of importance for neuroimmunopathology, the complex agencies or
mechanisms such as autoimmunity or viral infection which may modify
antigen and antibody, and the experimental models and clinical
disorders which are thought to partake wholly or in part of immune-
mediated processes.

In 1969, brain-specific Ags, including the S-proteins, were
described. Nerve growth factor (NGF) was known and studied but
was not thought to be intrinsic to the nervous system. The separation of protein and proteolipid components of myelin and the analysis of immunoglobulins (Ig) of the cerebrospinal fluid (CSF) as Ags
was under way. It was clear that infection, particularly by viruses,
could induce new antigens by modification of the cell surface by
viral action including the defined mechanism of transformation
(e.g., by SV 40). A primary viral infection could also have this
effect. It should be recalled that a viral etiology for multiple
sclerosis (MS) was not new, having been proposed by Pierre Marie
in the 19th century.

The role of antibody was poorly understood. At the experimental level antibodies were made to subcellular components such as
synaptosomes or synaptic vesicles, while a number of laboratories
had shown that in certain immunological disorders there were
specific or characteristic abnormalities and patterns of CSF Ig
content. Occasionally, the CSF was known to reflect in its Ig
composition systemic infections which did not specifically ·involve
the nervous system. This was usually manifested as a relative or
absolute increase in IgM. However, it was thought conservatively
that at least 50% of MS patients, regardless of disease stage,
had elevated IgG in the CSF. Abnormalities of the β/γ ratio
were also demonstrated. It was suggested that the brain was an
immunological organ, not capable perhaps of synthesizing IgG from
its own components, but being the site of important immunological
reactivity. In MS there may be gradients of IgG from areas of
perivascular inflammation or sclerotic plaque formation; such
gradients reflected the activity of IgG synthesizing cells,
presumably lymphocytes. Ab was also of importance (in a relatively negative sense) in immunodeficiency states with neurological
sequellae, of which the prime example was ataxia telangiectasia (AT)
as described in the ARNMD by McFarlin and others. A complex
recessive defect with variable deficiencies in IgA and occasional
increases in IgG and IgM, AT was also known to have defects in
cellular immunity and disordered architecture of thymus and lymph
nodes. The combined elements of humoral and cellular immune defect,

coupled with disordered mesenchyme and neuroectoderm, suggested an early and primary stem cell defect.

"Autoimmunity" (more correctly "autoallergy," after von Pirquet) was considered as a hypothetical mechanism for MS, myasthenia gravis (MG), polymyositis and the neurological manifestations of dermato-myositis. Cases with MG were described having concomitant thyroid-itis and Coomb's positive hemolytic anemia, both evidence for presumed autoimmune processes. The relationship of dermatomyositis to cancer was established and the analysis of the humoral and cellular immunity in patients with these disorders was in progress.

As an introduction to the later more detailed discussion of current concepts in a number of experimental and clinical neurolo-gical disorders, it will be helpful to review the major concepts existing in the "historical" time frame we have selected. MS and other demyelinating diseases were of central importance in the attempts to relate immune processes to neuropathology. Related major human demyelinating disorders which were included, at least implicitly, with MS as candidates for a complex immunological patho-genesis were diffuse sclerosis, Schilder's disease, acute dissemi-nated encephalomyelitis, Balo's disease, neuromyelitis optica (Devic's), metachromatic encephalopathies, post-infectious encephalo-myelitis and possible amyotrophic lateral sclerosis (ALS). It was known that some of these disorders, particularly MS, had typical but unexplained geographical distribution and were thought to be auto-allergic processes involving components of genetic susceptibility to an inciting event, which was widely hypothesized as a viral infection. The existence of the so-called "model" of MS, namely experimental allergic encephalomyelitis (EAE) was a further reason for the widespread attention to MS as an immune-mediated disorder. It can be added that epidemiological statistics revealed at least 100,000 known cases of MS in the United States, with another conservatively estimated 25,000 undiagnosed cases. Therefore, the disease assumed major proportions among entities commonly seen by neurologists. Interest in MS was also linked to a clear trend in the evaluation of virus infections, namely the recognition of viral infection as a prelude or concomitant to previously unexplained pathological processes. One example is a thoroughly studied viral CNS disorder, lymphocytic choriomeningitis (LCM), a disease of rodents widely studied in mice and caused by Arena virus, an enveloped RNA of 120 nm diameter. It was shown by various investigators that LCM infection existed in numerous possible states at the systemic and cellular levels and so-called phenotypic expression varied widely (35). One manifestation was cerebellar necrosis, accompanied by an acute, often necrotizing, inflammatory response. At certain stages, the disease was described as being primarily a result of host response with the active infection no longer present. A carrier state was also shown to exist in which immune complexes

were present, although without overt signs of systemic pathology,
therefore suggesting the immune complexes were not pathogenetically
significant. It was found that transmission of LCM required theta-
bearing lymphocytes and that the transfer of cell mediated immunity
to LCM was restricted by histocompatibility locus of the mouse, H2.

    The importance of cell-mediated immunity was further demonstrat-
ed by the ability of Cytoxan (CTX) treated animals to survive. Acute
infection in CTX-treated animals, depending on dose, also yielded
a number of complex intermediate states. Among these was a smol-
dering form of infection in which virus Ab or inflammatory response
were present for a number of months. This was not a true "slow
virus" infection as the slow virus infections are typically
unaccompanied by inflammation. The persistence of CNS lesions
without external clinical sign was explained as being due to virus-
specific immune tolerance. An animal viral infection, visna, (not
experimental) was known to be of great economic importance among
sheep in Iceland and elsewhere. Visna was shown to consist of a
vascular and diffuse angry inflammatory and round cell infiltration
in brain; inflammatory regions were accompanied by severe demyeli-
mination of brain and cord. In addition, a generalized viral
cytopathic effect, budding virions, etc., were observed as parts
of the acute infection, which generally has a brief course. Interest
in this disorder and its relation to EAE and human demyelinative
disorders have prompted other investigators to develop EAE in the
sheep (see below). Other disorders of unknown etiology of a postulated immune mediation were as follows: MG, for which antineuronal
Abs had been isolated, and in connection with which thymectomy was
known to improve selected patients; carcinomatous neuromyopathy or
the obscure "remote effect" of cancer, thought possibly to be due
to an unknown immune mechanism; and Guillain-Barré syndrome (a
polyradiculopathy), for which experimental allergic neuritis (EAN)
was a model. Finally, and importantly, EAE as a model disease was
developed following the observations in 1870-1900 of post-rabies-
inoculation encephalomyelitis in patients. It is known to exist
in a number of forms, including hyperacute, and was known to be
primarily of cellular immune origin being transferable passively
by cells from an affected animal to an unsensitized host. A
continuing controversy existed as to the adequacy of EAE as a model
for MS. Important in its own right, there have now been new
developments to be described below which makes the continued
investigation of EAE most important. The time table of significant
events in the history of EAE must include, for the orientation of
those not familiar with this experimental disease, the following
facts. After Waksman and others demonstrated, between 1950-1960,
that autoimmune experimental disease existed, it was found that
heterologous brain with Freund's adjuvant could produce the acute

and rapidly fatal disorder in a number of animals. Other investigators in 1959 pointed out the pathological distinctions between
post-vaccinal encephalomyelitis, MS and EAE. The cell-associated
delayed hypersensitivity was proven by others, and they showed
demyelination in normal brain cultures by the alpha-2 globulin fraction of EAE serum, and also cytopathic effects produced by lymph
node or T-cells of EAE animals. Others demonstrated globulins in
glia cells and in myelin sheaths by fluorescent Ab techniques.
It was shown that the sera of MS patients but not those with optic
neuritis, ALS, or encephalitis could demyelinate cultured rat
cerebellar cells (45) and reversible inhibition of sulfatide
synthesis was found in the demeylination model. At the time of
the ARNMD it was shown in preliminary work that a progestational
steroid (melengestrol acetate) could reverse or prevent the clinical
and histological EAE in rats.

These introductory comments and considerations lead to the
question whether sufficient knowledge of Ag-Ab immune complexes and
cell-cell interactions would be adequate to explain the phenomenology
of neuro-immunopathology. The astounding proliferation of knowledge
and methodology in all of the fields bearing on the above facts
require considerable oversimplification. Emphasizing that the
immediate requirement for immunopathology is to move more and more
into analysis of molecular interactions, we can in the case of cell-
mediated immunity, still organize these according to orderly
sequences. We can conveniently consider the afferent limb or sensitization phase where specific receptors on lymphocytes recognize
immunogens, a central phase of lymphocyte differentiation and
proliferation, and the efferent limb of the response where mediation of the responses occurs through specifically sensitized effector
cells, primarily T-cells. While some of the experimental techniques
through which these phases can be analyzed will be described, it is
considered more important to indicate the conceptual problems for
which existing techniques or required techniques will then become
obvious. Among the conceptual problems should be included the
following: a) the responses of which the CNS is capable and the
range of interactions, cytophathic, cytotoxic, etc., possible
between immunocompetent and CNS cells, b) the protein effects of
virus and other infections, for example oncogenicity in one species
and teratogenicity in another from the same virus (6), c) complex
disorders involving a relative or absolute immunodeficiency in
either cellular or humoral lines, d) the induction of CNS tumors
by carcinogens, viruses and other agents with concomitant activity
of blocking Ab, all relegated through the discipline imposed by
the unfolding of new facts in cellular and humoral immunity (17).
Aspects of immunopharmacology bearing on these questions will not
be taken up in detail.

Current Concepts of Neuro-Immunopathology

In discussing the repertoire of CNS responses to immune injury
we assume knowledge of the requirements for cooperating cells
beginning with macrophage and including helper T-cells in the
immune response, the existence of a suppressor T-cell population,
the production of lymphokines by challenged sensitized cells, and
the possibility of "macrokines" as elements in immune injury.
Although reactions of immediate hypersensitivity are not common
in the nervous system, these may be found in the future to play a
role and involve reaginic Ab (IgE) and components of the complement
cascade, kinin systems, slow-reacting substance of anaphylaxis
(SRSA), histamine, and the reactions of such cells as basophils,
neutrophils, eosinophils, and mast cells. In reviewing new information it is convenient to continue with the format initiated in
the introduction, namely a consideration of the pertinent Ags,
Abs, autoimmune processes, viral infections, participating cell
types and their interactions and finally experimental models and
clinical entities of current interest. Since the inclusion of
neuro-immunopathology in a program dedicated to the reticuloendothelial system assumes some importance for the latter in
immunopathogenesis, some documentation will be given for the
spectrum of activity of monocytoid cells and their derivatives
within the nervous system.

Antigens of the Nervous System

Methodology and the results of investigations on new brain-
specific proteins and Ags have advanced rapidly. Some expansion
of the old S-100 series has occurred with the production of Abs to
these substances. Laboratories working in this area have given
code designations to a series of proteins assumed to be brain-
specific; for example, 14-6-2, 14-3-2; glial fibrillary acidic
(GFA) protein and specific astrocytic proteins (Bignami); Ags of
neurons expressed in tissue culture (N-series, Schachner). Although
the significance to immunology or differentiation of the nervous
system is not yet clear, Abs to such proteins have already had
their value proved through their usefulness in delineating the
integrity (or lack of same) of the structures on which the corresponding Ags are located in defined pathological states. Little is
known at present about the complex interactions of cell-specific
neural and glial Ags, or the extent to which they may undergo re-
arrangement on the cell surface, as is the case of immunoglobulins
shown by others. Space does not permit a detailed account of other
specific Ags except to mention those isolated from microtubules
(tubulin) and actin and other filamentous proteins from microfila-
ments, specific proteins from nuclear cytoplasmic and synaptic
components; lipopolysaccharides and complex carbohydrates in micro-

tubules and other structures; phospholipids, phosphoglycerides,
gangliosides and other complex lipids, ribonucleoproteins of
defined (questionably viral) and undefined, structure; carcinoembryonic and tumor-related specific Ags on induced neuroblastomas,
astrocytomas and melanomas; and unassigned possibly antigenic
factors (nerve growth factor, NGF, acts in special cases such as
rat gliomas but may originate in the nervous system). The interaction of myelin and constituents of myelin including lipoproteins,
myelin basic proteins and several subfractions isolated by various
techniques, with immunocompetent cells giving rise to cell-mediated
immune reactions, and evidence for sensitization to myelin components in experimental and clinical disorders, particularly MS,
will be reviewed below. Myelin fragments and specific peptides
were initially assumed and later proven to contain immunogenic
segments which were thought to be exposed to EAE-activated lymphocytes but several lines of evidence have shown the basic protein
fraction is buried and requires release of a soluble component to
come into contact with non-sensitized or sensitized immunocompetent
cells. Controversy exists over the extent and meaning of cell-
mediated immunity in MS to myelin basic protein and some of these
studies will be reviewed below. This area of research is now open
to more detailed physical organic chemical studies such as the
relationship of hydrophobic bonding to alteration of accessible
protein segments in cell surfaces. A distinct relationship and
correlation of specific HL-A and H-2 alloantigen with certain
diseases, particularly MS and MG, has been noted (26,27). Four loci
of special interest in connecting specific diseases to the histocompatibility system had been indicated. Some of these are
related to the La-D or lymphocyte-determined systems. The DW-2
locus of special interest in association with MS has been described.
Reifert and Allen were the first to describe the existence of thymus-
derived lymphocyte theta-Ag in the mouse system of CNS cells,
probably neurons. Specific Yc globulin may exist in the nervous
system but its significance as an Ag is unknown.

Neuroviral Ags and transformation-induced Ags comprise fields
too vast to discuss comprehensively but certain findings must be
mentioned. Viruses of specific interest in connection with MS
include paramyxoviruses, especially measles and its three antigenic
determinants. Koprowski believes that macrophages in MS patients
contain paramyxo-like particles. This and related topics have
been reviewed (29). LCM has already been discussed in a previous
section of this paper. The transmission of Creutzfeldt-Jakob
disease from human brain to high primates has been accomplished
by Gadjusek and Gibbs and other laboratories (28). The other so-
called slow virus infections of great importance include scrapie,
visna (see above) and maedi of sheep, Kuru of the Fore people of
New Guinea highlands, subacute sclerosing pan-encephalitis (SSPE)
and progressive multifocal leukoencephalopathy (PML) in which viral

particles of the papova type have been shown (50). A number of
other candidate disorders have been included but at this time
insufficient evidence precludes their mention. Hadfield's report
of PML with paramyxovirus-like structures, Harano bodies and neuro-
fibrillary tangles raises the question of the etiological role of
slow viral infections in aging. Interaction among viruses in a
single host may affect the outcome of a disorder such as SSPE
(measles-induced) in which the presence of papova virus was found
either to exacerbate or ameliorate the encephalitis caused by the
SSPE agent (Yamamoto). PML is known to occur in immunosuppressed
patients and those with advanced lymphomas and this raises the
question of interaction of the agent with suppressed lymphocyte
populations. The papova variants have been found intracellularly
and on the oligodendrocytes specifically, but the mechanism of
demyelination which this may produce is yet unknown. Cerebellar
ataxia and its congenital transmission in cats has been reported
produced by feline panleukopenia virus. Schmidt-Rappin strain of
Rous sarcoma virus produces brain tumors in dogs; patients have
been reported with hereditary neurological syndromes combined with
a fatal varicella and generalized immunopathies. Again, the
precise interaction between the viral involvement and the immune
system is unknown. In Herpes-T encephalitis in monkeys the cell
of interest appears to be the eosinophil. In SSPE one of the
problems is the persistence of Ag in relation to immune responsiveness with the question of partial immunity being actually partial
tolerance. Assuming that the Ag is rejected, complexes are
eliminated by the normal individual. Chronic infection would then
be like a chronic graft-versus-host reaction with high Ab levels
(25). This raises the question of clonal mutants among competent
immune cells in the CNS compartment with an increase in brain Ig
but the question of the anti-measles specificity of the brain Ab
is unknown. Burnet's concept of split tolerance may also play a
role. The tremendous complexity of the relation between virus
infection and immunity precludes further discussion except to say
that once in the brain, viruses may affect many cell types including ependyma, choroid plexus-epithelium, neurons, astrocytes and
microglia. Highly selective actions are evidenced by the parvo-
virus infections which involve dividing cells in the external
germinal layer of the cerebellum, the infection of mouse ependyma
but not choroid plexus by Ross river virus and vice versa in the
case of Sendai; and the previously mentioned specific infection
of oligodendroglia by papova viruses in PML. Resultant demyelination may be due in part to the inability of infected oligodendroglia
to carry out their normal function of synthesizing central myelin.
Immunopathological lesions may result from the interaction of Ab
or immune cells with extracellular Ag liberating inflammatory
mediators and causing local cerebral edema. Direct cytolytic
effects cannot be discussed here. The long term effects of cytopathic effects from viruses, some of which may depend on altered

immune reactions include hydrocephaly, prencephaly and degree of aqueductal stenosis (34).

Antibodies of Interest in Immunopathology of the Nervous System

To each of the Ags mentioned above it is of course possible that Abs are developed, some of which have pathogenetic significance. The regulation of surface immunoglobulin and alloantigen on lymphocytes (Uhr) is beyond the scope of this presentation. It has been reported that anti-measles immunoglobulin in MS patients may be elevated without correlation with phases of the disease. New methods are available for concentration and resolution of the protein components in the CSF and serum. These components are for the time being assumed to be synthesized by non-nervous cells of immune origin which could possibly arise from progenitors within the nervous sytem but are more commonly thought to arrive there through the peripheral circulation. The cellular origin or oligoclonal Ig, shown to be present in certain cases of optic neuritis and MS, is unclear. Another researcher has indicated the presence of myelin fragments in CSF of certain MS patients; again, the relationship with stage of disease is unknown. The existence of specific anti-myelin Ab has not been demonstrated.

The role of immune complexes in CNS neuropathology is largely unknown. Others have demonstrated visceral immune complexes in SSPE but these are not thought to be of significance intracerebrally. Elevated levels of the amino acid serine in CSF of MS patients could represent a fragment of immunoglobulin or be the result of a separate process. It has been shown in a cytotoxicity assay depending on sensitized T-cells and labeled targets that serum and CSF of some SSPE and MS cases contain factors which block the cytotoxic reaction. Similar effects have been demonstrated in certain tumor systems, notably neuroblastoma. Specific antisera to cell surface Ags or subcellular components of immunocompetent cells within the nervous system have not been reported. Circulating anti-brain (possibly iso-) Abs have been shown to be present after brain surgery or other CNS trauma and in the case of certain brain tumors. A large literature exists on cytotoxic factors, some of which may be Abs in the Serum and CSF of MS and ALS patients. The specific Ab associations with disorders such as SSPE and MS have been mentioned. Some cases of Guillain-Barré appear to be associated with elevated Ab to cytomegalovirus (CMV) and the EB virus of Burkitt's lymphoma.

It has been shown that the form of amyloid found in myeloma primary amyloidosis or senile amyloid disease is different from that found in infection, rheumatoid arthritis or familial amyloidosis. The former probably consists of the variable end of Ig light

chain, whereas the latter consists of a B-pleated sheet complex
protein. Amyloidosis has also been found in gamma-3-heavy-chain
disease by Pruzanski. Other workers have shown that mouse amyloidosis may be suppressed by thymosin. There is no information
concerning the chemical structure of cerebral amyloid which occurs
in senile dementia, Alzheimer's Disease and the neuritic plaque.
Work in progress is aimed at determining the nature of the amyloid
demonstrated in the walls of blood vessels and in the cores of
plaques in these conditions. If it is determined that an Ab is
present it will be of great immunological interest. Areas of
interest whose relationship to nervous system immunopathology is
unclear are the soluble or cell-bound mediators of delayed and
immediate hypersensitivity, cytophilic Abs, antiviral Abs or unknown
significance and reaginic Ab (IgE). In considering the role of
Ab generated either within the CNS or acquiring CNS presence by
traversing the blood-brain barrier it is important to note the blood-
brain interface and the importance in nervous immunopathology of
perivascular cuffing with mononuclear cells (see further below)
and the deposits, not only of amyloid, but also of Ag and several
different Igs in the walls of blood vessels as demonstrated by
histochemical and immunochemical techniques. A final comment must
be made on neurological manifestations of immunodysregulation, for
example, ataxia telangiectasia (AT) as mentioned briefly above.
It has been suggested that co-culture of peripheral lymphocytes
from patients with immunodeficiency will depress Ig production by
peripheral lymphocytes from normal individuals. In the case of
AT no such studies have been done. Recent analysis of CSF
protein abnormalities in MS has indicated that some Ig moieties
are present in subnormal concentrations raising the question of
relative immunodeficiency or hypogammaglobulinemia for those
specific molecules. In the next section of this paper, a further
discussion of the cellular immune defect in AT will be presented.

    The complex genetics of the immune process and its relation-
ship to pathogenesis of specific disorders in animals has been
adequately covered in the immunological literature. In connection to neuro-immunopathology, it has been possible to selectively
breed sheep for high susceptibility to scrapie. Other investigators have shown that back-crossed animals from susceptible Lewis
and LxBN F_1 animals had less severe clinical and histological
disease compared to the homozygous litter mates. MLR and serological histocompatibility typing strengthened the suggestion that
this may be an Ir-gene-dose effect or the presence of disease-
suppressive effects controlled by genes linked to the compatibility
locus of the non-susceptible strain. Sensitivity to normal human
white matter was correlated with HL-A7 in patients with MS and
Jersild (27) and others have strong evidence in favor of an association between Ags HL-A 3, 7 and MS. Increased anti-measles virus
Ab titer of MS patients reflects the existence of an immune response

locus at 3,7 and W18. Further findings have stressed that the
possible link between Ir-gene products and HL-A 3,7. HL-A 7 was
significantly less common in patients showing a hypersensitive
response to brain Ag, and a strong correlation has been demonstrated
with LD-7A, a mixed-lymphocyte culture determinant closely linked
to HL-A 7, in patients with MS. A significant proportion of MS
patients showed diminished responsiveness to myxovirus Ags in
leukocyte migration tests. The situation in myasthenia (MG) is
less clear but there appears to be an association between MG and
certain histocompatibility frequencies. The same laboratory,
determining HL-A phenotypes of unrelated Caucasians with ALS, found
an increased frequency of HL-A 3. However, variations may be expected in other human populations (e.g., Mexican, Japanese). An increased frequency of HL-A W29 Ags, but not HL-A 3 or 7, has been
found in patients with SSPE.

New information on genetic factors in MS draws attention to
chromosome number 6 and H-LA types 3-7 (haplotype). Other investigators suggest that the greater the "gene dose" of HL-A 3, the more
MS is observed (nine families). Difficulties in this work include
the problems of segregation in families relative to incidence and
the possibility of existence of a "susceptibility gene" in connection with which the specific determinant for a given patient may
differ from one patient to the next or from one family to the next.
The power of continued research in this area will be the probability
of finding the extent to which the HL-A system marks for the
susceptibility gene. Summarizing the genetic information, it may
be hypothesized that sensitized immune cells respond to virus-
induced specific changes in Ags on infected cells which are
controlled by genes in a histocompatibility complex. These genes
may code for primate specificities or genes closely linked to them.
Doherty and Zinkernagel (16) and others have suggested that this
situation may or may not involve Ags specified by the viral genome
as part of the antigenic determinants, or that virus-specified
Ags alone are unimportant. Thus sensitized T-cells, produced in
a donor of one histocompatibility haplotype, would only recognize
virus-infected foci in target organs with the recipient sharing that
haplotype. Alternatively, T-cells in virus-infected cells may
interact in two separate ways; first, by a receptor for Ag and
second, by a product of genes in the histocompatibility complex.
Information from the Australian group on the biosynthesis of the
light chain component of Ig suggests it is not synthesized directly
as such but that a large precursor molecule is formed with modification to give the final product. Arrest at various stages of the
degradation of such a large precursor could account for unusual
Ab structures found in pathological processes. Others have drawn
attention to animal evidence that chronic stimulation of the RES
can result in the development of a single clone of paraprotein-
producing plasma cells (32). Hyposensitization with allergen

vaccines may predispose to this, although the labeled Ags have not
been found to bind to the paraprotein, indicating the specificity
of the latter for the allergen. The relationship between skin-
fixing Abs and neurological disorders has been reviewed. Little
evidence has been found for IgE reactivity or immediate hypersensitivity within the CNS. This may relate in some way to the
blood-brain interface. Anti-brain Abs were found to change tissue
respiration, content of SH-groups and histamine in brain tissue
of intact dogs, in patterns which were recognizable, but altered
after the induction of EAE. Reproducible patterns were changed
in animals sensitized to an encephalitogenic emulsion giving rise
to EAE. Many forms of Ab-mediated cytotoxicity exist   Investigators have described Ab-mediated cell-dependent immune lympholysis
which depends on specific Ab but can be produced with non-sensitized
lymphocytes. It has been found in idiopathic inflammatory myopathy,
granular deposits of IgG, IgM, and C_3, suggesting the existence of
soluble immune complexes as possible sources of vascular and muscle
damage. Pneumococcal meningitis patients with immunity to certain
strains of Pneumococcus may have specific Ab in the CSF; this may
have a protective function as a component of host defense. Other
investigators have studied the reactivity of antisera in rabbits
to human myelin by immunofluorescence and have found that whole
myelin induces Abs reacting with myelinated CNS tissue in a homogenous pattern, whereas basic protein of myelin (MBP) induced a
speckle pattern with cross-reactivity between MBP and histones.

Cellular Components of Immunopathology of the Nervous System

     We will divide the cellular components into normal cellular
members of the nervous system (neurons, glia, etc.) and immunocompetent cells which are present in the nervous system. The
fundamental reactions of neurons to multiple forms of injury
include a series of membranous , cytoplasmic and nuclear events
which have been widely described. The concept of the multiplex
neuron (Waxman) and the advances due to electron microscopy will
continue to broaden the understanding of specific neuronal changes
secondary to immune-mediated reactions. Differential staining
of neuronal and glia nuclei and the observations on axonal
changes such as those occurring in neuro-axonal dystrophy have not
yet been integrated into a comprehensive immunopathology of the
nervous system. It is clear that neurons contain histocompatibility
Ags as indicated by the rejection of vagal nodose neurons transplanted from BN to LE rats, and the survival in trophic function
of homografted neurons in the presence of adequate immunosuppression.

     The glial cells are the supporting structures of the nervous
system; in processes such as demyelination there are changes in
glia which are not only reactive but possibly causative. In an

experimental demyelinating lesion produced by injection of diphtheria
toxin into the spinal cord, zones of complete regeneration around
the lesion or massive demyelination of axons surrounded by demyelination of paranodal regions are accompanied by changes in oligodendrocytes and astrocytes with astrocytic proliferation and occupation
of enlarged extracellular spaces by astrocytic processes. This,
of course, is a terminal pathological process which does not give
any sign of the mechanism of regional demyelination. There is
some controversy over the extent to which regeneration of demyeli-
nated axons occurs either in the periphery or centrally, although
the situation in the periphery seems clearer. Ogata and Feigin
(36) have shown that Schwann cells arise in central nervous tissues
in demyelinating lesions of the brachium conjunctivum of the pons
(known to contain peripheral myelin). They suggest that the
appearance of cells having characteristics of Schwann cells represents selected maturation of multipotential primitive reticular
cells, consistent with the view that Schwann cells are mesenchymal.
The interactions between Schwann cells and axons suggest a complex
transmission of information, most of which is determined by axonal
factors. Space does not permit the detailed accounting of Schwann
cell reactions to nerve injury including Wallerian degeneration,
responses to localized nerve lesions, segmental demyelination or
primary and secondary demyelination. In each case the response
of the existing Schwann cells and the alteration in reactivity of
the axon-Schwann cell system as evidenced by later changes, are
subjects for active investigation. The oligodendrocyte, as the
central homolog of the Schwann cell in terms of biosynthesis of
myelin, has been shown to be affected specifically by serum and
lymphocytes from patients with MS. The reactions of glia include
hypertrophy, hyperplasia, swelling, focal swelling, degeneration,
phagocytosis, lysosomal storage, and focal degeneration ("dying
back"). These rather limited reactions become more complex in
view of interdependence of certain structures and processes on
the integrity of other cells. Biochemical reactions of astrocytes and oligodendrocytes undergoing immune injury are not
catalogued and represent work for the future. It is assumed by
some that the reactions revealed in the pathological states in
some way represent accentuations or representations of normal
glial functions. Oligodendraglia probably do not synthesize DNA
under normal conditions and have a limited capacity for repair,
often losing their differentiation under conditions of reactivity.
Astrocytes, on the other hand, may be replenished in young animals
from the subependymal region by mitosis and react in this way
usually within a period of two-three days following injury. The
oligodendrocyte, astrocyte, and microglia (to be considered
separately below) have been reviewed by Ribadeau-Dumas. The bio-
physical properties of cultured human glial cells have been
studied. Bornstein has reviewed the immunological factors
affecting neuroglia in EAE and MS. Field (17) has studied the

reactivity of astrocytes in sheep scrapie (see above), a disease
throught to be due to a filtrable agent which has resisted viral
identification. This condition emerged in sheep in Iceland after
inoculation of brain material from a patient dying of acute MS.
Although the question of development of an Ag associated with
reactive astrocytes could not be resolved, the fact remains that
hypertrophied astrocytes develop a factor to which lymphocytes
become sensitized. This was shown in lymphocytes of MS patients,
GPI of tertiary syphilis and glioma, with greater sensitization to
scrapie infected mouse brain than so-called "normal" lymphocytes.
It is stressed that the reaction was nonspecific. The detailed
role of the blood-brain barrier in immunopathologic reactions
undergone by glia is not clear except to emphasize that astrocytes
in particular provide the cytoplasmic bridge between endothelial
cells of brain parenchymal capillaries, and nueronal membranes.
Reversible breakdown of the blood-brain barrier is produced by
numerous substances including electrolytes and non-electrolytes
having little or no lipid solubility but differing in chemical and
ionic properties. Osmotic shrinkage of vessel cells may occur.

      The importance of these reactions is yet to be fully appreci-
ted, but it is clear from the numerous pathological responses at
the vascular-brain interface (perivascular cuffing with round cells;
amyloid angiopathy, IgM and IgG deposition about cerebral vessels)
that this is a crucial site in the pathogenesis and maintenance of
immune-mediated reactions (20). In moving to a consideration of
the microglia and their relationship to astrocytes and mononuclear
leukocytes it is convenient to cite the study of other investigators
who found that the main cellular defense against brain injury is
a migration of mononuclears with a limited astrocyte response.

      The phagocytic microglia have been cells of mystery in the
nervous sytem since Pio del Rio Hortega separated them from oligodendroglia by metallic impregnation techniques during the 1920's
(38). Having been identified as possibly mesodermal, it has been
shown that these cells probably arise at the end of fetal life from
lymphocytoid cells whose characteristics change on entry into the
CNS through the choroid plexuses and basal leptomeninges (44). It
has been suggested that labeled marrow cells injected from one
closely inbred animal into another appeared in labeled microglia-
like nuclei in the brain, not only in choroid plexus but beneath
the ependymal lining of the ventricles. Microglia-like cells in
this subependymal zone synthesized DNA and were found both on and
within choroid plexus. In this site, so-called Kolmer cells
having features of macrophages appeared among other derivatives of
blood-borne mononuclear cells. There is a variable turnover, in
traffic in and out of brain substance, of circulating monocytes
which are normally non-proliferative. It is possible that they may
synthesize DNA and divide under circumstances of injury (especially

well-observed in relationship to motoneurons after nerve transection).
Other investigators have contributed to understanding of the
microglia by demonstrating their relationship to CNS capillaries
and to the pericytes adjacent to vascular endothelium, possibly
representing a transitional cell type. Reactive cells bearing some
features of microglial have been found in degenerative CNS disorders. Following injury microglial clusters appear. It is
believed that there is a specific histochemical marker for
microglia. Transformation of blood monocytes into microglia and
pericytes has been studied. The participation of monocytoid cells
including microglia in pathological processes involves unknown
details of their phagocytic capability. The role of the pericytes
may be much greater than previously imagined. Endothelial cells
can become phagocytic, undergoing fine structural changes in sub-
arachnoid locations. Monocytoid cells upon contacting this tissue
appear to assume the properties of macrophages by an unknown
mechanism. The reticulohistiocytic cell which differentiates into
histiocytes and microglia, also participates in the formation of
primary cerebral reticulosis which may assume neoplastic
proportions. Intracranial reticulum-cell sarcoma may be associated
with IgA deficiency, and solitary intracerebral tumors of lymphoreticular cells (so-called microglioma, reticulum-cell sarcoma)
may be associated with a serum paraprotein. The link between the
evolution of monocytoid-microglial cells and their participation
in immunopathology is more obvious in the light of new information
on the role of macrophages in sarcoidosis and other granulomatous
disorders of the nervous system, new information on macrophage
specificity and characteristics of receptors for Igs and other
substances on macrophage membranes, the altered interactions
between lymphocytes and monocytes in immune deficiency syndromes,
and new techniques for demonstration of specific macrophage immune
reactions such as immune cytolysis (19), the use of substrates of
CNS origin such as myelin or whole brain to test the cytotoxic
reactions of monocytes,macrophages and microglia and the question
of receptors on microglia or monocytoid cells in the CNS for Abs
specific to CNS Ags. Monocytes within the CNS carry and multiply
virus (Monjan); others have studied the development of fibroma-
myxoma virus complexes within immune and non-immune macrophages.
Infection of alveolar macrophages with a neurotrophic vaccinia
has been documented. Herndon (23) presented EM evidence that a
cell resembling a macrophage may be involved in the process of
demyelination of axons affected in mouse hepatitis recurrent
demyelination. Ultrastructural observations in Guillain-Barré
syndrome by Hart et al., (22) have demonstrated involvement of
phagocytes in the process of demyelination and resorption of
intact myelin sheaths. The existence of techniques for the
centrifugal isolation of neurons, glial cells and immunocompetent
cells and their subsequent maintenance in tissue culture will
assist in developing experiments by which their separate functions

in immunopathology can be determined.

The cellular components of the immune system which are active
in the nervous system include lymphocytes, monocytes and their
derivatives and other leukocytes and mast cells. In these
Proceedings, some investigators covered the essentials of new
information on the functional differences among T and B cells and
the growing understanding of the complexity of surface Ags and
receptors. It is neither appropriate nor possible to review these
facts in detail but the major topics can be categorized as follows:
B and T lymphocyte differentiation and immunological tolerance;
lymphocyte membranes and in vitro methods for studying B-cell
triggering and thymus differentiation; blockading effects of
multivalent Ags including Ag-Ab complexes on effector cells;
ultrastructural aspects of lymphoid physiology and methods for
isolation of receptor-specific immunocytes; animal models and the
ramifications of the T-cell receptor problems. The controversy
over the synthesis of Ig by T-cells (Putnam) has been partly
elucidated by the work of others on the B-cell dependency of T-cell
surface IgG. It has been suggested that T-cells must synthesize
the receptors which bind IgG but this does not solve the problem
of the control or regulation of receptor synthesis. Emerging
problems include development of lymphocytes in bone marrow or
other peripheral stage, the development of subclasses of T-cells
and the cytokinetics and fate of sensitized lymphocytes (2). New
findings not yet applied by those interested in immunopathology
of the nervous system include the Ly-Ags of mouse lymphocytes
representing five sets of alleles with specificities as helper,
killer, suppressor and idiotype-determinants. The appearance of
B-cell markers in bone marrow rosette-forming-cells is apparently
dependent on thymic hormone. Problems of Ag recognition include
the development of reagents such as specific alloantisera by
which inhibition of the immune response can be studied. Ig and
alloantigens interact on lymphocytes in specific ways. Lymphocytes fail to re-express Ag receptors after interaction with a
tolerogenic amino acid polymer. This may have significance in
the nervous system where small peptides and polypeptide co-
polymers are known to have neurotransmitter activity. Ag-activated
T-cells recruit, via chemotaxis, mononuclear and polymorphonuclear
leukocytes, thus involving them in inflammation dependent on
cellular immunity. Associated immunological phenomena include
delayed hypersensitivity reactions, homograft rejection, graft-
versus-host responses, and alteration of B-cell responsiveness to
Ag by facilitation or suppression. For example, T-cell deficient
mice may respond to Ag by producing IgM but not IgG. Although
considered specific for T-cells, rosette formation may be observed
in fibroblasts, macrophages, and tumor cells. Ag-Ab complexes may
function as blocking factors on T-cells. Of interest will be
further data on the replacement of T-cells by soluble or cell-bound

factors which may be eluted. Immune-regulation includes the
cellular parameters of unresponsiveness (state of tolerance;
related situations), suppression by T-cells, and the role of T-
cells in experimental models of autoallergy such as the NZB mouse.
It is assumed, but not proven, that T-cells within the CNS will be
able to be categorized according to currently described subtypes.
The presence of IgA-specific immunodeficiency in many patients
with ataxia telangiectasia (AT) raised hopes for application of
the finding by other investigators, that the addition of suppressor
T-cells to a culture of normal B-cells suppressed specific Ig
production. No suppression by T-cells has been thus far found in
peripheral lymphocytes isolated from AT patients and co-cultured
with normal B-cells. The application of T-cell suppressor theory
to AT is still viable, however, in view of the tumors which occur
in certain AT patients (presumably due to defective surveillance),
the observation of chromosome-14-lymphocyte translocation, altered
levels of alpha-fetoprotein, immaturity of the thymus, etc.
Further problems in the regulation of the immune response relate
to the interaction of T- and B-cells in humoral Ab responses and
the observation in a lympho-proliferative disorder ("idiopathic
splenomegaly with associated pancytopenia") that surface markers
of both T- and B-lymphocytes may be present. Fluctuation or
oscillation of interacting cell types during the immune response
might result from two principal mechanisms; negative feedback and/
or interaction between populations. These possibilities may depend
on the action of unrecognized external factors. Techniques for
studying T-cell immunity in vitro allow investigation of the development of specific tolerance by blocking T-cell responses at the
same time these cells are stimulated to make enhancing Ab. In the
mixed lymphocyte reaction (MLR), one of the powerful methods in
use, the assay of differential response of lymphocyte subpopulations
to antigenic stimulation and cooperating cell interactions, may
allow quantitation of cell-mediated immunity to specific Ags, or
the extent to which synthesized cells react to cells with which
they share specificities (F_1 hybrid versus parental, for example).
B-cell helper activity may exist in the MLR; the inhibition test
may be used to select non-stimulator cells as a form of cellular
typing. This has been applied by the Dutch transplantation group
in studies of histocompatibility. In studies on MS patients the
MLR showed no difference between responsiveness of MS and non-MS
cells. However, in unpublished work two siblings with AT, tested
by MLC, demonstrated significant reductions in reactivity in
the one-way MLC as compared with controls or a phenotypically
unaffected sister. Although variability in interpretation of the
role of macrophages in the immune response is considerable, a
number of cells defined generally as the "adherent population" or
accessory cells ("A" cells, probably macrophage-like or monocytoid
cells) are known to not only participate in, but be required by
certain immune responses. MLC responses have been enhanced by

removal of adherent cells. It is conceivable that some cells in
the adherent population may have a suppressor function and there
is separate evidence for this. It is believed that most Ags are
dependent on some form of "A" cell participation (these
Proceedings). Since few immunopathological processes within the
nervous sytem proceed in the absence of monocytoid cells it is
predicted that further delineation of their function in the
nervous system will have far-reaching implications. To a limited
extent, this reaction might depend on the increase in Ag concentration and the specific infiltration of mononuclear cells also seen
in delayed hypersensitivity reactions secondary to hapten-specific
sensitization. The use of the method of blast-transformation or
similar radioisotopic assays has raised important questions about
the specificity of such studies. The existence of a population
of lymphocytes in the peripheral blood which will "turn on" to a
specific Ag thought to be the cause of sensitization must be
carefully interpreted. It is suggested that among the controls for
interpretation of such results must be absorption experiments
which help narrow the specificities involved. Since a defective
lymphocyte transformation and delayed hypersensitivity are
characteristics of well-known syndromes such as Wiskott-Aldrich,
one must also interpret negative clinical responses with care and
ascertain, by establishment of a profile of immunocompetency, the
extent to which the subject or participating cells are capable of
appropriate responses.

     It will be noted that up to this time little has been said
about specific interactions between immunocompetent cells within
the nervous system and the subclasses of cells described as
components of the nervous system. It has been observed that
microglia laden with lipids and adherent to neurons. Lampert and
Lampert (31) have requested great care in the interpretation of
paramyxovirus-like particles in association with macrophages and
neurons whose nuclear material appears to be undergoing proliferation of chromatin fibers. They have suggested that these reactions might be artifacts and that the macrophage activity in
particular might simply represent enhanced phagocytic responses.
It has been suggested that monocytes release immune complexes in
the neighborhood of Ag in the CNS. Other investigators have
performed adoptive transfer of cells from Sindbis-virus-infected
mice to immunosuppressed animals and have shown that the viral-
induced perivascular inflammatory reaction is immunologically
specific for the viral Ag. This reaction, although not mediated
by Ab,is observed to be associated with a significant reduction in
CNS virus content, greater than that achieved by passive transfer
of immune serum alone. In connection with the foregoing experiment
it should be pointed out that host cells exposed to Ab producing
donor cells by long-standing close contact do not acquire specific
immunological memory. In experimental ectromelia virus infection,

other researchers have underscored the striking similarity between
cell-mediated cytotoxic mechanisms evoked by cytopathic viruses,
non-cytopathic viruses, oncogenic viruses, histocompatibility Ags
on allogeneic tumor cells, erythroid cells and perhaps tumor-
associated transplantation Ags. They suggest the T-cells are a
rapidly mobilizable force which respond in host defense against
a broad range of foreign Ags. Finally, although little is known
about the role of soluble mediators within the CNS, the specific
activities of endotoxin, endogenous pyrogens (Greisman and
Hornick (21)), vasoactive hormones, neurotransmitters, products of
non-immunological leukocytes, the chalones, and B-cell stimulated
factors from monocytes, or chemotactic substances elaborated by
primary elements of the neurophil, all of which are presumably
soluble or cell-bound smaller molecules, must be mentioned as
subjects for active investigation in the immunopathology of the
nervous system. It should be clear that the examples given above
apply also to the development of neoplastic cells, an area comprising too large a topic for comprehensive discussion in this
presentation. The experimental tumor produced by ethylnitroso-
urea, the properties and characteristics of sublines of cells
developed from human astrocytomas and other primary cerebral
neoplasms and drug-induced differentiation of rat brain tumors
are examples of experimental approaches in this area. Yet
undiscovered cellular interactions may depend on expansion of the
techniques exemplified by mitogenesis, involving new methods of
cell activation.

Animal and Clinical Observations on Immunopathology of
the Nervous System

The classical problems still relate to recognition, discrimination and specificity in immunopathology. When Ag speaks, who
interprets, who answers, who obeys? Ag-Ab reactions in the rat
brain have been evaluated with evidence, after introduction of
rabbit Ab to serum albumin along with the Ag that implantation
through the cannula induces the same effect as the administration
of norepinephrine. This effect stimulated the release of amines
and local anaphylaxis, alternatively blocking resorption of
norepinephrine released by normal means. The results were
summarized in EAE of experiments on demyelination, remyelination
and sclerosis in cultured mammalian CNS tissue. This body of
work puts much of the phenomenology of experimental autoallergic
CNS disease, beginning with the observations on rabies post-inocula-
tion encephalomyelitis in 1870, on an in vitro biological-experi-
mental basis. Newer findings in EAE include the demonstration
that MBP is in a perivascular situation early in the course (by

fluorescent Ab technique). Different authors outlined the requirements for production of hyperacute EAE in rats by using whole nervous
tissue of β-pertussis as adjuvant. Guinea pig BP appears to be the
only purified protein [(residues 45-89), or mid-peptide as the active
region] capable of inducing HEAE. It was suggested that the immunogenic determinant unique to guinea pig mid-peptide may act as a
secondary encephalitogenic determinant or as a non-encephalitogenic
carrier, preferentially inducing helper T-cells to recruit IgE secreting B-cells in response to the encephalitogenic determinant. A
spinal cord protein in rats found by some investigators is anti-
encephalitogenic. Work by others has established that a basic (A_1)
protein of 170 amino acid residues is the myelin factor responsible
for the induction of EAE. The encephalitogenic peptide is Phe-Ser-
Trp-Gly-Ala-Glu-Gly-Pln-Lys. The nonapeptide is active in rabbits
and guinea pigs and residues 44-89 contain a region encephalitogenic
in rabbits. The terminal 34 residue segments of the A_1 molecule is
active in monkey. Antigenic regions in the A_1 protein, in addition
to the Trp containing region, elicit cell-mediated responses in
guinea pigs. HMB-modification of A_1 protein results in peptides
which produce a delayed skin reaction. X-ray diffraction patterns
and structural characteristics suggest that the A_1 protein conforms
to a classic Davson-Danielli picture, unfolded and electrostatically
adsorbed on the outer surface of myelin. This speculation concludes
that "a purpose of the protein may be to organize lipids during
membrane biosynthesis according to the requirements for myelin
structure." As stated previously in this review, it has since been
found that the structure of MBP is largely buried (51). Animals
which have developed the full clinical symptoms of EAE by administration of A_1 can be restored symptomatically and histologically
by administration of A_1. This may occur through initial activation
of a clone of lymphocytes which have become sensitized to brain and
produce Ab there to, following which re-exposure to the sensitizing
Ag results in release of degradative factors. Second administration of the same protein may result in an Ag (protein)-Ab (on the
circulating lymphocytes' surfaces) reaction prior to their gaining
access to the CNS. The mode of action is thus specifically cytotoxic. Del Canto, by analogy, found cells in EAN which bind horseradish peroxidase conjugate of MBP. Some insight into the location
of free antigenic sites in a well known immunopathologic response is
the location of fluorescein-labeled IgG from sera of patients con-
valescing from acute post-streptococcal glomerulo-nephritis. The
streptococcal nephritogen localizes in the glomeruli. Further
experimental studies on EAE must now move to the level of isolation
of specifically interacting cells such as those in the perivascular
cuff to evaluate their possible transformation by MBP; the dissection of cell types at the injection site; and the extent to which
local anti-brain-specific Ig production is produced in each location.
Modulation of EAE (inhibition) has been produced by intrathymic

injection of encephalitogen, presumably through a mechanism of
tolerance. An immunosuppressant (L-asparaginase) is also capable
of inhibiting EAE, as is heparin in guinea pigs. Some of these
reactions have been reviewed by Cendrowski (9). Because of the
intriguing pathological relationships of sheep visna and MS on
the one hand versus EAE on the other, it has been established
that EAE in sheep with whole brain and adjuvant was found to be
reversible with anti-sheep thymocyte serum whose effect on visna
is currently under investigation. EAE sheep peripheral lymphocytes were found to respond significantly less in MLC than normals;
however, the unresponsiveness varied at different stages of the
induction of EAE. Studies in the EAE and EAN models of transformation in the presence of mitogens and specific CNS fractions
are being carried out by numerous laboratories. It is important
that EAE more closely resembles post-vaccinal encephalomyelitis
(such as rabies or disseminated encephalitis) than MS which is
characterized by early inflammatory lesions (not unlike those of
EAE) which, however, progress to severe demyelination and localized
sclerotic plaques. The course of EAE appears to be progressively
downhill and is generally too short to observe the fluctuations
seen in MS; a chronic relapsing form of EAE has not been confirmed.
The observation that some lymphoid cells infiltrating the brain
in EAE are potentially reactive with brain Ags (Lennon) points to
the feasibility of applying this technique to post-mortem sections
of human brain to ascertain whether infiltrating cells in fresh
MS plaques are reactive with brain Ags. Diseases which may be
intermediate between MS and EAE (recurrent acute necrotizing
hemorrhagic encephalopathy) have been reported in patients which
have cell mediated hypersensitivity to an encephalitogenic myelin
basic protein (5).

Etiologic considerations of MS are still centered around the
virus theory. Inoculation of mice intracerebrally with 6-94 virus
(para-influenza type I virus isolated from cultured MS patient
brain cells) produces selective degeneration of cerebral white
matter preceded by mononuclear infiltration of the ependyma. The
intranuclear filamentous inclusions found in MS have also been
reported in nerves, uremic neuropathy, muscle from carcinomatous
neuropathy and polymyositis, cerebellum from PML, brain stem from
Bickerstaff encephalitis, cerebrum from SSPE, and CMV encepahlitis
and Creutzfeldt-Jakob disease (37). Advances in the investigation
of MS were reported.   Areas of greatest interest included the
following: relative risk for MS among individuals with HL-A 3 and
7 including MLC studies identifying individuals with LD 7A, family
studies of MS, treatment of MS with transfer factor and immunosuppressants (no uniform results), evaluation of serum factors in the
7 S-globulin range which block lymphocyte activity against chronic
measles infection in target cells, interfering particles which, as
deletion mutants of virus, replicate along with the virus but

interfere with its complete growth, the existence of virus
particles in non-neural tissues of MS patients, the relationship
to measles Ab and other virus Abs of oligoclonal IgG in CSF of
patients with MS and other possible or non-viral neurological
diseases, the spectrum of activities of the candidate paramyxoviruses, and new immunological techniques, particularly immuno-
electron microscopy for the demonstration of specific localization
of IgG. A synthesis of some of the factors involved in the viral
etiology of MS is as follows: para-influenza or paramyxoviruses
are the current candidate viruses (25,39,46). At the tissue level,
a gradient of anti-measles Ig from sclerotic plaques, reflecting
possible perivascular Ab synthesis, was shown. This might
participate in a loop of toxic Ab myelinoclasis, macrophagocytosis,
and lymphocyte-stimulation. Ab or cell-mediated cytotoxicity
might require Ag-Ab complexes plus cooperating hematogenous (A?)
cells, linking the peripheral and CNS immune compartments. There
is some evidence that oligodenroglia may become phagocytic. Direct
viral infection could induce new cell-surface Ags. If these were
viral-specific, then C^1-dependent, Ab-mediated lysis or lysis by
sensitized T-cells could occur. At what stage reduction in systemic
immunocompetence occurs is unknown. Specifics of lymphocyte subpopulations are unavailable. The enhanced enzymic degradability
of MBP at sclerotic plaque margins could produce immunogenic or
mitogenic peptides. A similar action might result from peripheral
infection with a virus having cross-reactivity with myelin.
Encephalitogenic sites on MBP are internally consistent but only
partially species-specific. Structural characteristics of
Eylar's A_1 protein and active nonapeptide are beyond the scope of
this talk, except to say that the Trp-containing latter may induce
or suppress EAE. A_1 protein produces EAE in guinea pig or rhesus,
but if given as CNS tissue or pure A_1 prior to heterologous
sensitization, it suppresses induction, requiring that an immune
mechanism be coupled, through sensitization, to an Ag or antigenic
site which cross-reacts with A_1 protein in CNS tissue (14,15, 30).
Heterologous MBP would be expected to suppress this response. A
hypothetical virus-MBP complex would be immunogenic; analogous
suppression might occur, offering a therapeutic possibility.
Exacerbations or remissions in MS might depend on complex
variations in Ab level and lymphocyte reactivity (40,41,42,49).
Other CNS Ags, vulnerable to virus modification, may be implicated
in the same scheme.

Persistence of virus or a "slow virus" concept is obviously
circumstantial and although consistent with the clinical course,
does not explain pathological differences and spongiform
encephalopathy (SSPE) versus plaque formation (MS) (48). Viruses
whose nucleocapsids contain single-stranded RNA might depend on
a reverse transcriptase, as shown for C-type RNA viruses or oncorna
viruses such as Rous sarcoma or Rasucher leukemia. The C-viruses

causing visna and progressive pneumonia of sheep resemble the
oncorna viruses. Incorporation of viral information for replication into the chromosome produces a repressible or suppressible
virogene. Impaired feedback control of the immune response,
plausible in measles infection, would result in a low level of
lymphocyte-transformation, with Ab and possibly a T-cell sub-
population as regulatory agents. An IgG with high affinity for
Ag would not sustain the process but the process might be
sustained for production of a low-affinity, possibly IgM Ab with
secondary lymphoid hyperplasia related to defective T-cell control
of B products, some of which might be cytotoxic (47).

Herpes simplex produces latent infection of spinal ganglia
in mice, a model which would be useful for investigations on the
state of the virus in latent infection. Maintenance of latency
or the "slow" state may depend on non-immunologic mechanisms.
The techniques of nucleic acid hybridization and ultracentrifugal
isolation are central to progress in this work. The Weizmann group
has shown that MS patients whose cells do not transform to BP
generally improve. Sheramata and colleagues (41) have demonstrated
hypersensitivity to fractions of myelin in MS patients during acute
attacks using migration inhibition techniques.

The immunological status of patients having well-characterized
neurological diseases is under vigorous study (3,7,10). Several
patients with Guillain-Barré syndrome have been reported in
association with Hodgkin's disease. Some of these failed to
express delayed hypersensitivity to a test panel of standard Ags.
Their peripheral lymphocytes may also be defective in other tests
of cell-mediated immunity (13). The relation of this to an auto-
allergic disease of the nervous system would depend on suppression
of immune surveillance. A CNS disease of neonates with increased
levels of IgM was reported. Those with increased cord IgM often
had CSF pleocytosis and elevated protein. Two of these had
specific infections (toxoplasma and CMV). Patients with a gene
for myotonic dystrophy, a dominant condition including distal
myopathy, myotonia, mental changes, cataract, hypogonadism and
possible neuropathy, manifest deficiency of IgG with an abnormally
fast catabolism of IgG and abnormalities of IgM and IgA. The
relationship of these immunoglobulin disorders to the gene defect
is under investigation. A case of late-onset combined immunodeficiency with progressive encephalopathy was reported, in which
serum IgM was undetected but CSF IgM was present in normal amounts.
Evidence supported the contention that the CSF IgM is produced by
lymphoid cells in the brain. Adenosine deaminase (ADA) has been
suggested as one cellular marker for conversion of monocytes to
active macrophages. In addition, a measurable defect in erythrocyte ADA has been found in two patients with severely impaired
cellular immunity (Giblett). Although few at present, these

clinical observations relating CNS involvement to immunodeficiency
disease point the way toward discovery of other CNS disorders of
immune etiology.   Other investigators have discussed a congenital
immunological defect with progressive brain disorder.   The relation
between aging and immunodeficiency is in this category as are
carcinomatous neuromyopathy, familial white cell defects with
recurrent bacterial infections and systemic disease, and the protean
immune-related nervous system manifestations of connective tissue
disorders, including the lupus-polyarteritis-giant cell arteritis-
dermatomyositis spectrum.   Many of these disorders are characterized
by mixed cryoglobulinemias.

     The role of the thymus in EAE has been mentioned in connection
with the inhibition of EAE by intrathymic injection of encephalitogen.   Research on MG in addition to further characterization of
the receptor for acetylcholine at the neuromuscular junction and
the effects of humoral Abs to such receptor from patients with MG
on experimental animals has also led to detailed investigations of
the thymic hormone based on the hypothesis that MG is a consequence
of thymic disease with a release of thymic hormone producing
neuromuscular block.   It has been suggested that a synthetic
peptide from bovine thymic hormone causes a neostignine
responsive neuromuscular block in mice, resembling the neuromuscular block of MG.   The effects of prednisone administration
to MG patients and immunological studies using mitogens and other
indices of immunoregulation support the concept that T-cell defects
and autoantibodies are present in an increased percentage of
patients with MG.

     A separate treatise is in preparation on the pharmacology of
immune-mediated nervous pathology.   This will include considerations of transfer factor in cellular immunity and mechanisms of
action of specific immunosuppressants (antimetabolites, alkylating
agents, anti-cellular antibodies, and coricosteroids) at various
levels of the immune response.

REFERENCES

1.   Bartfeld, H., Atoynatan, T., Int. Arch. Allergy Appl. Immunol.,
     39 (1970) 361.
2.   Bartfeld, H., Atoynatan, T., Brit. Med. J., 2 (1970) 91.
3.   Behan, P.O., Behan, W.M.H., Feldman, R.G. and Kies, M.W.,
     Arch. Neurol., 27 (1972) 145.
4.   Bertrams, J., Kuwert, E., Liedtke, U., Tissue Antigens, 2 (1972)
     405.
5.   Brockman, J.A., Stiffey, A.V., Experientia, 26 (1970) 413.
6.   Caspary, E.A., Field, E.J., Eur. Neurol., 4 (1970) 257.
7.   Caspary, E.A., Field, E.J., Brit. Med. J., 2 (1971) 613.

8. Cendrowski, W., Neurol. Neurochir. Pol., 5 (1971) 133.
9. Cendrowski, W., Neurol. Neurochir. Pol., 4 (1970) 465.
10. Chuprikov, A.P., Lab. Delo., 9 (1969) 551.
11. Ciongoli, A.K., Lisak, R.P., Zeiman, B., Koprowski, H. and
    Waters, D., Neurology, 25 (1971) 891.
12. Coates, A., Mackay, I.R., Crawford, J., Cell Immunol., 12
    (1974) 370.
13. Currie, S., Knowles, M., Brain, 94 (1971) 109.
14. Dau, P.C., Peterson, R.D.A., Arch. Neurol., 23 (1970) 32.
15. De Boer, W.G., Lancet, ii (1971) 96.
16. Doherty, P.C., Zinkernagel, R.M., Transplant. Rev., 19 (1974)
    89.
17. Field, E.J., Caspary, E.A., and Carnegie, P.R., Nature, 233
    (1971) 284.
18. Finkelstein, S., Walford, R.L., Myers, L.W. and Ellison, G.W.,
    Lancet, i (1974) 736.
19. Gallily, R., Ben-Ishay, Z., J.R.E.S., 18 (1975) 44.
20. Gregory, M.C., Hughes, J.T., J. Neurol., Neurosurg., Psychiat.,
    36 (1973) 769.
21. Greisman, S.E., Hornick, R.B., J. Infect. Dis., 128 (1973) s265
22. Hart, M.N., Hanks, D.T., and Mackay, R., Arch. Path., 93
    (1972) 552.
23. Herndon, R.M., Griffin, D.E., McCormick, U. and Weiner, L.
    P., Arch. Neurol., 32 (1975) 32.
24. Iwasaki, Y., McMichael, J. and Koprowski, H., Acta Neuropath.,
    31 (1975) 315.
25. Jabbour, J.T., Roane, J.A., Sever, J.L., Neurology, 19 (1969)
    929.
26. Jersild, C., Fog, T., Hansen, G.S. and Ammitzbøll, T.,
    Tissue Antigens, 3 (1973) 243.
27. Jersild, C., Svejgaard, A., Fog, T. and Ammitzbøll, T.,
    Tissue Antigens, 3 (1973) 243.
28. Johnson, R.T., Gibbs, C.J., Arch. Neurol., 30 (1974) 36.
29. Johnson, R.T., Herndon, R.M., In: Progress in Medical Virology,
    (Ed. J.L. Melnick) S. Karger, Basel, Switzerland (1974) 214.
30. Lamoreux, G., Carnegie, P.R., McPherson, T.A., Mackay, I.R.
    and Bernard, C., Ann. Inst. Pasteur, 118 (1970) 562.
31. Lampert, F., Lampert, P., Arch. Neurol., 32 (1975) 425.
32. Lambert, C.D., Trewby, P.N., J. Neurol., Neurosurg., Psychiat.,
    37 (1974) 835.
33. McDevitt, H.O., Bodner, W.F., Lancet, ii (1974) 1269.
34. Mims, C.A., In: Progress in Medical Virology, (Ed. J.L.
    Melnick) S. Karger, Basel, Switzerland, (1974) 1.
35. Nathanson, N., Cole, G.A., Adv. Virus Res., 16 (1971) 397.
36. Ogata, J., Feigin, I., Neurology, 25 (1975) 713.
37. Payne, C.M., Sibley, W.A., Acta Neuropath., 31 (1975) 353.
38. Poirier, J., Goutieres, F., Ribadeau Dumas, J.L., La Nouvelle
    Presse Medicale, 1 (1972) 1513.
39. Prineas, J., Science, 178 (1972) 760.

40. Sheremata, W., Colby, S., Karkhanis, Y. and Eylar, E.H.,
    Can. J. Neurol. Sci., 2 (1975) 87.
41. Sheremata, W., Colby, S., Lusky, G. and Cosgrove, J.B.R.,
    Neurology, 25 (1975) 833.
42. Sheremata, W., Cosgrove, J.B.R., Eylar, E.H., Arch. Neurol.,
    30 (1974) 1.
43. Simon, J. and Anzil, A.P., Acta Neuropath., 27 (1974) 1.
44. Stenwig, A.E., J. Neuropath., Exp. Neurol., 31 (1972) 696.
45. Tabira, T., Webster, H. DeF. and Wray, S.H., Arch. Neurol.,
    32 (1975) 1.
46. Tanaka, R., Iwasaki, Y. and Koprowski, H., Arch. Neurol., 32
    (1975) 80.
47. Vuia, O., Internat. Res. Commun. Sys., (1973).
48. Wattre, P., Path.-Biol., 20 (1972) 793.
49. Webb, C., Teitelbaum, D., Abramsky, O., Lancet, ii (1974)
    66.
50. Weiner, L.P., Narayan, O., Progress in Medical Virology,
    S. Karger, Basel, Switzerland, 18 (1974) 229.
51. Westall,F.C., Immunochem., 11 (1974) 513.

ALTERED NEUTROPHIL FUNCTION INDUCED BY SERUM FROM PATIENTS WITH SYSTEMIC LUPUS ERYTHEMATOSUS

G. W. NOTANI, A. J. KENYON and R. B. ZURIER*

Memorial Sloan-Kettering Cancer Center
New York, New York (USA)

Systemic lupus erythematosus (SLE) is a disease characterized by immunoglobulin deposition and inflammation in several organs (1). It is not clear whether these deposits, presumably in the form of immune complexes, are ingested and/or degraded by phagocytic cells. Uptake and "processing" of antigen by phagocytic cells appear necessary for lymphocytes to respond to antigen (14). Thus defects in cell mediated immunity which have been demonstrated in patients with SLE (1,9,11) may be due, in part, to defects in phagocytosis. Although leukocytes from patients with SLE exhibit impaired phagocytic activity (4), other factors have been implicated to explain the decrease in this cellular function. Thus a decrease in IgM antibody titers to bacterial antigens (2), leukopenia (13) and low levels in serum of critical complement components (16) might account for impaired phagocytosis. Recent in vitro studies have indicated the presence of a factor(s) in sera from patients with SLE which suppresses the phagocytic function of normal macrophages (18)· and neutrophils (20). In the neutrophil studies, only some of the SLE sera were tested in the presence of autologous serum. The studies described in this paper confirm and extend the observation that suppression of lysosomal enzyme release from phagocytic cells is due to presence in SLE serum of an inhibitor of phagocytosis rather than solely the lack of factors which support particle uptake. All of the SLE sera were tested in the presence of normal serum, so that optimal conditions for phagocytosis were

*Present address: University of Connecticut School of Medicine, Farmington, Connecticut (USA)

147

provided.   In addition, suppressor activity was studied temporally in the serum of a patient with SLE, and the factor was partially characterized.

METHODS AND MATERIALS

Preparation of Polymorphonuclear Leukocytes (PMN)

Blood (heparinized) was obtained from normal fasting individuals. Leukocytes were separated by flotation on dextran as previously described (19). Leukocyte suspensions contained approximately 85% PMN. In some experiments leukocyte preparations containing 96% PMN were obtained by means of a Hypaque-Ficoll cushion (3).

Test System for Inhibition of Lysosomal Enzyme Release

Cells were suspended in a medium (pH 7.4) of 10 mM phosphate buffered saline to which had been added calcium and magnesium at final concentrations of 0.6 and 1 mM, respectively. The buffer was designated PiCM. Cell suspensions were incubated (37 C, 30 min) in autologous plus homologous or autologous plus SLE serum (13.3% v/v), then exposed to zymosan for 30 min. The amounts of autologous sera used were always sufficient for optimal enzyme release; that is, further addition of autologous serum in experiments did not increase the enzyme release. Throughout the 60 min incubation tubes were turned by hand every 10 min. At the end of experiments, tubes were centrifuged at 100 g at 4 C for 5 min. The cell-free supernatants were removed for β-glucuronidase determination (5). Reduced enzyme release reflected a corresponding reduction in particle uptake (20). Determination of enzyme release is consistently more reproducible than assay of phagocytosis. Comparisons are therefore made in terms of enzyme release. Sera which reduced enzyme release greater than 25% (p < 0.01 vs controls) in each individual experiment were considered to be inhibitory.

Patients

The diagnosis of SLE in all patients studied was based on the criteria set forth by the American Rheumatism Association (7). The patient whose serum was studied over time is a 15-year old girl (U.S.) who presented with polyarthritis, fever, malar rash, vasculitic rash on upper extremities and oral mucosal ulcers. There was an associated anemia (Hgb. 9.7) leukopenia (WBC 2400) and LE cells were observed in the preparation from peripheral

blood. Antinuclear antibodies (ANA) were present in serum to a
dilution of 1:512 with a peripheral pattern on immunofluorescence.
Determination of anti-DNA antibody titer (Farr test) showed DNA
binding of 61.2% (normal < 30%). There developed hematuria and
proteinuria and a renal biopsy confirmed the diagnosis of membrano-
proliferative glomerulonephritis. The patient's serum was studied
at the time of disease onset before therapy was begun, and 1 and
2 months later during treatment with 40 mg per day of prednisone.

Ammonium Sulfate Fractionation

Equal volumes (0.5 cc) of serum and neutral saturated ammonium sulfate were stirred together for 20 min, then centrifuged
at 10,000 rpm for 20 min (Sorvall RC2B centrifuge). Supernatant
was discarded, the pellet resuspended in 0.5 ml PiCM and dialyzed
overnight against PiCM buffer.

Precipitation with Polyethylene Glycol (PEG)

PEG (MW 6,000, 50% v/v in borate buffer, pH 8.5) was added
to 0.5 cc SLE serum to final concentrations of 7.5% and 10%.
Suspensions were stirred for 20 min, then centrifuged at 3,000 rpm
(Sorvall RC-3) for 10 min. Supernatants were discarded, pellets
were resuspended in PiCM buffer.

RESULTS

Release of lysosomal enzymes from normal human leukocytes
exposed to zymosan was reduced significantly by 36 of 44 (81.8%)
SLE serums tested. Release of enzymes was selective: lysosomal
(β-glucuronidase) but not cytoplasmic (LDH) enzymes were released.
The effects on lysosomal enzyme release of serums from 44 SLE
patients and homologous serum from 10 normal individuals are documented in Table I. Results obtained using purified (96%) preparations of PMN were comparable to those obtained with leukocyte suspensions containing 85% PMN.

Serum from patient U.S. was studied on 3 separate occasions
(Table II). During this time the patient improved clinically
(arthritis, mucosal ulcers and skin rash resolved) and biochemically
(C3 complement component nearly normal by second month after therapy). During this period of therapy there was a five-fold decrease
in suppressor activity of the serum (Table II). The patient had
elevated levels of serum IgG and IgA and normal levels of IgM.
Concentrations of serum immunoglobulins did not change appreciably
during the period studied.

TABLE I

Inhibition of Lysosomal Enzyme Release from Human Leukocytes*

Serum	No.	Percent inhibition of enzyme activity released into supernatant**
		β-Glucuronidase
Normal homologous	10	12.0 + 11.4
SLE	44	40.2 + 14.0
Inhibitory	36	45.0 + 10.2
Noninhibitory	8	18.6 + 6.5

*Normal human leukocytes (1×10^6) incubated (37 C, 30 min) with autologous serum and the test serum (homologous or SLE), then incubated (37 C, 30 min) with zymosan

$$**\text{Percent Inhibition} = 100 - \frac{\text{Control (autologous serum only)} - \text{Test (autologous and SLE}}{\text{Control enzyme release}}\frac{\text{serum) enzyme release}}{}$$

Values expressed as mean + SD for number of sera tested.

TABLE II

Inhibition of Lysosomal Enzyme Release
by Serum from SLE Patient U.S.

Duration of Treatment	C3 Complement	Hgb	WBC	Dilution	Percent Inhibition
0 (onset of disease)	27	9.7	2400	None 1:5 1:25 1:125	63.1 27.3 7.5 0
1 month (40 mg prednisone daily)	46	10.6	6400	None 1:5 1:25 1:125	59.3 6.7 0 0
2 months (40 mg prednisone daily)	85	12.4	9700	None 1:5 1:25 1:125	30.1 0 0 0

TABLE III

Inhibition of Lysosomal Enzyme Release
by Serum from SLE Patient U.S.

Serum Treatment	Inhibition of β-Glucuronidase Release
None	58.2%
Dialysis	31.3%
7.5% PEG	0
10% PEG	27.7%
50% Ammonium Sulfate	24.5%

The suppressor activity in the pretreatment serum of patient U.S. was characterized partially. It was stable to 30 min incubation at 56 C. It did not precipitate after treatment with 7.5% polyethylene glycol (PEG). Similar amounts of suppressor activity were recovered after precipitation with 10% PEG, and with 50% ammonium sulfate saturation and after dialysis of the serum (Table III).

DISCUSSION

These studies suggest that sera from a substantial proportion of patients with SLE contain factor(s) which suppress(es) phagocytosis by –and enzyme release from– normal human neutrophils. The suppression by SLE serum of enzyme release despite the presence in cell suspensions of fresh autologous serum suggests that suppression is due to presence of inhibitor rather than solely to lack of factors which support phagocytosis. Suppressor action does not appear due to anticomplementary activity in SLE serum (20).

The presence in cell suspensions of altered gamma globulin (IgG) is associated with reduced opsonic activity (17). It is possible that immune complexes or altered immunoglobulins block opsonin binding sites on the leukocyte plasma membrane. Svennson has in fact demonstrated (18) that cryoproteins (containing IgG) from several patients with SLE inhibit phagocytic activity of macrophages in a manner similar to the SLE sera. However, removal of cryoprecipitates from serum of several patients in our study did not remove suppressor activity (Hicks, M.J. and Zurier, R.B., unpublished observations).

The presence of suppressor activity does not appear to be related solely to disease activity or to treatment (20) but serial studies have not been done to determine whether changes in disease activity and/or treatment alter suppressor activity in individual patients. In the patient we studied over time, the degree of suppressor activity in the serum fell progressively as the patient improved clinically and serologically. Horwitz and Cousar (12) have demonstrated suppression of lymphocyte transformation by SLE serum. They found the suppressor activity to reside in the IgG fraction and to correlate directly with disease activity.

Immune precipitation of antigen-antibody complexes is increased in PEG (10), a compound which can be used to fractionate plasma proteins according to size of the molecules (15). PEG has been used at a concentration of 7.5% to precipitate soluble antigen-antibody complexes; under such conditions free antigen or free antibody remain soluble (8). In PEG precipitation studies presented here suppressor activity could not be recovered in the 7.5%

PEG pellet. This finding suggests that immune complexes in the serum of patient U.S. are not responsible for inhibition of phagocytosis and lysosomal enzyme release. Since 50% of suppressor activity was lost following dialysis, it is possible that factors other than gamma globulins are responsible for suppressor activity. In studies of sera from other patients with SLE, however, total suppressor activity appears to reside in the IgG fraction (Hicks, M.J. and Zurier, R.B., unpublished observations). It is not unlikely that a variety of proteins (altered or native) present in SLE serum can function as inhibitors of leukocyte function. It will be necessary to characterize further suppressor factors in patients with SLE in attempts to gain greater understanding of mechanisms which regulate phagocytic function.

ACKNOWLEDGEMENTS

The authors thank Dr. Naomi Rothfield who made available for study sera from patients with SLE. These studies were aided by grants from the National Cancer Institute CA 15604 and Program Project Grant CA 16599-01, by USPHS Grants AM-17309 and AI 12225 and a Clinical Center Grant of the Arthritis Foundation.

REFERENCES

1. Abe, T. and Homma, M., Acta. Rheumatol. Scand., 17 (1971) 35.
2. Baum, J. and Ziff, M., J. Clin. Invest. 48 (1969) 758.
3. Boyum, A., Tissue Antigens, 4 (1974) 269.
4. Brandt, L. and Hedberg, H., Scand. J. Haematol., 6 (1969) 348.
5. Brittinger, G., Hirshorn, R., Douglas, S. and Weissmann, G., J. Cell Biol., 37 (1968) 394.
6. Cochrane, C.C. and Koffler, D., Adv. Immunol., 16 (1973) 185.
7. Cohen, A.S., Reynolds, W.E., Franklin, E.C., et al., Bull. Rheuma. Dis., 21 (1971) 643.
8. Creighton, D.W., Lambert, P.H. and Miescher, P.S., J. Immunol., 111 (1973) 1219.
9. Hahn, B.H., Bagby, M.K. and Osterland, C.K., Am. J. Med., 55 (1973) 25.
10. Harrington, J.C., Fenton, J.W. and Pert, J.H., Immunochemistry, 8 (1971) 413.
11. Horwitz, D.A., Arth. Rheuma., 15 (1972) 353.
12. Horwitz, D.A. and Cousar, J.B., Am. J. Med., 58 (1975) 829.
13. Mittal, K.K., Rossen, R.D., Sharp, J.T., Lindsky, M.D. and Butler, W.T., Nature, 255 (1970) 1255.
14. Mosier, D.A. and Coppelson, J., Proc. N.A.S., 61 (1968) 542.
15. Polson, A., Potgeiter, G.M., Largier, J.F., Mears, C.E.F. and Jombert, F.J., Biochem. Biophys. Acta., 82 (1964) 463.
16. Rosen, F.S., Seminar Hematol., 8 (1971) 22.

17. Rowley, D., Thoni, M. and Isliker, H., Nature, 207 (1965) 210.
18. Svensson, B.O., Scand. J. Immunol., 4 (1975) 145.
19. Weissmann, G., Zurier, R.B., Spieler, P.J. and Goldstein, I.M., J. Exp. Med., 134 (1971) 149s.
20. Zurier, R.B., Arth. Rheum., 19 (1976) 21.

PRIMARY IMMUNODEFICIENCY DISEASES

R. A. Good and M. A. Hansen

Memorial Sloan-Kettering Cancer Center

New York, New York 10021

Rare individuals, experiments of nature who suffer the often devastating consequences of a congenital defect within the immune system, have served to teach us much of what we know today about the complex immune response of man (36,39,54). From a careful study of them and their diseases, together with related studies in the laboratory, we have learned, for example, that there are two major and quite separate arms making up the immune system of man and animals. These two basic components are now known as (1) cell-mediated immunity (CMI), or the T-cell (T for thymus derived) system, and (2) humoral immunity, or the B-cell (bone marrow or bursa-derived) system.

Humoral immunity provides for primary defense against high grade encapsulated bacteria such as pneumococci, streptococci, Hemophilus influenzae, meningoccocci, and Pseudomonas aeruginosa. The factors mediating humoral immunity are known as antibodies, and they are comprised of serum proteins known as immunoglobulins (17,45). The B cells of humoral immunity synthesize and, in their fully differentiated form as plasma cells, secrete antibody molecules which make up the five major immunoglobulin (Ig) classes. These are:

IgM. The IgM molecules account for 5 to 10% of the total serum immunoglobulin. Large in size itself, and with a high molecular weight, IgM offers front line protection against microbes and other large-sized immunogens with repeating antigenic determinants on their surface (i.e. capsules of pneumococci, the cell wall of gram negative bacilli, bacterial flagella, and viruses). IgM is produced early in the humoral immune response and works so

effectively that only one IgM molecule, combining with an appropriate antigen, can activate complement and lyse a cell. (IgG
lysing, on the other hand, requires the presence of two or more
IgG molecules working together). In normal people, IgM is distributed predominantly in the bloodstream. It is, of the five immunoglobulin classes, the least thymic dependent (41). It has a half-
life of 5 to 6 days in the serum.

IgG. A full 85% of all the immunoglobulins in man are of the
IgG class. These are distributed throughout the extracellular
fluids, including the blood, lymph, and interstitial tissues of
normal adults. Combining with a wide variety of antigens of all
shapes and sizes, it is particularly responsible for antibody
action against gram positive pyogenic bacteria, certain viruses,
and antitoxins in human adult serum. The Fc portion of the IgG
molecules enables them to cross the placenta (they are the only
immunoglobulin to do so), and thus provide the newborn infant with
passive immunity for the first several weeks of life. The Fc portion also enables the IgG molecule to fix to complement, and to
receptors on skin, B cells, and macrophages. IgG is involved in
opsonic activity, making bacteria more susceptible to phagocytes.
It has a half-life in the serum of 18 to 23 days.

IgA. Approximately 10% of the total immunoglobulin in human
serum is recognized as IgA. It is found in large quantities in
parotid saliva, colostrum, and in tears, all secretions which
sponsor only trace amounts of IgG. A smaller amount of IgA can be
found in bile, secretions of the intestinal glands, and in pros-
tatic fluid. Since IgA is selectively concentrated in body secretions, it is thought to play a protective role at mucous surfaces.
IgA antibody can be active against a wide spectrum of antigens,
including Brucella suis, diphtheria bacilli, bacterial toxins,
viruses, and ragweed pollen. Selective deficiency of IgA is
associated with frequent respiratory infections and increased incidence of autoimmune diseases. The IgA group of immunoglobulins
is the most thymic dependent of the five (41). It circulates in
the serum and, in a special secretory form, seems to be a major
force in local antibody systems at the body surfaces, thus becoming
a first line of defense.

IgE. The reaginic factor responsible for the immediate
hypersensitivity skin reactions is a special immunoglobulin now
known as IgE. Trace amounts of IgE are present in serum, with
elevated levels noted in certain patients with pollen allergies
and helminth infections. Its function in the immune response is
not yet known.

IgD. The most recently discovered class of immunoglobulin,
IgG has a half life of approximately 3 days. Only a small amount

is found in the plasma of a healthy person, where it demonstrates
antibody activity against penicillin, insulin, milk proteins,
diphtheria toxoid, and nuclear thyroid antigens. IgD also appears
on the surface of B lymphocytes, usually with IgM, and may be
important as a B cell antigen receptor, perhaps involved in B cell
differentiation.

Cell-mediated immunity enables a person to develop delayed
allergic reactions (as in the tuberculin skin test), and to reject
skin allografts. It enables cells to initiate graft vs. host
(GvH) disease (as in bone marrow transplantation) and to respond
in vitro by blast transformation and replication when stimulated
by: certain plant mitogens such as phytohemagglutin (PHA) and
concanavalin A (Con A) in solution; by allogeneic cells, as in the
mixed lymphocyte culture test (MLC); and by different antigens to
which the host has been previously exposed. The T cells of CMI
provide for primary defense against virus, fungi, and low grade
bacterial pathogens, such as the tubercle bacillus. In addition,
CMI provides the host with an intricate immune memory which can be
of lifetime help in fighting off infection and in responding to
antigens which the body recognizes as foreign.

At least three kinds of T cells have been identified and
studied (60). These are helper cells, killer cells, and suppressor
cells. Helper cells are so-named because they cooperate with the
B cell system, and are an absolutely necessary presence in the
making of antibody against a whole group of so-called T-dependent
antigens. Killer T cells are capable of direct cytotoxicity
against certain antigens found on body cells or against foreign
cells, while suppressor cells, capable of inhibiting antibody production by B lymphocytes, seem to be an important regulatory influence within the immune system (24,48). Shou et al. (61) in
our laboratory, for example, have recently reported the existence
of a population of· cells present in healthy volunteers which can
be induced by Con A to manifest suppressor functions. We see this
as indicative of an important but latent immunologic control.

Currently, suppressor cells are the focus of much attention
because laboratory and clinical reports have begun to link them
with the immunodeficiencies in a number of disease states which,
for a long time, have been poorly understood, i.e., common variable
hypogammaglobulinemia (a primary immunodeficiency disease) (74),
Hodgkin's disease (73) and multiple myeloma (both diseases with a
significant secondary immunodeficiency) (12), and in animals, with
autoimmune disease (30).

It is Waldmann et al. (74) who have recently shown that the
common variable immunodeficiency (CVID) is associated with the
presence in circulation of T lymphocytes which suppress synthesis

and secretion of immunoglobulins. Siegal et al. (64) and Schwartz
et al. (58) in our laboratory have confirmed and extended this
finding and have been able to show that suppressor cells are
present in the blood not only in patients with CVID, but also in
those with X-linked infantile agammaglobulinia, thymoma deficiency,
and with some cases of severe combined immunodeficiency (SCID), in
which acceptance of the B cell population from the donor has been
inhibited by suppressor cells (64).

        The immunodeficiency diseases of man can be considered as
either primary or secondary in nature. The primary immunodeficiencies are generally genetically determined and may in polar
forms involve deficient function of either the B cell or the T
cell system. Many of the primary immunodeficiency diseases,
however, are characterised by defects in both systems (19,39,36,38,
54).

        Manifestly important to effective, incisive treatment of
these diseases is an exact knowledge of the disorder involved. It
is therefore to be hoped that what we understand on a cellular
level today may tomorrow be comprehended at the molecular level.
Already this is, in some cases, beginning to happen.

        Among those primary immunodeficiency diseases listed by the
WHO expert committee are: (25)

        Bruton-type X-linked Infantile Agammaglobulinemia. (13,11,34,
26) This disorder represents a polar form of primary immunodeficiency disease, with the defect occurring in the B-cell (humoral)
system only. While the affected male babies seem normal at birth,
they begin at 6-12 months to be troubled by frequent infections
due to high grade encapsulated pyogenic bacteria such as pneumococcus, Hemophilus influenza bacilli, streptococcus, meningococcus,
and pseudomonas aeruginosa. Normally, such organisms are checked
by the B cell system.

        These patients are, on the other hand, able to withstand T-
cell mediated infections due to many viruses, fungi, tubercle
bacilli, and atypical acid fast organisms. Interestingly, they
are also not much troubled by gram negative bacterial pathogens.
In the lymph node there is a complete absence of germinal centers
and a cell population in the far cortical areas which ordinarily
sponsor a B cell population, and a lack of plasma cells in the
medullary cords, but otherwise the lymph node seems quite normal.
Usually the thymus is normal. Touraine et al. (69,70) and Incefy
et al. (42) found that bone marrow cells of agammaglobulinemic patients could, under the influence of thymic extract factors,
differentiate into T lymphocytes, thus establishing the presence
of T precursors. Indeed, the entire T cell system of these pa-

tients is quite intact.

In patients with agammaglobulinemia, fully developed B lymphocytes and plasma cells are absent from the blood and lymphoid
tissues, yet they have normal or near normal numbers of lymphocytes in the blood. The circulating T cell population seems to be
increased in number, and there is also present an as yet undefined
population of lymphocytes. Perhaps they are K cells, or possibly
imperfectly developed B cells, since they carry other markers
associated with B cells, such as Fc receptors, complement receptors, and an inability to form rosettes with sheep red blood cells
(64).

The lack of fully developed B lymphocytes makes these patients
incapable of producing immunoglobulin antibodies, except infre-
quently, and in very low numbers. IgA, IgD, and IgE, and usually
IgM have not been found in the X-linked agammaglobulinemic patient,
though extremely low levels of IgG have been reported. Diagnosis
is therefore aided by analysis of quantitative levels of immunoglobulin in the patient's serum. Individual immunoglobulins can
be enumerated by several means, one of which is the Mancini radio-
immunodiffusion assay. As early as the second week of life, the
patient may demonstrate low serum IgM (25).

Rather surprisingly, Siegal et al. (64) in our laboratory
found an excess of suppressor cell activity in these patients,
leading us to surmise that (1) it is possible that even in X-
linked infantile agammaglobulinemia that the primary pathogenesis
relates to an X-linked genetically determined presence of suppressor cells, which could act early in the differentiation of B cells
by interfering with development and capacity to generate cell
surface immunoglobulin, or (2) (the preferred hypothesis) it is,
rather, a genetically determined defect of B cell differentiation
which would, in fact, lead to the generation of suppressor cells
as a secondary event. This idea is further supported by the
finding of a suppressor population in a patient with thymoma and
agammaglobulinemia.

No way has yet been found to permanently correct the X-linked
Bruton type agammaglobulinemic defect. For successful episodic
treatment, early detection is important. Patients usually respond
well to appropriate drug therapy for the frequent bacterial infections which are inevitable. As prophylaxis, intramuscular injection of gammaglobulins at regular and frequent intervals is
recommended. On such a regimen, patients have reached adult life.
Injections, however, are painful and may be associated with many
unpleasant reactions, e.g., the syncopal anaphylactoid reactions
associated with absorption of aggregated gamma globulin.

DiGeorge Syndrome. (23) In counterposition to X-linked
infantile agammaglobulinemia is the DiGeorge syndrome, characterized by a lack of T cell number and' function due to absence of a
normal thymus gland. This is a rare disease, the cause of which
remains unknown. Unlike most of the primary immunodeficiency diseases, no genetic basis as yet has been ascertained, although
athymic nude mice, an experimental counterpoint, are known to have
an autosomal recessive defect associated with their malady.

Humoral immunity in the DiGeorge patients seems to be relatively intact. The far cortical areas of the lymph node are normal,
and in some cases poorly developed germinal centers are present (B
cell areas). There is, furthermore, a fairly well-developed plasma
cell population in the medullary cords. DiGeorge plasma cells
produce antibody when stimulated by thymus independent antigens,
but those antigens which require T helper cell cooperation cannot
elicit a response. Thymic dependent areas of the lymph node are
abnormal and severely depleted, and no T cells are to be found in
the blood. Chest reontgenograms should reveal absence of thymic
shadow, although absence of thymus is not always complete. (Such
patients are recognized as suffering from the partial DiGeorge
syndrome.) Incefy et al. (42), and Touraine et al. (69), exploring
a new technique for analyzing immunodeficiency diseases, have
found that patients with the partial DiGeorge syndrome possess
marrow cells which were readily differentiated into cells with T
markers after a short period of incubation with thymic extracts,
suggesting that the stem cells with T potential are somehow made
possible by the partial thymus.

Troubled by more than immunodeficiency, DiGeorge patients
sometimes die early because of congenital abnormalities in the
heart vessels. Frequently they have no parathyroid glands, and
they can present an abnormal appearance, with a prominent fore-
head, enlarged ears, and bowed mouth.

Diagnosis is made by testing for presence of thymus, for proper function of cell-mediated immunity, and enumeration of T cell
lymphocytes. An allogeneic fetal thymus transplant has, in several
instances, completely corrected the immunological disorders associated with the DiGeorge syndrome. Fetal thymus, generally inserted
into the abdominal muscle, is used because the fetal thymic epithelium is capable of differentiating into a mature thymus, while the
young thymocytes are as yet not mature enough to initiate a GvH
reaction. Transplantation brings a rapid restoration of T cell
function, and karyotypic analyses reveal that the newly immunocompetent lymphocytes are of host origin. Soon after transplantation,
B cell numbers, which are often elevated, tend to return to normal
(6,15,2,28).

Treatment of endocrinopathy due to absence of parathyroids requires appropriate use of Vitamin D or equivalent sterol therapy in order to raise the serum calcium level.

Severe Combined Immunodeficiency (SCID) (40). SCID is a particularly lethal form of primary immunodeficiency, because it encompasses total or near absence of both cell-mediated and humoral immunity. Genetically determined, it can manifest considerable variation in phenotypic expression, such that a degree of T or B cell development can occur in some affected family members, while in others there may be no sign of immune activity at all. The WHO expert committee has set up categories for X-linked, autosomal recessive, and sporadic SCID, each a distinct form of the disease (25). It can occur in some patients in conjunction with a specific enzyme deficiency (adenosine deaminase).

Children with SCID have an epithelial thymus, but it is underdeveloped, as are the lymph nodes and spleen, which are almost completely devoid of lymphoid cells. There is no in vitro response to mitogens, but cells of some children have responded to allogeneic cells in MLC reactions. Lymphocyte numbers are low, but not absent, while antibody production is either absent or grossly deficient.

If not treated, patients with SCID die in their first or second year of some sort of infection, to which they fall universal prey. Live virus immunization often leads to lethal infection, and blood transfusions can lead to death by GvH. From birth they are bothered by persistent and invasive infections with Candida, BCG, polio and measles virus, fungus infection, and high and low grade pathogens.

Knowing that multipotential and committed stem cells were concentrated in the bone marrow and that patients with SCID lacked, or had an impairment of, the T and B stem cell population, we decided as recently as ten years ago that bone marrow transplantation might be a means of reconstituting the faulty cells of such patients (3,27).

The process sounds deceptively simple. Needles and syringes are used to aspirate normal red bone marrow from cavities inside the bones of the donor. Suspensions of living bone marrow stem cells are then made and injected directly into the blood stream or peritoneal cavity of the recipient. The donor stem cells circulate through the blood of the patient to lodge in bone cavities where they will, if all prospers, grow and replicate (68).

From the first tentative experiments in animals and then humans, it was found that matching of donor and recipient at the

major histocompatibility system (H_2 in mice or HLA in man), in
addition to testing for donor and recipient lack of response in
the MLC test, were necessary in order (1) to avoid a lethal GvH
reaction and (2) to promote normal development of lymphoid and
hematopoietic cells (78). Siblings have, in almost every successful case, been used as donors, there being a 1 in 4 chance of HLA
compatibility among sibs.

Even with HLA matching between siblings, GvH has remained a
clinical problem for bone marrow transplantation. GvH occurs when
immunocompetent cells contained in the graft material recognize
the tissue of the host as "non-self" and attack. Major sites of
injury in GvH reactions are lymphatic tissues, hematopoietic
tissues, skin, intestines, and liver of the host. GvH, however,
has not produced lethal disease when siblings matched at HLA and
in the mixed leukocyte culture have been employed for the cellular
engineering to correct SCID. When less perfect matches have been
used, GvH has been lethal, or a major disease problem for these
patients. Some methods have been devised to mitigate the vigor of
GvH, most notably: (1) a separation of stem cells in donor marrow
from immunocompetent lymphocytes. (Density separation in albumin
gradients can to some degree accomplish this); (2) Use of fetal
liver, instead of fully developed bone marrow. (Fetal liver
contains stem cells, but not T cells mature enough to initiate a
GvH reaction); (3) Following the lead of Müller-Ruchholtz et al.
(51,52,53), we are working now on a serological technique, using
an antisera specific for T lymphocytes (but nonreacting with stem
cells), which could be the most effective way of separating the
potential aggressor cells from the stem cells.

In our own experiments with Yunis et al. (79), the fact that
fetal liver, spleen from strains of mice which develop T lymphocytes relatively late in life, and spleens from more mature neonatally thymectomized mice can be used to achieve full hematopoietic reconstruction after fatal irradiation, even across major
histocompatibility barriers without producing GvH, indicates that
it is the postthymic T cell which must be eliminated if GvH is to
be avoided (72,80).

Positive evidence of a successful transplant can be gleaned
from genetic markers, such as those found on red blood cells and
granulocytes in the marrow and blood, from chromosomes on dividing
cells in the marrow, as well as from the complete reconstitution
of humoral and cell-mediated immunity. Incefy et al. (43,44) are
studying marrow of patients with SCID before and after transplantation, using both relatively crude thymic extracts and the highly
purified and defined peptide derived from the thymus. These studies have shown that patients with several different genetic forms
of SCID have stem cells after successful marrow transplantation

that can now be differentiated into cells with T and B markers, whereas, prior to the transplant, such cells were lacking. We hope that such studies will lead us to ways of correcting SCID through a macromolecular manipulation of cells, eliminating the need for marrow transplantation, which is difficult and sometimes in itself a life-threatening mode of treatment, but these studies are still in their infancy.

In our recent suppressor cell study, Siegal et al. (64) had two patients with SCID, one with an associated adenosine deaminase (ADA) deficiency. Both children demonstrated suppressor activity in vitro after bone marrow transplantation. T cell engraftment had been achieved in each case, but at the time of the study, they both remained hypogammaglobulinemic. Here the degree of suppression seemed to correlate with the in vitro cell yield, which had not been true for patients with other immunodeficient syndromes. This phenomenon may be related to the failure of B cell "take" in one of the children.

Severe Combined Immunodeficiency with Adenosine Deaminase Deficiency (31,49,56,16). Similar to the SCID described above, this particular disorder is complicated further by the apparent absence of the enzyme adenosine deaminase in red blood cells, lymphoid cells, and elsewhere in the body. While the amount of functional and cellular deficiency varies more in this disease, it is in most other ways like the X-linked and autosomal recessive forms of SCID. Some lymphoid tissue development does occur, however, and the thymus in these patients resembles an atrophic, involuted organ, as opposed to the underdeveloped, fetal-like thymus of the other SCID's.

The relationship between ADA deficiency and this form of SCID has not yet been defined, but it is possible that (1) some toxic product resulting from the enzyme deficiency could cause the immunodeficiency, or that (2) this enzyme could be essential to the development of lymphoid cells. Parents of these children, presumed to be carriers of the disease, were found to have ADA levels intermediate between the patients and the controls (49).

Some patients with the ADA form of SCID have been immunologically reconstituted by bone marrow transplantation, after which lymphoid cells were found to possess normal amounts of ADA, whereas red cells continued to lack the enzyme completely. The question remains whether or not ADA deficiency is pathogenic in this disease, or merely represents a closely linked marker of the SCID deficiency gene.

In our most recent studies, Trotta et al. (71) have discovered that, in our patients with apparent lack of ADA, the enzyme is,

indeed, present but coupled in the cell with an inhibitor of ADA
action. Whether or not this inhibition interferes in some way
with lymphocyte development needs to be determined. Recently
Giblett et al. (32) described a patient in whom a nucleoside
phosphorylase deficiency was associated with a profound deficiency of T cells and cell-mediated functions.

Thymoma with Immunodeficiency (33,47). This particular immunodeficiency, accepted as an experiment of nature, led to laboratory
investigations which were instrumental in teaching us about the
thymus and its important role in the development and maintenance
of proper immune function. (36). Patients with thymoma and immunodeficiency are generally hypo- or agammaglobulinemic. With deficiencies in T and B cell number and function, they are susceptible
to a broad range of bacterial, viral, and fungal infections. They
may also be troubled by anemia. Removal of the thymus tumor,
which is composed of stromal epithelial cells, has not corrected
the immunodeficiency.

Although the pathogenesis of this disease is not yet clear,
Siegal et al. studied one thymoma patient who had demonstrable
circulating B cell precursors which, when provided with normal
serum, could differentiate into apparently normal B cells (62).
It was shown in this study that plasma factors were required early
in the B cell differentiation process.

Siegal et al. (62,63,64) discovered that, like patients with
SCID, patients with thymoma and agammaglobulinemia have suppressor
cells in their marrow, a finding which has been confirmed by
Waldmann et al. (75). In Siegal's study of the thymoma agammaglobulinemia syndrome, removal of neither rosetting cells nor
adherent cells from the mononuclear population of the patient
permitted plasma cell generation after stimulation by pokeweed
mitogen. Because technical difficulties may, however, have been
responsible for the negative findings, the question remains a very
open one, but at the moment it seems that the thymoma agammaglobulinemia patients may differ from at least some of the common
variable immunodeficiency patients, in that suppressor cells are a
secondary occurrence, rather than pathogenetic. In this particular matter, thymoma patients seem similar to those with Bruton's
agammaglobulinemia.

Immunodeficiency Associated with Ataxia-Telangiectasia (5,9,
55). Ataxia-telangiectasia is a multi-faceted familial disease,
of which the immunological disorder is only one of many problems.
The endocrine and central nervous systems, lymphoid tissues,
liver, and skin are all abnormal in these patients, but an underlying, connecting cause has heretofore eluded investigators. From
an immunological point of view, the disease presents a variety of

demonstrable immune deficiencies from patient to patient, and
from time to time in the same patient. Some form of both humoral
and cell-mediated immune deficiencies, however, are generally
present, with IgA and IgE absent or decreased (IgE is frequently
also absent from circulation in healthy family members). Antibody responses to a broad spectrum of bacterial and viral antigens are impaired.

     Lymph nodes are often small, with reduced numbers of follicles
and germinal centers in the far cortical regions. Plasma cells
are normal or diminished, but there is generally evidence of
reticulum cell hyperplasia. Thymus dependent areas of the lymph
node are depleted, and the lymphoid tissues along the gastro-
intestinal tract are hypocellular and underdeveloped. The thymus
appears to be embryonic in development. Lymphoid tissues thus
reflect deficiencies of both cellular and humoral immunity.
Often, however, the cell-mediated immunity in these patients seems
slight at first, but becomes progressively deficient. Delayed
hypersensitivity is usually diminished, and rejection of skin
allografts prolonged (5). Boumsell et al. (10), however, found
that marrow cells from one child with ataxia-telangiectasia,
complicated with lymphosarcoma, could be readily induced to
differentiate into T lymphocytes. Inadequate local immunity is
especially apparent when these patients are stimulated by viral
antigens seeking ingress through oral or intranasal routes.

     Patients with ataxia-telangiectasia suffer from frequent
sinopulmonary infections, which seem to be linked to the deficiency
of IgA. When challenged by antigens, they are often able to mount
an antibody response, but their response falls short of that
produced by age-matched controls.

     Neither the origin nor the cause of ataxia-telangiectasia is
known. Some have considered the possibility of a maternal-fetal
immunological incompatibility as a causative factor. A basis in
isoimmunization has also been proposed as cause, since isoimmunization with a homologue of the T cell specific theta antigen might
account for both the central nervous system and the immunological
abnormalities. A third possibility is the presence of a chronic
progressive viral infection, the origin of which would be related
to the host's genetic makeup.

     Waldmann (76) recently has discovered that these patients
harbor alpha-feto protein in their blood, an interesting finding
in view of the fact that more than 10% of the patients with A-T
develop some sort of malignancy. This usually takes the form of
reticulum cell sarcoma, lymphosarcoma, leukemia, or cancer of the
gastro-intestinal tract. The elevated level of fetal proteins,
especially that of hepatic origin, lends credence to the hypothesis

that the liver, as well as the thymus, has not matured completely, and that there is in this disease a general failure of tissue differentiation. Especially affected are those tissues requiring endodermal-mesodermal interactions or mesenchymal induction for maturation. Such a defect could explain the immunological mal-development in patients with A-T and may serve also to contribute to the high incidence of neoplasia.

Unfortunately, no corrective treatment of a permanent or long-lasting nature has yet been found for patients with A-T. They generally succumb in early or middle adolescence to broncho-pulmonary infections, lymphoreticular malignancy, or both.

The Wiskott-Aldrich Syndrome (WAS) and Immunodeficiency (46,7,8,18). Patients with the Wiskott-Aldrich syndrome (WAS) present a typical clinical triad of symptoms, including eczema, purpura due to thrombocytopenia, and severe recurrent infections. There is variation in severity of symptoms from patient to patient, but all experience bloody diarrhea during the first year of life and eczema within the first four months.

A disorder of the X-linked inheritance, the basic polypeptide or enzyme deficiency which underlies the WAS remains undefined. It is, however, a particularly lethal form of immunodeficiency, and the affected male children die in infancy or early childhood from infection due to virus, bacteria, or fungus; from bleeding caused by an abnormality of platelet number and structure; or from cancer (10%). Usually, but not always, malignancy takes the form of lymphoreticular sarcoma.

Abnormality within the humoral immune system is evidenced by elevated levels of IgA and IgE, and a reduced level of IgM. IgG levels are generally normal. Low IgM is thought to be responsible for an inability in these patients to produce antibody to certain antigens, especially polysaccharides. Patients with the WAS demonstrate a progessive failure of CMI. There is often a reduced number of lymphocytes in peripheral blood, and a progressive decline of lymphocyte responses in vitro. Changing morphology of lymphoid tissue reflects the progressive deterioration of immune function.

According to WHO guidelines (25), there are two forms of this disease, one associated with a deficiency in monocyte IgG receptors and the other not. Also, in some WAS patients, platelet metabolism is normal, while in others, it is not, yet patients in the same family are consistently similar in either all having or not having the platelet defect. Spitler et al (66) concluded after studying nine patients with the WAS (and selected family members), that al-though there are great clinical similarities found in the disease, the underlying immunopathologic abnormalities were many and varied. They argue that the basis for the combined humoral and cell-mediated deficiencies could rest at different levels of the immune response in different patients.

Until recently, standard therapy for patients with the WAS provided supportive care only. Now, however, three additional modes of immunotherapy are available, i.e., plasma infusions, bone marrow transplantation, and transfer factor therapy.

Patients involved in plasma therapy receive infusions from a single donor approximately every three weeks. Although not of proven efficacy, this is done to counteract the impaired humoral immunity, and many investigators believe it to be of some help. A small number of bone marrow transplantations have now been success-fully performed on patients with the WAS, and clinical improvement has been attained, but the difficulties inherent in bone marrow transplantation itself, (i.e., the need for a properly matched donor and the use of immunosuppressive agents, in addition to the risk of a fatal GvH disease), curb its potential usefulness for the moment. Transfer factor, a dialyzable extract of sensitized leukocytes, apparently can in some patients with the WAS, bring about clinical improvement and partial reconstitution of cell-mediated immuntiy. This mode of therapy has, however, been helpful only for those patients with a defect in the monocyte IgG receptor. Blaese links the immunodeficiency in these patients to a defect of mononuclear cell function (8).

Complement Component Immunodeficiency Diseases (22,39). There are nine different complement components in man (Cl–C9), which together account for 10% of the human serum globulin. These components are made up of eleven different proteins. Antigen-antibody complexes set complement components into motion, and, once stimulated, they act in a specific sequence that has become known as the complement cascade. That classical sequence begins with Cl, going on to subsets Clq, Clr, Cls, then C4, C2, C3, C5, C6, C7, C8, and finally, C9. The total cascade can be measured in vitro by the lysis of sensitized red blood cells.

The complement cascade, with its several by-products, serves immune function by acting as mediator of the inflammatory process, altering vascular permeability, causing leukocyte chemotaxis, smooth muscle contraction, and histamine release from mast cells.

Disorders within the complement system generally are heredi-tary and may cause a wide variety of immunological disorders. It is of interest to note, for example, that the Clq is somewhat diminished in patients with X-linked infantile agammaglobulinemia, and almost absent in some patients with SCID. It is, however, nearly normal in patients with the DiGeorge syndrome, suggesting

that there is some connection between synthesis of C1q and proper
development and/or function of immunoglobulin molecules. See
Table I for a list of specific diseases, most of them highly
lethal, which have been associated with different complement
deficiencies.

TABLE I

Complement Deficiency   Associated Diseases

C1q   severe combined immunodeficiency, re-
                             current severe infections

C1r   ecthyma gangrenosum, recurrent bacter-
                             ial infections, lupus-like syndrome,
                             frank lupus erythematosus, progressive
                             glomerulonephritis, arthritis

C1s   lupus erythematosus

C4   lupus, arthritis, anaphylactoid pur-
                             pura

C2   lupus, fatal dermatomyositis, ana-
                             phylactoid purpura, lupus-like syn-
                             drome, progressive glomerulonephritis,
                             anaphylactoid purpura, recurrent in-
                             fections

C3   marked increased susceptibility to in-
                             fection

C5   lupus, diarrhea, and wasting

C5 dysfunction   Leiner's syndrome, persistent skin
                             infection with yeast, and gram nega-
                             tive bacteria

C6   Recurrent gram negative bacterial in-
                             fections, polyarticular gonorrheal
                             arthritis, recurrent meningitis

C7   Raynaud's syndrome, lupus?

C8   Recurrent gram negative infections,
                             gonorrheal meningitis

C9

Exactly how a particular complement abnormality results in the different disease symptoms noted in the table is still an open question. Since complement, however, is but one of many biologic mediators of disease, each being extremely complex and interrelated, it apears that answers, when they come, will also be complex. Ballow et al. (4) have shown a reconstitution of deficient Clq in SCID patients following bone marrow transplantation. Kidney transplantation used to treat end stage renal disease consequent to Clr deficiency increased levels of CH50 (total hemolytic complement), Cl, and Cls. This patient had a genetically determined deficiency of Clr. The presumption is that the kidney could produce sufficient Clr to correct, by cellular engineering, this inborn error of metabolism (21).

Selective IgA Deficiency (35,1,14,20,57). Although it is possible to have a selective deficit of any of the immunoglobulin classes or subclasses, it is IgA which most frequently is found lacking. Patients with IgA deficiency lack both serum and secretory IgA. Under normal circumstances, the secretory IgA is associated with a molecule called the secretory piece, which facilitates transportation of the IgA molecule across the mucosal barrier into body secretions. It has generally been found that people with an IgA deficiency, although lacking serum and secretory IgA, have an abundance of the free secretory piece in saliva and jejunal fluids. The genetic background of this disorder is unclear. Both autosomal recessive and autosomal dominant cases have been noted, but most often it seems to occur sporadically.

Under normal circumstances the secretory form of IgA is the predominant immunoglobulin found in body secretions, i.e., colostrum, tears, saliva, and hence, at body surfaces, where there is maximal exposure to foreign antigens. Secretory IgA is therefore extremely important as a first line of defense. While, unexpectedly, many people lacking IgA seem to experience no ill consequences at all, others, the majority, suffer from one or many of the following:

Atopy. Soothill et al. (65) have recently learned that classic atopy, caused by IgE, is more prevalent in patients who develop IgA more slowly than those in the normal population. It has also been noted that IgE levels are sometimes elevated in the IgA deficient patients.

Recurrent respiratory infections. Secretory IgA is thought to play a role in preventing bacterial agents and viruses from forming colonies on mucous surfaces. Increased numbers of respiratory infections would then seem a natural result of insufficient IgA. The presence of IgE in normal or elevated amounts has been associated with the respiratory problems of these patients.

Autoimmune disorders. The most common autoimmune diseases encountered in IgA deficient patients are rheumatoid arthritis and
lupus erythematosus, but a whole galaxy of autoimmunities have
been linked by Ammann and Hong to IgA deficiency (1,35). The
autoimmune phenomena in these patients have not been satisfactorily explained in toto as yet, but several theories have been
advanced. One possibility is that autoimmunity could result from
a thymus deficiency, which sometimes occurs in conjunction with
IgA deficiency (DiGeorge patients, for example, often have low
levels of IgA). Autoimmunity could also be fostered by the immune
imbalance naturally caused by the absence of IgA, or by the presence of a virus infection permitted by inadequate host defense.
Inadequate immunity, in fact, might allow many antigens, formerly
forbidden, to enter the body (across the gastro-intestinal epithelium, for example), where they may stimulate chronic inflammation
or autoimmune activity.

Gastrointestinal disease. Patients with insufficent amounts
of IgA suffer from many gastro-intestinal disorders. It is not
yet known whether this is due to lack of protection normally
administered in that area by secretory IgA, or whether it is
instead a consequence of autoimmune activity.

Malignancy. Carcinomas observed in these patients have
almost always occurred in places where IgA is normally found,
i.e., the colon, lung, esophagus, stomach, and lymphoid tissues.

Evidently the IgA deficiency is caused by a lack of IgA
secretion, not synthesis, for the majority of these patients
exhibit normal numbers of circulating B lymphocytes with IgA
molecules at their surfaces. The reason these cells do not mature
into IgA secreting plasma cells may be due to suppressor T cell
activity. It seems plausible that the immunodeficiency present in
these patients would be corrected by a process enabling the impaired cells to differentiate into IgA secreting bodies. Wu et
al. (77) and Waldmann et al. (74) have recently had some success
in reversing this process in vitro.

Very recently Strober et al. (67) have described one patient
with a unique form of IgA deficiency. The young boy, troubled by
chronic diarrhea, demonstrates normal levels of IgA in the serum,
but has no secretory IgA and no secretory piece. Further analysis
has revealed normal or elevated levels of surface IgA bearing
lymphocytes in the blood, which in vitro, have responded vigorously
to pokeweed mitogen, secreting IgA.

In vitro studies of biopsied jejeunal mucosa, however, revealed absence of IgA synthesis, therefore, indicating that,
although normal IgA bearing and secreting cells were present in
the serum, these cells were not to be found as expected in the

mucosa. The authors postulated that the absence of the secretory IgA piece, which is produced by an epithelial, not a lymphoid cell, has resulted in a deficient mucosal microenvironment that is incapable of providing the normal homing, differentiative, or proliferative signal to the circulating IgA precursor cells.

Common Variable Immune Deficiency (74,59,25). The majority of immune deficiency disorders still cannot be specifically defined, and thus have been lumped together under the rather vague heading of common variable immune deficiencies. Most prevalent of the primary immunodeficiency states, the particular immunodeficiency manifested (decreased numbers of all or several immunoglobulins, feeble antibody production, impaired cell-mediated immunity) can vary significantly from time to time in the same patient and from patient to patient in the same family. It is expected that, as we learn more about this heterogeneous group, several distinct disease entities will eventually appear.

Family studies have given evidence for autosomal recessive and autosomal dominant modes of inheritance. It is also, because of late onset, sometimes thought to be acquired. Patients in this group may suffer from a wide variety of problems, including pneumonia, sinusitis, otitis media, gastro-intestinal infection, particularly infestation by Giardia lamblia, and autoimmune disorders. Pernicious anemia, histamine-fast achlorhydria, and gastric cancer are encountered in these patients more frequently and at an earlier age than normal.

Immunoglobulin studies have revealed that patients with common variable immune deficiency may have: (1) no circulating B cells at all, (2) B cells incapable of synthesizing immunoglobulins, or (3) B cells incapable of secreting immunoglobulins. The B cell defect can thus occur at each level of the differentiation process. Touraine et al. and Incefy et al. (69) have found that the problem in these patients does not involve absence of a T cell precursor, since in their experiments, bone marrow cells could be induced under the influence of thymic factors to differentiate into T cells.

Waldmann et al. (74) have shown, however, that the immune defect in common variable immunodeficiency is not to be found solely in the B cell system, but is rather due to a complex interaction between abnormally active T lymphocytes and potentially normal B lymphocytes. His studies suggest that in many patients with this disease, there is an abnormality of regulatory (supprsssor) T cells, which inhibit B cells from maturing and producing antibody. In our laboratory, Siegal et al. (64) have confirmed this finding. All the common variable immune deficient patients we studied demonstrated some degree of suppressor activity.

We must now find a way to counteract the abnormal suppressor
cells; then, perhaps, a substantial number of patients with common
variable immunodeficiency syndrome could be given relief.

The current treatment for the common variable immunodeficiency patient is as yet restricted to supportive and prophylactic
measures. As with other immunodeficiency diseases, early recognition is important. Health problems should be carefully diagnosed
and treated specifically with appropriate therapy. For prophylaxis, large doses of gamma globulin have been administered at
frequent intervals, but this has not proven adequate. Gamma
globulin concentrates that could be given intravenously might well
be of greater help in the future. Meanwhile, infusions of plasma
(from donors free of hepatitis, Australian antigen and antibody to
Australian antigen) selected from a small pool (the buddy system)
have given these patients most help in the buttressing of an
impaired immune system.

REFERENCES

1. Ammann, A.J. and Hong, R. Selective IgA deficiency; presen-
    tation of 30 cases and a review of the literature. Medicine
    50:223, 1971.
2. August, C.S., Berkel, A.I., Levey, R.H., Rosen, F.S., and
    Kay, H.E.M. Establishment of immunologic competence in
    child with congenital thymic aplasia by graft of fetal thymus.
    Lancet i:1080, 1970.
3. Bach, F.H. and Good, R.A. Bone marrow and thymus transplants:
    cellular engineering to correct primary immunodeficiency.
    In: Clinical Immunobiology, Vol 2, ed. by F.H. Bach and
    R.A. Good, Academic Press, New York, 1974. pp. 65-115.
4. Ballow, M., Day, N.K., Biggar, W.D., Park, B.H., Yount, W.J.
    and Good, R.A. Reconstitution of C1q following bone marrow
    transplantation in patients with severe combined immuno-
    deficiency. Clin. Immunol. Immunopathol. 2(1):28, 1973.
5. Biggar, W.D. and Good, R.A. Immunodeficiency in ataxia-
    telangiectasia. In: Immunodeficiency in Man and Animals;
    Proceedings. (D. Bergsma, ed., R.A. Good, and J. Finstad,
    scientific eds.), Sunderland, Mass., Sinauer Associates,
    Inc., 1975. pp. 271-276 (Birth Defects: Original Article
    Series, Vol. XI, No. 1, 1975).
6. Biggar, W.D., Park, B.H., Stutman, O., Gajl-Peczalska, K.J.,
    and Good, R.A. Fetal thymus transplantation: Experimental
    and clinical observations. In: Immunodeficiency in Man
    and Animals; Proceedings. (D. Bergsma, ed., R.A. Good and
    J. Finstad, scientific eds.), Sunderland, Mass., Sinauer
    Associates, Inc., 1975. pp. 361-366, (Birth Defects: Original
    Article Series, Vol. XI, No. 1, 1975).

7. Blaese, R.M., Strober, W., Brown, R.S., and Waldmann, T.A. The Wiskott-Aldrich syndrome. A disorder with a possible defect in antigen processing or recognition. Lancet i:1056, 1968.

8. Blaese, R.M., Strober, W., and Waldmann, T.A. Immunodeficiency in the Wiskott-Aldrich syndrome. In: <u>Immunodeficiency in Man and Animals</u>; Proceedings. (D. Bergsma, ed., R.A. Good and J. Finstad, scientific eds.), Sunderland, Mass., Sinauer Associates, Inc., 1975. pp. 250-254, (Birth Defects: Original Article Series, Vol. XI, No. 1, 1975).

9. Boder, E. Ataxia-telangiectasia: Some historic, clinical and pathologic observations. In: <u>Immunodeficiency in Man and Animals</u>; Proceedings. (D. Bergsma, ed., R.A. Good and J. Finstad, scientific ed.,), Sunderland, Mass., Sinauer Associates, Inc., 1975, pp. 255-270. (Birth Defects: Original Article Series, Vol. XI., No. 1, 1975).

10. Boumsell, L., Incefy, G.S., Bernard, A., Schwartz, S., Smithwick, E., and Good, R.A. Differentiation in vitro in ataxia-telangiectasia associated with lymphosarcoma. J. Pediatr. 87:435, 1975.

11. Bridges, R.A. and Good, R.A. Connective tissue diseases and certain serum proteins in patients with agammaglobulinemia. Ann. N.Y. Acad. Sci. 86:1089, 1960.

12. Broder, S., Humphrey, R., Durm, M., Blackman, M., Meade, B., Goldman, C., Stroker, W., and Waldmann, T. Impaired synthesis of polyclonal (nonparaprotein) immunoglobulins by circulating lymphocytes from patients with multiple myeloma. Role of suppressor cells. New Engl. J. Med. 293:887-92, 1975.

13. Bruton, O.C. Agammaglobulinemia. Pediatrics, 9:722-728, 1952.

14. Buckley, R.H. Clinical and immunologic features of selective IgA deficiency. In: <u>Immunodeficiency in Man and Animals</u>; Proceedings. (D. Bergsma, ed., R.A. Good and J. Finstad, scientific eds.), Sunderland, Mass., Sinauer Associates, Inc., 1975. pp. 134-140. (Birth Defects: Original Article Series, Vol. XI, No. 1, 1975).

15. Cleveland, W.W., Fogel, B.J., Brown, W.T., and Kay, H.E.M. Foetal thymic transplant in a case of DiGeorge's syndrome. Lancet ii:1211, 1968.

16. Cohen, F. Adenosine deaminase and immunodeficiency. In: <u>Immunodeficiency in Man and Animals</u>; Proceedings. (D. Bergsma, ed., R.A. Good and J. Finstad, scientific eds.), Sunderland, Mass., Sinauer Associates, Inc., 1975. pp. 124-127. (Birth Defects: Original Article Series, Vol. XI, No. 1, 1975).

17. Cohen, S. and Milstein, C. Structure and biological properties of immunoglobulins. Advan. Immunol. 7:1, 1967.

18. Cooper, M.D., Chase, H.P., Lowman, J.T., Krivit, W., and Good, R.A. Wiskott-Aldrich syndrome: an immunologic deficiency disease involving the afferent limb of immunity. Amer. J. Med. 44:449, 1968.

19. Cooper, M.D., Faulk, W.P., Fudenberg, H.H., Good, R.A., Hitzig, W., Kunkel, H.G., Roitt, I.M., Rosen, F.S., Seligmann, M., and Soothill, J.F. Meeting report of the second international workshop on primary immunodeficiency Diseases in man. Clin. Immunol. Immunopathol. 2:416-445, 1974.

20. Cooper, M.D. Cells producing IgA: Comment. In: Immuno-deficiency in Man and Animals; Proceedings. (D. Bergsma, ed., R.A. Good and J. Finstad, scientific eds.), Sunderland, Mass., Sinauer Associates, Inc., 1975. p. 143. (Birth Defects: Original Article Series, Vol., XI., No. 1, 1975).

21. Day, N.K. and Good, R.A. Reconstitution of C1 after kidney transplantation in a patient with chronic glomerulonephritis. (Unpublished observation.)

22. Day, N.K. and Good, R.A. Deficiencies of the complement system in man. In: Immunodeficiency in Man and Animals; Proceedings. (D. Bergsma, ed., R.A. Good and J. Finstad, scientific eds.), Sunderland, Mass., Sinauer Associates, Inc., 1975. pp. 306-311, (Birth Defects: Original Article Series, Vol. XI, No. 1, 1975).

23. DiGeorge, A.M. Congenital absence of the thymus and its immunologic consequences. In: Immunologic Deficiency Dis-eases in Man. (R.A. Good and D. Bergsma, eds.), The National Foundation Press, New York, 1968. pp. 116-121.

24. Dutton, R.W. and Scavulli, J. Suppressor T cells in the regulation of the immune response. J. Reticuloendothel. Soc. 17:187-189, 1975.

25. Fudenberg, H.H., Good, R.A., Goodman, H.C., Hitzig, W., Kunkel, H.G., Roitt, I.M., Rosen, F.S., Rowe, D.S., Seligmann, M., and Soothill, J.R. Primary immunodeficiencies. Bull. WHO 45:125-142, 1971.

26. Gajl-Peczalska, K.J., Park, B.H., Biggar, W.D., and Good, R.A. B and T lymphocytes in primary immunodeficiency disease in man. J. Clin. Invest. 52:919-928, 1973.

27. Gatti, R.A., Meuwissen, J.J., Allen, H.D., Hong, R., and Good, R.A. Immunologic reconstitution of sex-linked lympho-penic immunologic deficiency. Lancet ii:1366-1369, 1968.

28. Gatti, R.A., Gershanik, J.J., Levkoff, A.H., Wertelecki, W., and Good, R.A. DiGeorge syndrome associated with combined immunodeficiency. J. Pediatr. 81:920, 1972.

29. Geha, R.S., Schneeberger, E., Merler, E., and Rosen, F.S. Heterogeneity of "acquired" or common variable agammaglo-bulinemia. New Engl. J. Med. 291:1-6, 1974.

30. Gerber, N.L., Hardin, J.A., Chused, T.M., and Steinberg, A.D. Loss with age in NZB/w mice of thymic suppressor cells in the graft-versus-host reaction. J. Immunol. 113:1618-1625, 1974.

31. Giblett, E., Anderson, J., Cohen, F., Pollara, B., and Meuwissen, H.J. Adenosine deaminase deficiency in two patients with severely impaired cellular immunity. Lancet ii:1067, 1972.

32. Giblett, E.R., Ammann, A.J., Wara, D.W., Sandman, R., and
     Diamond, L.K. Nucleoside-phosphorylase deficiency in a child
     with severely defective T-cell immunity and normal B-cell
     immunity. Lancet i:1010-1013, 1975.
33. Good, R.A. and Varco, R.L. Clinical and experimental study
     of agammaglobulinemia. J. Lancet 75:245, 1955.
34. Good, R.A., Kelly, W.D., Rötstein, J., and Varco, R.L.
     Immunological deficiency diseases--agammaglobulinemia, lympho-
     gammaglobulinemia, Hodgkin's disease and sarcoidosis. Progr.
     Allergy 6:187, 1962.
35. Good, R.A. and Rodey, G.E. IgA deficiency, antigenic barriers
     and autoimmunity. Cellular Immunol. 1:147, 1970.
36. Good, R.A. Immunodeficiency in developmental perspective.
     In: "The Harvey Lectures, Series 67". Academic Press, New
     York, 1973. pp. 1-107.
37. Good, R.A. and Bach, F.H. Bone marrow and thymus transplants.
     cellular engineering to correct primary immunodeficiency.
     In: Clinical Immunobiology, Vol. 2., (F.H. Bach and R.A.
     Good, eds.), Academic Press, Inc., New York, 1974. pp. 65-115.
38. Good, R.A. The primary immunodeficiency diseases. In:
     Textbook of Medicine, 14th Ed., (P.B. Beeson and W. McDermott,
     eds.), W.B. Saunders, Philadelphia, 1975. pp. 104-109.
39. Good, R.A. and Finstad, J. Adaptive immunity. In: Patho-
     physiology, (E.D. Frohlich, ed.), J.B. Lippincott, Philadel-
     phia (in press).
40. Hitzig, W.H., Landolt, R., Muller, G., and Bodmer, P.
     Heterogeneity of phenotypic expression in a family with Swiss
     type of agammaglobulinemia: Observations on the acquisition
     of agammaglobulinemia. J. Pediatr. 78:968, 1971.
41. Horowitz, S. and Hong, R. Selective IgA deficiency--some
     perspectives. In: Immunodeficiency in Man and Animals; Pro-
     ceedings. (D. Bergsma, ed., R.A. Good and J. Finstad,
     scientific eds.), Sunderland, Mass., Sinauer Associates, Inc.,
     1975. pp. 129-133. (Birth Defects: Original Article Series,
     Vol. XI, No. 1, 1975).
42. Incefy, G.S., Touraine, J.L., Touraine, F., L'Esperance, P.,
     Siegal, F.P., and Good, R.A. In vitro studies on human T-
     lymphocyte differentiation in primary immunodeficiency dis-
     eases. Trans. Assoc. Am. Physicians 87:258-262, 1974.
43. Incefy, G.S., Boumsell, L., Touraine, J.L., L'Esperance, P.,
     Smithwick, E., O'Reilly, R., and Good, R.A. Enhancement of
     T-lymphocyte differentiation in vitro by thymic extracts
     after bone marrow transplantation in severe combined immuno-
     deficiencies. Clin. Immunol. Immunopathol. 4:258-268, 1975.
44. Incefy, G.S., Boumsell, L., Kagan, W., Goldstein,G.,DeSousa,M.,
     Smithwick, E., O'Reilly, R., and Good, R.A. Enhancement
     of T lymphocyte differentiation in vitro by thymic extracts
     and purified polypeptides in severe combined immunodeficiency
     diseases. Trans. Assoc. Am. Physicians (In press).

45. Janeway, C.A., Rosen, F.S., Merler, E., and Alper, C.A. The Gamma-Globulins. Little, Brown and Company, Boston, 1967.

46. Krivit, W. and Good, R.A. Aldrich's syndrome. A.M.A. J. Dis. Child. 97:137-153, 1959.

47. MacLean, L.D., Zak, S.J., Varco, R.L., and Good, R.A. Thymic tumor and acquired agammaglobulinemia--a clinical and experimental study of the immune response. Surgery 40:1010-1017, 1956.

48. Marx, J.L., Suppressor T cells: Role in immune regulation. Science 188:245-247.

49. Meuwissen, H.J., Pickering, R.J., and Pollara, B. Adenosine deaminase deficiency in combined immunological deficiency disease. In: Immunodeficiency in Man and Animals; Proceedings. (D. Bergsma, ed., R.A. Good and J. Finstad, scientific eds.), Sunderland, Mass., Sinauer Associates, Inc., 1975. pp. 117-119. (Birth Defects: Original Article Series, Vol XI, No. 1, 1975).

50. Miller, M.E. Introduction: complement diseases in man. In: Immunodeficiency in Man and Animals; Proceedings. (D. Bergsma, ed., R.A. Good and J. Finstad, scientific eds.), Sunderland, Mass., Sinauer Associates, Inc., 1975. pp. 299-300. (Birth Defects: Original Article Series, Vol. XI, No. 1, 1975).

51. Müller-Ruchholtz, W., Müller-Hermelink, H.K., and Sonntag, H.G. Specific inactivation of lymphoid cells in bone marrow grafts. Transplant. Proc. 5:877-880, 1973.

52. Müller-Ruchholtz, W., Wottge, H.U., and Müller-Hermelink, H.K. Selective grafting of hemopoietic cells. Transplant. Proc. 7:859-862, 1975.

53. Müller-Ruchholtz, W., Wottage, H.U., and Müller-Hermelink, H.K. Bone marrow transplantation in rats across strong histocompatibility barriers in selective elimination of lymphoid cells in donor marrow. Transplant. Proc. (in press) March, 1976.

54. Park, B.H. and Good, R.A. Principles of Modern Immunobiology, Lea and Febiger, Philadelphia, 1974.

55. Peterson, R.D.A., Kelly, W.D., and Good, R.A. Ataxia-telangiectasia: its association with a defective thymus, immuno-logical-deficiency disease, and malignancy. Lancet i:1189, 1964.

56. Pollara, B., Moore, J.J., Pickering, R.J., Gabrielsen, A.E., and Meuwissen, H.J. Combined immunodeficiency disease: an inborn error of purine metabolism. In: Immunodeficiency in Man and Animals; Proceedings. (D. Bergsma, ed., R.A. Good and J. Finstad, scientific eds.), Sunderland, Mass., Sinauer Associates, Inc., 1975. pp. 120-123. (Birth Defects: Original Article Series, Vol. XI, No. 1, 1975).

57. Polmar, S.H., Waldmann, T.A., and Terry, W.D. The relation-
     ship of IgA and IgE deficiency. In: Immunodeficiency in
     Man and Animals; Proceedings. (D. Bergsma, ed., R.A. Good
     and J. Finstad, scientific eds.), Sunderland, Mass., Sinauer
     Associates, Inc., 1975. pp 147-150. (Birth Defects: Original
     Article Series, Vol. XI, No. 1, 1975).
58. Schwartz, S.A., Choi, Y.S., and Good, R.A. (in preparation).
59. Seligmann, M., Fudenberg, H., and Good, R.A. A proposed
     classification of primary immunological deficiencies. Am. J.
     Med. 45:817-825, 1968.
60. Shiku, H., Kisielow, P., Boyse, E.A., and Oettgen, ·H.F.
     Immunogenetic identification of functional T-cell subsets.
     Trans. Proc. (in press).
61. Shou, L., Schwartz, S.A., and Good, R.A. Suppressor cell
     activity after Con A treatment of lymphocytes from normal
     donors. (manuscript submitted).
62. Siegal, F.P., Pernis, B., and Kunkel, H.G. Lymphocytes in
     human immunodeficiency states: a study of membrane-associated
     immunoglobulins. Eur. J. Immunol. 1:482-486, 1971.
63. Siegal, F.P., Wernet, P., Dickler, H.B., Fu. S.M., and Kunkel,
     H.G. B lymphocytes lacking surface Ig in patients with immune
     deficiency: initiation of Ig synthesis in culture of cells of
     a patient with thymoma. In: Immunodeficiency in Man and
     Animals; Proceedings. (D. Bergsma, ed., R.A. Good and J.
     Finstad, scientific eds.), Sunderland, Mass., Sinauer Associates,
     Inc., 1975. pp. 40-44. (Birth Defects: Original Article
     Series, Vol. XI, No. 1, 1975).
64. Siegal, F.P., Siegal, M., and Good, R.A. Suppression of in
     vitro differentiation of plasmacytoid cells by leukocytes
     from hypogammaglobulinemic patients. (in preparation).
65. Soothill, J.F. Interactions in immunodeficiency. In:
     Immunodeficiency in Man and Animals; Proceedings (D. Bergsma,
     ed., R.A. Good and J. Finstad, scientific eds.), Sunderland,
     Mass., Sinauer Associates, Inc., 1975. pp. 50-52. (Birth
     Defects: Original Article Series, Vol. XI, No. 1, 1975).
66. Spitler, L.E., Levin, A.S., Stites, D.P., Fudenberg, H.H.,
     and Huber, H. The Wiskott-Aldrich syndrome. Immunologic
     studies in nine patients and selected family members.
     Cell. Immunol. 19:201-218, 1975.
67. Strober, W., Krakauer, R., Kaleveman, H.L., Reynolds, H.Y.,
     and Nelson, D.L. Secretory component deficiency, a disorder
     of the IgA immune system. New Engl. J. Med. 294:351-356,
     1976.
68. Thomas, E.D., and Storb, R. Technique for human marrow
     grafting. Blood 36:507-515, 1970.
69. Touraine, J.L., Touraine, F., Incefy, G.S., and Good, R.A.
     Effect of thymic factors on the differentiation of human
     marrow cells into T-lymphocytes in vitro in normals and
     patients with immunodeficiencies. Ann. N.Y. Acad. Sci.
     249:335-342, 1975.

70. Touraine, J.L., Touraine, F., Incefy, G.S., Goldstein, A.L., and Good, R.A. Thymic factors and human T lymphocyte differentiation. In: The Biological Activity of Thymic Hormones; Proceedings (D.W. van Bekkum, ed.), Kooyker Scientific Publications, Rotterdam, 1975. pp. 31-35.

71. Trotta, P.P., Smithwick, E.M., and Balis, M.E. Restoration of normal adenosine deaminase activity in the red cell lysates of carriers and patients with severe combined immunodeficiency disease. (submitted for publication).

72. Tulunay, O., Good, R.A., and Yunis, E.J. Protection of lethally irradiated mice with allogeneic fetal liver cells: influence of irradiation dose on immunologic reconstitution. P.N.A.S. 72:4100-4104, 1975.

73. Twomey, J.J., Laughter, A.H., Farrow, S., and Douglass, C.C. Hodgkin's disease. An immunodepleting and immunosuppressive disorder. J. Clin. Invest. 56:467-475, 1975.

74. Waldmann, T.A., Broder, S., Blaese, R.M., Durm, M., Blackman, M., and Strober, W. Role of suppressor T cells in pathogenesis of common variable hypogammaglobulinaemia. Lancet ii: 609-613, 1974.

75. Waldmann, T.A., Broder, S., Durm, M., Blackman, M.,Krakauer,R., and Meade, B. The role of suppressor T cells in the pathogenesis of hypogammaglobulinemia with a thymoma. Clin. Res. 23:447A, 1975.

76. Waldmann, T.A., discussant in: Immunodeficiency in Man and Animals; Proceedings (D. Bergsma, ed., R.A. Good and J.Finstad, scientific eds.), Sunderland, Mass., Sinauer Associates, Inc., 1975. p. 275. (Birth Defects: Original Article Series, Vol. XI, No. 1, 1975).

77. Wu, L.Y.F., Lawton, A.R., Greaves, M.F., and Cooper, M.D. Evaluation of human B lymphocyte differentiation using pokeweed mitogen (PWM) stimulation: In vitro studies in various antibody deficiency syndromes. In: Proc. Seventh Leukocyte Culture Conference, (F. Daguillard, ed.), Academic Press, N.Y. pp. 485-500, 1973.

78. Yunis, E.J., Hilgard, H.R., Martinez, C., and Good, R.A. Studies on immunologic reconstitution of thymectomized mice. J. Exp. Med. 121:607-632, 1965.

79. Yunis, E.J., Good, R.A., Smith, J., and Stutman, O. Protection of lethally irradiated mice by spleen cells from neonatally thymectomized mice. P.N.A.S. 71:2544-2548, 1974.

80. Yunis, E.J., Fernandez, G., Smith, J., and Good, R.A. Long survival and immunological reconstitution following transplantation with syngeneic or allogeneic fetal liver and neonatal spleen cells. Trans. Proc. (in press).

MALNUTRITION, THE THYMOLYMPHATIC SYSTEM AND IMMUNOCOMPETENCE

R. L. GROSS and P. M. NEWBERNE

Massachusetts Institute of Technology
Cambridge, Massachusetts (USA)

Abundant evidence has accumulated in recent years demonstrating that adequate nutrition is essential to normal immunological function of the host. The close relationship of malnutrition, morbidity and mortality with experimental evidence of depressed immune responsiveness was initially elucidated in a number of extensive studies of children with protein-calorie malnutrition, including both Kwashiorkor and marasmus (4,7,12). Histological evidence of depleted thymus-dependent areas in lymphoid tissue correlated well with studies of cell-mediated immunity, such as skin testing, T cell rosette assay and mitogen stimulation. These parameters appeared to reflect impairment secondary to nutritional deficiency because delayed hypersensitivity returned to normal in a significant number of patients after dietary protein supplements. These cases however, represented far-advanced and severe nutritional deficiencies not permitting the assessment of the possible contribution of concomitant nutrient deficiencies such as vitamin A, iron and folic acid.

The lipotropic factors choline, methionine, folic acid and vitamin B_{12} play a central role in cellular metabolism through their regulation of the transfer and utilization of 1-carbon moieties Early studies in our laboratory revealed that rats deprived prenatally of choline and methionine or vitamin B_{12} have a decreased resistance to infection with Salmonella typhimurium in adult life (5,11). Since these preliminary studies we have extended our investigations to more detailed evaluations of immune function, particularly in folic acid deficiency.

The methylation of deoxyuridylate represents the rate-limiting

step in DNA synthesis. In the absence of folic acid the de novo
synthesis of thymidylate ceases and megaloblastic changes appear in
actively replicating cells. Other important metabolic functions of
folic acid include the de novo synthesis of purine bases; intercon-
version of glycine and serine; synthesis of choline; methylation of
homocysteine to methionine; and finally a postulated role in the
initiation of protein synthesis at the ribosomal level via the
formylation of methionyl t-RNA and m-RNA. Folic acid thus plays
a fundamental role in cell growth and reproduction via its involvement in nucleic acid, protein, amino acid and phospholipid metabolism. It seems reasonable to expect that the generation of an immune response requiring antigen recognition, blastogenesis, cellular transformation and production of lymphokines, all of which are
associated with intense metabolic activity including DNA synthesis,
might be impaired in folate deficiency.

Our studies (6) on the immune function in humans with megaloblastic anemia due to folic acid deficiency led us to the development of a model in the rat which allowed more extensive investigations of cellular immune function in experimentally induced deficiency.

MATERIALS AND METHODS

Experimental Human Subjects

Clinical studies were performed in South Africa where a high
incidence of megaloblastic anemia due to nutritional deficiency is
seen among the Bantu people. The experimental subjects in this
study were divided into 5 groups: group I consisted of 11 patients
with folate deficiency on a nutritional basis alone; group II included 7 obstetric patients with megaloblastic anemia of pregnancy,
although deficient dietary intake also contributed to the deficiency; group III consisted of 5 patients with combined megaloblastic
and iron deficiency anemia; group IV comprised 5 patients with iron
deficiency anemia; and group V consisted of 13 normal age-matched
volunteers, as well as age-matched hematologically normal obstetric
patients.

The following procedures using serum were performed for each
group of patients: folate by the Lactobacillus casei method of
Waters and Mollin (10), vitamin B_{12} by the Lactobacillus leichmannii
method of Spray (8), iron by the method of Bothwell and Mallet (1),
and total iron binding capacity (TIBC) by the method of Dacie and
Lewis (3).

All patients were evaluated prior to and after institution of folate supplementation. Delayed hypersensitivity was assessed by response to skin sensitization with dinitrochlorobenzene (DNCB), a contact allergen. In addition phytohemagluttinin (PHA) mitogen studies were performed on peripheral blood lymphocytes.

Animal Model

Weanling Wistar-Lewis rats were placed on either a control or folate-deficient diet. After 8 weeks on these diets, a battery of immunologic tests was performed on each animal. In vivo testing consisted of intradermal PHA skin testing. Responses were graded histologically by a pathologist on the basis of degree of mononuclear cell infiltration. T-cell levels in peripheral blood, spleen and thymus were assessed by an autoradiographic method (5). Spleen cells were tested for their ability to respond to PHA using ^{3}H-thymidine uptake as a measure of response. Lymphocyte-mediated cytotoxicity was assayed by a ^{51}Cr-release method. Animals were sensitized to allogeneic thymocytes from Brown-Norway rats. Allogeneic thymocytes were then labelled with ^{51}Cr and incubated with spleen cells from the sensitized experimental animals. The amount of ^{51}Cr released into the supernatant after the incubation period was taken as a measure of T cell-mediated cytotoxic killing.

RESULTS

The hematologic status of the patients included in each of the 5 groups is shown in Table I. A highly significant difference in skin reactivity to DNCB after a challenge dose was demonstrated between the 3 folate-deficient groups when compared to both the iron-deficient and control groups. Indeed only 1 of 23 patients with folate deficiency responded to DNCB as compared to 17 of 18 positive responses in the other 2 groups. After therapy with folate supplements was instituted, the experimental groups were rechallenged with DNCB and 80% of the folate deficient patients converted to a positive response at a mean time after folate supplementation of 14 days.

Similar lymphocyte depression was noted in the PHA studies. Prior to folate treatment, the deficient patients showed significantly decreased ability to undergo transformation after PHA stimulation when compared to control and iron deficient groups. Patients were followed sequentially with transformation studies after folate treatment was started. We found that in these patients [^{3}H-6] thymidine uptake returned to normal rapidly, as soon as 2 days after folate was started, rose to levels as high as 4 to 5 times normal and then returned to the normal range. This rapid improvement after institution of folate supplementation strongly suggests that the abnormalities noted on initial testing were directly related to folate deficiency.

TABLE I

Hematological Status of Patients in the Study

Group*	Number Tested	Age	Hemoglobin g/100 ml	Serum Folate** ng/ml	Serum B_{12}*** pg/ml	Serum Iron**** µg/100 ml	TIBC***** µg/100 ml	Transferrin Saturation % Fe/TIBC
I	11	28.5 (12–58)	4.8 (2.0–8.5)	4.0 (0.4–9.6)	256 (120–480)	145 (67–240)	274 (186–395)	54.3 (25.0–93.9)
II	7	22.5 (16–32)	9.1 (7.5–10.4)	5.2 (3.4–7.6)	204 (100–300)	150 (80–360)	399 (262–617)	36.3 (22–58.4)
III	5	28.6 (22–38)	5.5 (2.8–8.2)	3.3 (0.0–4.8)	275 (105–420)	53 (46–68)	286 (236–310)	18.5 (16.0–21.9)
IV	5	34.6 (21–60)	6.8 (4.4–9.2)	8.2 (6.0–10.0)	430 (200–890)	53 (5.0–88)	451 (392–571)	11.8 (1.2–19.8)
V	13	27.8 (22–38)	12.3 (10.5–14.6)	11.4 (8.2–20)	254 (190–310)	162 (102–327)	397 (311–600)	40.8 (31.0–56.2)

Values are given as means and ranges.
*Group I=nonobstetric, folate deficient; II=obstetric, folate deficient; III=iron and folate deficient; IV=iron deficient; V=normal.
**Assayed by the Lactobacillus casei method of Waters and Mollin (11).
***Assayed by the Lactobacillus leichmanii method of Spray (9).
****Assayed by the method of Bothwell and Mallett (1).
*****Total iron binding capacity, assayed by the method of Dacie and Lewis (3).

As shown in Table II, there was no significant difference between groups with regard to body weight or hematocrit. Histologic examination in the folate group revealed mild megaloblastic changes in the bone marrow and small intestine. Serum folate levels revealed a significant difference. All control animals had folate levels greater than 100 ng/ml whereas the mean for deficient animals was only 24 ng/ml.

TABLE II

Hematologic Status of Control and Folate-Deficient Rats

Group	Weight (g)	Hematocrit (%)	Serum Folate (ng/ml)
Control	484 + 8.2 SE (458-520)	45.4 + 1.3 SE (40.2-49.1)	100
Deficient	454 + 12.9 SE (410-502)	44.5 + 1.6 SE (39.1-49.8)	24 + 7.9 SE (10-53)

As shown in Table III, responses graded histologically by a pathologist on the basis of degree of mononuclear cell infiltration revealed that there was considerable difference in degree of response among animals of both groups but control animals exhibited the more marked response in each instance.

TABLE III

Histological Response on the Basis of the
Degree of Mononuclear Cell Infiltration

Group	Skin Test Histological Response*
Control	3.8 + 0.37 SE (Range: 3 to 5)
Deficient	1.6 + 0.37 SE (Range: 0 to 3)

*Histological grading: 0=no response; 1=minimal; 2=mild; 3=moderate; 4=marked; 5=severe.

Kinetic and labelling experiments by Hollingsworth, et al.
(5) had previously demonstrated that in the rat, T cells selectively
take up ^{3}H-uridine in both the peripheral blood and spleen (Table
IV). This suggests that in folate deficiency there is a paucity of
functional T-cells in lymphoid tissue and the circulation.

TABLE IV

^{3}H-Uridine Labelling of T-Cells

| Group | Percent T-Cells (Labelled with ^{3}H-Uridine) | | |
	Spleen	Thymus	Blood Lymphocytes
Control	70.3 + 2.4 SE (61-78)	81.8 + 1.5 SE (76-88)	67.0 + 1.7 SE (63-71)
Deficient	41.6 + 1.9 SE (35-52)	73.0 + 2.0 SE (68-83)	43.6 + 2.8 SE (35-51)

Spleen cells were then tested for their ability to respond to
PHA using ^{3}H-thymidine uptake as a measure of response (Table V).
The deficient animals showed significantly depressed response to
PHA with no deficient animal having a response in the normal range.
The observed depression of mitogenic response may simply represent
deficient numbers of thymic-dependent spleen cells in the deficient
animals, or may reflect a normal numerical population of T-cells
that are functionally incapable of responding to the mitogen.

TABLE V

Uptake of ^{3}H-thymidine by Spleen Cells
Stimulated by Phytohemagglutinin (PHA)

| Group | PHA Stimulation (CPM) | | |
	Control (no PHA) Cultures	Stimulated Cultures	Stimulation Index (Con/Stim)
Control	1315 + 312 SE (877-1701)	19,398 + 1014 SE (14,501-23,412)	14.8
Deficient	1405 + 102 SE (910-1966)	4263 + 579 SE (1789-8248)	3.0

The lymphocyte-mediated cytotoxicity of spleen cells from
sensitized animals is shown in Table VI. When compared to controls,
the folate-deficient animals exhibited negligible killing. There
seemed to be some correlation between killing capacity and degree
of folate deficiency as the highest levels of killing among the
deficient animals were in those animals with the higher serum
folate levels.

TABLE VI

Lymphocyte-Mediated Cytotoxicity by the ^{51}Cr-Release Assay

Group	Cytotoxicity Percent Killing*
Control	29.1 ± 3.7 SE (Range: 16.3 to 40)
Deficient	5.2 ± 1.5 SE (Range: 0 to 13.9)

*The amount of ^{51}Cr released into the supernatant was taken as a
measure of T-cell mediated cytotoxic killing.

DISCUSSION

The results of these studies strongly suggest that folic acid
is important to the normal function of the cell-mediated immune
system. Although the exact mechanism responsible for the blunted
immune response is unknown in folate deficiency, results from the
rat model indicate that all measured parameters of the cellular
immune response are depressed under these conditions; actual T
cell numbers, antigen recognition, blastogenesis and effector cytotoxic function. The most likely metabolic dysfunction resulting
in immune unresponsiveness would involve DNA synthesis; however,
any or all of metabolic functions of folic acid may be involved.
Further studies are required to more fully define the exact defect.

Another aspect of folic acid deficiency that may have even
more important clinical relevance is that of marginal folate deficiency. Preliminary results in our laboratory indicated that
rats which are made marginally deficient in adult life have measurable differences in mitogenic response and cytotoxic killing
ability. These differences have been observed as early as 3 weeks

after institution of a marginal folate diet. In addition, T cell-
dependent B cell function seems to be affected as the number of
plaque-forming cells in the spleen after sheep red blood cell immunization is reduced in both severe and marginal deficiencies,
compared to controls. If more extensive studies corroborate these
early findings, one could suggest that the immune system is extremely sensitive to inadequate folic acid levels.

     The finding that folate deficiency interferes with normal im-
mune function may have important clinical implications. Total body
folate stores are rapidly depleted during periods of inadequate
dietary intake, with megaloblastic changes detectable as early as
6 to 8 weeks later. Because of the lability of body stores, deficiency of folate is not uncommon even in developed nations. A
recent study among the Zulu populations in South Africa revealed
that 44% of pregnant women, 32% of non-pregnant women, and 19% of
adult men were folate deficient (2). In industrialized society,
the major contributing factor to folate deficiency is pregnancy.
Temperley, et al.(9) in Great Britain showed that the mean serum
folate level in a large cohort of pregnant women at the time of
delivery was below normal, and remained so for up to 3 months.
The effects of this marginal deficiency on the developing fetus remain to be determined. Other factors which contribute to folate
deficiency and which may involve large segments of the population
include intestinal disorders (sprue), mechanical bowel disorders
(blind loops), alcohol, and commonly used medications. The latter
include most importantly oral contraceptive agents and anticonvul-
sants, both classes of drugs interfering with intestinal folate
absorption. Because of the large segments of the population at
risk and the potential contribution of folate deficiency to morbidity and mortality in affected groups, it is clear that more in-
tensive investigations are required to determine the extent and
clinical significance of this deficiency.

REFERENCES

1. Bothwell, T.H. and Mallett, B., Biochem. J., 59 (1955) 599.
2. Colman, N., Barker, E.A., Barker, M., Green, R. and Metz, J.,
    Am. J. Clin. Nutr., 28 (1975) 471.
3. Dacie, J.V. and Lewis, S.M., Practical Haematology, Churchill,
    London (1968) 4th ed.
4. Geffhuysen, J., Rosen, E.U., Katz, J., Ipp, T. and Metz, J.,
    Brit. Med. J., 4 (1971) 527.
5. Gross, R.L., Reid, J.V.O., Newberne, P.M., Burgess, B., Marston,
    R. and Hift, W., Am. J. Clin. Nutr., 28 (1975) 225.
6. Hollingsworth, J.W. and Carr, J., Cell. Immunol., 8 (1973) 270.
7. Smythe, P.M., Schonland, M., Brereton-Stiles, G.G., Coovadia,
    H.M., Grace, H.J., Koening, W.E.K., Mafoyane, A., Parent, M.A.

and Vos, J., Lancet, 2 (1971) 939.
8.   Spray, G.H., Clin. Sci., 14 (1955) 661.
9.   Temperley, I.J., Meehan, M.J.M. and Gatenbey, P.B.B., Brit.
     J. Haematol., 14 (1968) 13
10. Water, A.H. and Mollin, D.R., J. Clin. Pathol., 14 (1961) 335.
11. Williams, E.A.J., Gross, R., and Newberne, P.M., Natr. Reports
     Intern., 12 (1975) 137.
12. Work, T.H., Ifekwunigwe, A., Jellife, D.B., Jellife, P. and
     Neumann, C.G., Ann. Int. Med., 79 (1973) 701.

ANTIGEN AND IMMUNE COMPLEX INDUCED SUPPRESSION OF DELAYED HYPERSENSITIVITY

E. R. HEISE, E. ROWLAND and N. BEATTY

Bowman Gray School of Medicine
Winston - Salem, North Carolina (USA)

Both cellular and humoral immune responses are known to be regulated by mechanisms that are incompletely understood, especially with respect to delayed hypersensitivity (DH). Injection of soluble or alum-precipitated antigen near the time of sensitization prevents the development of DH and the synthesis of the gamma-2 class of antibody, leaving the gamma-1 class unaffected (1,3). Moreover, established DH is suppressed by systemic administration of specific antigen (6). Antigen-induced suppression of the induction and expression of DH has been termed "immune deviation" and "de-sensitization", respectively. These experimental phenomena are thought to have counterparts in certain granulomatous diseases in which skin-test anergy is a prominent feature.

Hypotheses to explain antigen-induced suppression of DH include the following: a) the functional elimination of T lymphocytes by mechanisms such as receptor blockade or shedding of antigen receptors, and b) a feedback inhibition system operating through suppressor cells or inhibition factors (4,5).

Since skin-test anergy often develops in animals or individuals possessing circulating antibodies, we considered the possibility that antigen challenge might result in the formation of antigen-antibody complexes possessing immunosuppressive activity. A precedent for this possibility is the work of Axelrad (12) on suppression of DH with antigen and specific antisera.

The experiments to be described were designed to determine: a) whether or not immune complexes formed in antigen excess, i.e. soluble complexes, can suppress the induction and expression of DH,

b) the effect of desensitization on the proliferative capacity of
different lymphocyte populations, and c) whether or not antigen
stimulated lymphocytes release a suppressor substance in vitro that
is capable of inducing a state of desensitization.

MATERIALS AND METHODS

The experimental system involved sensitizing guinea pigs with
bovine serum albumin (BSA) in complete Freund's adjuvant (CFA).
Thus, BSA served as the deviating and desensitizing antigen and
purified protein derivative (PPD) served as the reference antigen
and specificity control. The footpad route was used for sensitization. The intraperitoneal route was used for injection of the
deviating or desensitizing antigen, immune complexes, or culture
supernatant fluid.

Rabbit anti-BSA hyperimmune serum was prepared and specific
antibody was isolated by affinity chromatography using BSA-conju-
gated Sepharose beads since rabbit and guinea pig globulins are
both cytophilic for guinea pig macrophages.

Immune complexes were prepared by addition of BSA to specific
antibody at 4 times equivalence (in moderate antigen excess) and
were incubated 30 min at room temperature prior to injection.

For sensitization, guinea pigs received 2 mg of BSA (Pentex)
5x crystallized in CFA, divided into 4 intradermal sites. Twenty-
one days later the animals were skin-tested with 20 µg PPD and 10
µg BSA. Skin thicknesses were measured before and 24 hr after antigen injection. The net increase in skin thickness was taken as a
measure of the induration produced in the DH reaction.

Lymphocyte culture fluids were prepared from 24 hr cultures of
sensitized lymph node, spleen and blood lymphocytes (10×10^6/ml
RPMI 1640 + 10% normal guinea pig serum (GPS), cell density 4×10^5/
cm^2) incubated with BSA (experimental) or without BSA, but subsequently reconstituted with an equivalent (1000 µg/ml) amount of BSA.
Culture fluids were centrifuged at 150 x g for 10 min, filtered
(0.45 µ pore diameter) and stored at -15 C until use.

Lymphocyte transformation was assessed by culturing lymphocytes
(1×10^6 spleen or lymph node cells, 0.5×10^6 white blood cells)
in 0.2 ml RPMI 1640 + 10% GPS with or without antigen for 66 hr at
37.5 C. The lymphocytes were collected 6 hr after the addition of
2 µC of tritiated thymidine with a multiple automatic sample harvest-
er. The radioactivity incorporated into cellular DNA was determined
in a Beckman LS-133 liquid scintillation spectrometer. The reduction of DNA synthesis in cultures of desensitized cells was calcu-

lated as follows:

$$\frac{\text{cpm (desensitized)} - \text{cpm (unstimulated control)}}{\text{cpm (sensitized)} - \text{cpm (unstimulated control)}} \times 100 = \% \text{ reduction}$$

RESULTS

Immune Deviation

The data shown in Table I indicate that immune complexes which were formed in moderate antigen excess significantly suppressed the development of DH. However, immune complexes were not significantly more suppressive than equivalent amounts of free BSA. In contrast, animals which received antibody alone developed marginally greater DH than did the untreated sensitized controls. As expected, suppression was specific for BSA, the deviating antigen.

Desensitization

Free BSA and BSA in the form of soluble antigen-antibody complexes were also compared for their ability to suppress established DH. The experimental groups were desensitized 3 weeks after sensitization. The time course of skin-test hyporesponsiveness to the desensitizing antigen is shown in Fig. 1. The results indicate that the magnitude and duration of the suppressed state produced following treatment with free BSA is comparable to that produced by an equivalent amount of BSA in the form of soluble immune complexes. Following a single desensitizing dose, the animals remained partially suppressed for a period of 3 weeks but recovered skin-test reactivity by the fourth week, a time at which sensitivity to BSA in control animals begins to wane. Results of skin tests to the unrelated antigen PPD are shown in Fig. 2. A transient period of nonspecific hyporesponsiveness lasting about 3 days can be noted. In other experiments (not shown) we determined that the multiple skin tests used in this protocol did not contribute significantly to the persistence of the skin-test anergy.

Prolongation of the desensitized state was produced by weekly administration of BSA or soluble complexes (Fig. 3). However, some animals died following repeated desensitizing treatments with BSA but not following repeated desensitization with soluble immune complexes. Death was attributed to shock of a delayed anaphylactic type. These experiments were interpreted to indicate that the desensitizing effects of exogenously formed soluble immune complexes is attributable to interaction between sensitized cells and antigen determinant groups and not through Fc receptor binding.

TABLE I

Suppression of the Development of Delayed Hypersensitivity to
BSA by Treatment with Specific Antigen and Immune Complexes*

| | | 24 HOUR SKIN TEST REACTION** | |
GROUP	TREATMENT	Induration $(mm^2, \bar{X}, SE)$	Erythema $(mm^2, \bar{X}, SE)$
I	None	1.2, 0.1	52, 4
II	Antibody-BSA (4x) Complex	0.6, 0.1	27, 5
III	BSA	0.6, 0.1	25, 8
IV	Antibody only	1.6, 0.3	74, 14

*All animals were sensitized on day 0 with 2 mg BSA in 0.4 ml CFA.
Experimental groups were treated on days -1, +1 and +3 with 2.8 mg
BSA + 3.4 mg specific antibody (Group II), 2.8 mg BSA (Group III)
or 3.4 mg antibody (Group IV). All animals were skin-tested on
day 21 with 3 μg BSA.
**Comparative P values (induration and erythema, respectively): I
vs II, 0.001, 0.001; I vs III, 0.001, 0.002; I vs IV, NS, 0.051;
II vs III, NS, NS.

Desensitization with Lymphocyte Culture Supernates

Experiments were designed to determine whether or not sensitized lymphocytes cultured with BSA release suppressor substance(s).
Administration of a 24 hr culture supernatant fluid obtained from
antigen stimulated lymphocytes resulted in specific unresponsiveness for a period of 3 weeks (Fig. 4). Supernatant fluid obtained
from cells cultured in the absence of antigen and subsequently reconstituted with BSA was also immunosuppressive. However, reconstituted culture fluid appeared to be less effective than the experimental culture fluid, suggesting that a factor in addition to
antigen was involved in desensitization. It was therefore important to distinguish between the effects of BSA present in the culture fluid and possible suppressor substance(s) released into the
medium. The first approach taken was to culture sensitized cells
with BSA for 2 hr, followed by washing the cells to remove antigen from the medium. The cells were then incubated for an additional 22 hr in the absence of additional antigen. Under these conditions pretreated cells proceed to undergo a blastogenic response.

Antigen free culture fluids prepared in this way did not suppress skin test reactivity (results not shown). A second approach was to fractionate experimental culture supernates with 50% saturated ammonium sulfate in order to separate BSA from immune complexes or other potential suppressor substance(s) that could be precipitated by half saturated ammonium sulfate. When the ammonium sulfate precipitable and soluble fractions were tested _in vivo_, significant desensitization occurred only in animals which received the 50% ammonium sulfate soluble fraction (Table II).

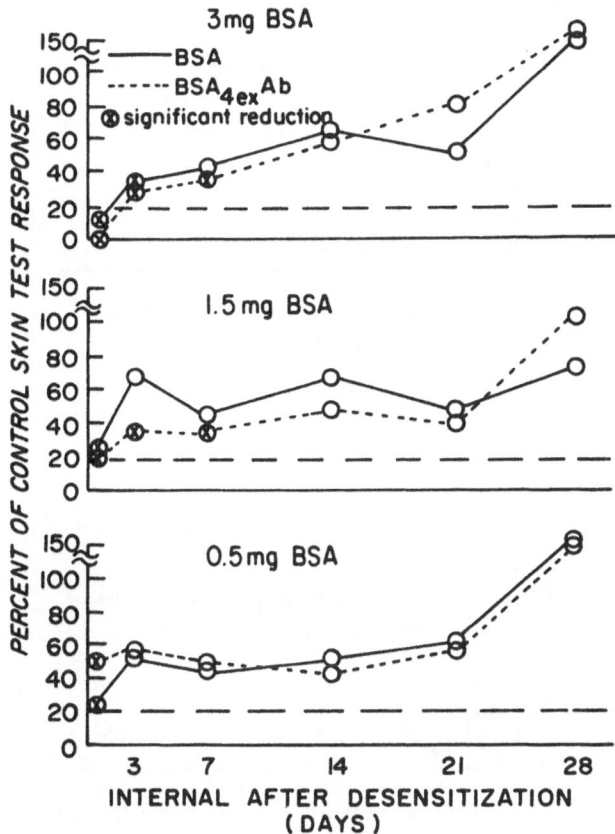

Fig. 1. Desensitization with specific antigen and immune complexes. Groups of animals sensitized to BSA 21 days previously received an intraperitoneal injection of BSA or an equivalent amount of BSA in the form of soluble complexes. At intervals, the animals were skin tested with 10 μg BSA.   --- = negative skin reactions; ⊗ = significant suppression (p < 0.05).

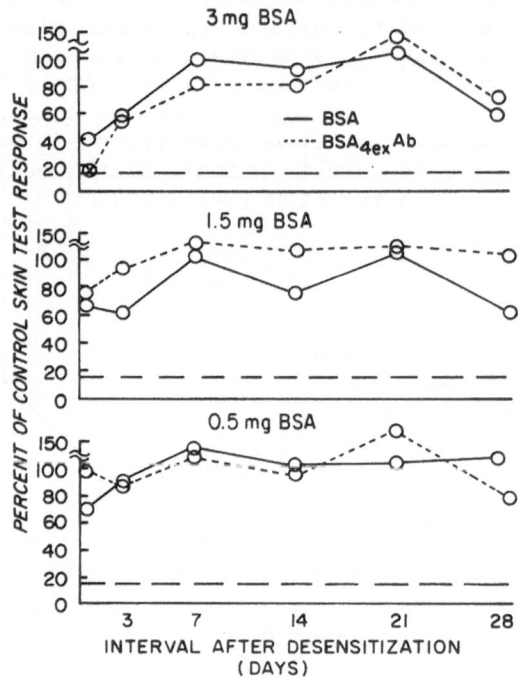

Fig. 2.   Specificity of desensitization.   Skin test reactions to PPD in animals treated with BSA or immune complexes.

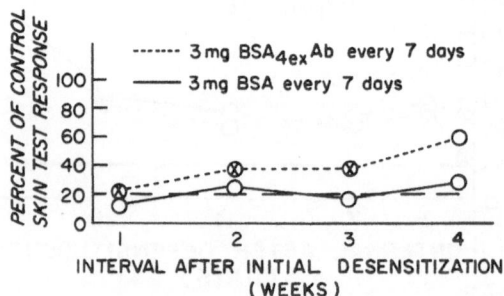

Fig. 3.   Prolongation of skin test hyporeactivity by weekly treatment with BSA or immune complexes.   Animals were skin tested 3 days after each treatment.   --- = negative skin reactions; ⊠ = significant suppression (p < 0.05).

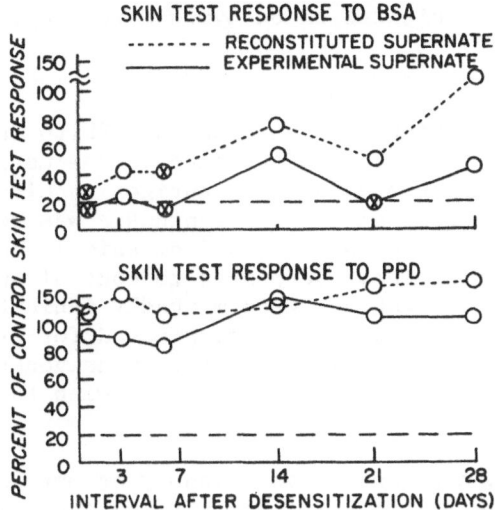

Fig. 4. Desensitization with BSA-stimulated lymphocyte culture
fluid. Groups of 4 sensitized guinea pigs with equal mean DH
reactivity received: a) 3 ml of experimental supernatant fluid containing 3 mg BSA, b) 3 ml of antigen reconstituted control culture
fluid, or c) 3 ml of aged culture medium (control group).

TABLE II

Desensitization with Lymphocyte Culture Supernatant
Fluid Fractionated with Half-saturated Ammonium Sulfate

Skin Test Interval	Test Antigen	Percent of Skin Test Response in Nondesensitized Animals	
		Soluble Fraction	Precipitable Fraction
12 hours	BSA	0	63
3 days		27	100
7 days		17	63
12 hours	PPD	19	74
3 days		43	82
7 days		61	75

Effect of Desensitization on Lymphocyte Proliferation

Peripheral blood lymphocytes, obtained from animals desensitized with 3 mg BSA 20 days after sensitization were cultured with appropriate concentrations of PHA, BSA and PPD to assess their ability to proliferate in vitro. DNA synthesis was measured by ^{3}H-thymidine incorporation 3 days after antigen or mitogen stimulation. As shown in Fig. 5, BSA-induced DNA synthesis was completely inhibited in cultures of cells derived from animals for at least 7 days after desensitization. Responses to control antigen (PPD) and to the nonspecific mitogen PHA were markedly inhibited 24 hr after desensitization and were partially suppressed for at least 7 days. Thus, suppression of the proliferative responsiveness of blood lymphocytes in vitro is correlated temporarily with the period of nonspecific anergy.

Spleen, lymph node and blood lymphocytes obtained from desensitized and control sensitized animals were cultured in vitro with BSA, PPD and PHA to determine whether the proliferative potential of other lymphocyte populations was also suppressed (Table III). Regardless of the stimulus, the DNA synthetic responses were most strongly suppressed in blood lymphocytes, less strongly in spleen cells and least inhibited in cultures of lymph node cells. These results indicate that desensitization is associated with marked inhibition of cell proliferation.

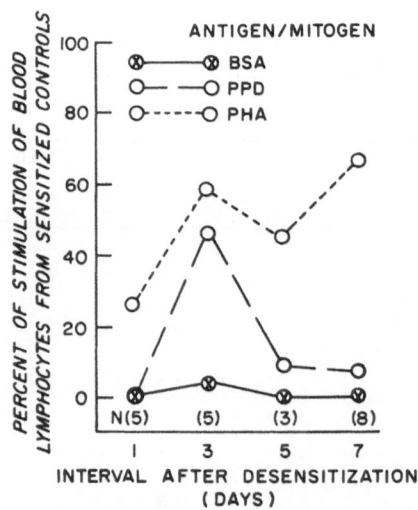

Fig. 5. Suppression of DNA synthesis in vitro in cultures of blood lymphocytes obtained from desensitized animals.

TABLE III

Suppression of In vitro DNA Proliferative Responses in
Lymphocytes Obtained from Desensitized Guinea Pigs

Lymphocyte Source	Percent of ^{3}H-thymidine Incorporated into Lymphocytes of Sensitized Control Animals		
	BSA	PPD	PHA
Peripheral Blood	7	12	14
Spleen	24	20	28
Lymph Node	26	30	50

DISCUSSION

The most important observations made during this investigation
are: a) immune deviation and desensitization can be induced with
either antigen or antigen-antibody complexes in which uncomplexed
antigen determinant sites are present, b) desensitization involves
a brief nonspecific anergic state and a more prolonged state of
antigen specific skin test hyporesponsiveness and lymphocyte nonreactivity, and c) convincing direct evidence for an antigen free,
cell-derived suppressor substance in either lymphocyte culture
supernatant fluids or in serum of desensitized animals was not obtained.

The data are of interest in relation to theories which seek to
explain how DH is regulated. The major possibilities at present
appear to be an antigen receptor blockade or a feedback mechanism
involving suppressor cells. Our results do not support a feedback
mechanism involving antibody or immune complexes. It is likely
that a humoral suppressor is rapidly bound to lymphocytes and may
therefore be difficult to detect in serum or culture fluids. The
present data taken together with published results permit the conclusion that desensitization is associated with a profound but temporary functional suppression of both dividing and lymphokine-pro-
ducing cells. This suggests that both long-lived and short-lived
lymphocytes are affected during desensitization. It is also clear
that proliferative cells in lymph nodes and spleen are also suppressed although not as completely as are blood lymphocytes.

Any hypothesis of desensitization must account for both the

antigen specific and nonspecific aspects of anergy. One possibility
is that the initial period of nonspecific suppression involves suppressor cells and that the persisting specific unresponsiveness is
the result of antigen receptor blockade or suppressor cells. A second possibility is that inactivation or removal of a critical cell
type, such as the monocyte, from the circulation prevents normal
lymphocyte function. The solution to these questions will have important implications for the pathogenesis and therapeutic approaches
to a number of important diseases.

ACKNOWLEDGEMENTS

    This investigation was supported by Grant No. HL 16769 from
the Heart and Lung Institute and by a Grant from the North Carolina
Lung Association.

REFERENCES

1. Asherson, G.L. and Stone, S.H., Immunology, 9 (1965) 205.
2. Axelrad, M.A., Immunology, 15 (1968) 159.
3. Dvorak, H.F., Billote, J.B., McCarthy, J.S. and Flax, M.J.,
    J. Immunol., 24 (1965) 966.
4. Kantor, F.S., Hall, C.B. and Lipsmeyer, E., Second Inter.
    Convocation of Immunol., Buffalo, New York, 1970, Karger, Basel,
    Switzerland (1971) 213.
5. Polak, L., Clin. Exp. Immunol., 19 (1975) 543.
6. Uhr, J.W. and Pappenheimer, A.M., Jr., J. Exp. Med., 108 (1958)
    891.

PLASMA CELL SURFACE ANTIGEN IN THERMAL BURNS, CARCINOGENESIS AND PRIMARY IMMUNODEFICIENCIES

N. S. HARRIS and P. D. THOMSON

Shriners Burns Institute and University of Texas
Medical Branch, Galveston, Texas (USA)

Previous studies have shown that human B-lymphocytes (B-cells) possess a receptor for the binding of the Fc portion of certain immunoglobulin (Ig) classes, or possess Ig as a surface marker (1, 5). Evidence has also been presented that T-lymphocytes (T-cells) may also bear Ig receptors (7), but more widely accepted is the fact that these T-cells form spontaneous rosettes (E-rosettes) when mixed with sheep erythrocytes in vitro (5,10). These surface markers have allowed the identification and enumeration of normal lymphoid cells, but may not be sufficient to fully identify the immunologic nature of certain disease states in man (9).

A more recent study has described a new surface marker on B-cells (3). This surface antigen has been shown to be absent on T-cells and on B-cells which bear either surface-Ig or Fc·receptors for Ig. Since this antigen was identified using an antiserum prepared from human plasmacytoma cells, it has been named the plasma cell (PC) antigen. Results from this previous study showed that in 75 normal individuals, 11.5% of the lymphocytes in their peripheral circulation had this plasma cell antigen.

In order to better define the clinical importance of this new surface marker and its relevance to the immunologic identification of certain disease processes in man, we have chosen 3 distinct groups of patients for initial study. These patients include those with acute thermal burns (greater than 40% body surface area), those with different types of tumors, and those patients with primary immunodeficiencies.

It is our purpose to identify this PC antigen on the surface of lymphocytes from these patients and to correlate these findings with their T-cell and Ig-bearing cell levels.

MATERIALS AND METHODS

Antisera

Human anti-plasma cell serum (HuAPS) was produced by injecting 8-week old New Zealand White rabbits i.v. with 2 weekly injections of 1 x 10^8 human plasmacytoma cells (3). Serum was harvested from blood which was removed by cardiac puncture 5 days after the last injection of cells. This serum was decomplemented at 56 C for 30 min and absorbed with the human T-cell line Molt-4 (8) and with the lymphoblast cell line RPMI 4098. This HuAPS was then sterilized by passage through a .22 μ Millipore filter.

All other antisera used were commercial preparations of fluorescein labelled Ig. These included goat anti-human IgG (Cappell), IgA (Hyland), and IgM (Hyland) which were extensively absorbed with cell line RPMI 4098 and with cultured human plasma cells.

Cells

Patients with primary immunodeficiencies, different types of tumors, and acute thermal burns were selected for study and 7-10 cc of whole blood were drawn from these patients. This blood was diluted 1:2 with Hepes buffered RPMI 1640 (ABS) and subjected to Ficoll-Hypaque density centrifugation. This cell separation technique yielded a relatively pure population of peripheral blood lymphocytes. In some cases, contaminating red blood cells were lysed with Tris-buffered ammonium chloride. These cells were washed 3 times in RPMI 1640 and stained with the appropriate fluorescein conjugated immunoglobulin, or were used for E-rosette assays.

Control cells were drawn from age matched donors where possible and were treated in the same manner. In selected patients, bone marrow needle biopsies were obtained and this bone marrow was also handled in the same manner.

Fluorescent Staining

Patient and control peripheral blood lymphocytes or bone marrow lymphocytes with membrane bound Ig or with PC antigen were detected by immunofluorescence. Cells with Ig markers were incubated with either goat anti-human IgG, IgA, or IgM diluted 1:2 in a

direct method for 30 min at 4 C. Those lymphocytes bearing PC antigen marker were detected indirectly by incubating them with HuAPS
at 4 C for 30 min. These cells were then washed 4 times with RPMI
1640 and incubated with goat anti-rabbit IgG for an additional 30
min at 4 C.

     After both the direct and indirect staining methods, all cells
were washed 4 times to remove any unbound fluorescein stain. These
cells were placed on clean glass slides under cover slips and were
counted with the aid of a Leitz Ortholux Microscope with HB200 mer-
cury lamp, BG38 and BG12 exiter filters, and 2 K495 barrier filters.
All cells were viewed under oil immersion with simultaneous phase
contrast and reflected fluorescence.

     At least 100 cells were counted on each slide and the percent-
age of cells with membrane fluorescence was determined.

                              E-Rosettes

     Human peripheral blood lymphocytes and sheep erythrocytes were
washed separately 3 times in RPMI 1640. A suspension of 0.5% sheep
erythrocytes was mixed with 4×10^6 lymphocytes and allowed to in-
cubate for 5 min at 37 C. This cell mixture was spun at 1100 rpm
for 5 min and incubated on ice at 4 C for 1 hr. The excess super-
nate was removed, leaving enough to resuspend the remaining cells
by gently rocking for 5-10 sec. A drop of the resuspended cells
was transferred to a clean glass slide and counted with the aid of
a Leitz Ortholux phase contrast microscope. Any lymphocyte with 4
or more sheep erythrocytes on its surface was considered a rosette.

                               RESULTS

     The lymphocyte surface markers of 120 normal individuals are
shown in Table I. These findings were compatible with earlier results (3) and showed that the PC antigen level remained near the
previous 11.5% mean. It should also be noted that this marker
seemed to be more consistent for B-cells than the Ig surface markers as reflected by the standard deviations.

     Table II categorizes various patients with primary immuno-
deficiencies. Those patients diagnosed as having severe combined
immunodeficiency exhibited low or absent levels of T-cells as determined by the E-rosette assay. At the time of study, they also
exhibited high levels of Ig-bearing cells as well as high levels of
cells with the PC antigen. Patients with Bruton's x-linked agammaglobulinemia had low levels of B-cells with surface Ig and had low
values for cells exhibiting the PC antigen. Another group of

patients, termed X-linked dysgammaglobulinemic, had a similar clinical picture as those patients with Bruton's disease. However,
they presented a different picture when compared to the Ig-bearing
B-cells of the Bruton's patients. The X-linked dysgammaglobulinemic
patients exhibited normal or higher than normal Ig-bearing cells
and higher levels of circulating cells with the PC antigen. A
similar pattern was observed for those patients classified as non-
X-linked dysgammaglobulinemias.

In acute thermal burns (Table III) it was found that at 24 hr
postburn, T-cell levels were dramatically decreased while PC antigen levels were only slightly lower than normal with respect to
the values as shown in Table I. At 3 days postburn, the PC antigen level and cells with surface IgG increased two-fold and did not
return to normal levels until 5 or 6 weeks postburn. The T-cell
level decreased after a brief return to normal at 3 days and did
not approach normal levels until the fifth week postburn.

Table IV shows data collected from a randomized group of tumor
patients. Those with malignant melanoma had near normal levels of
T-cells, but 1/3 of these patients exhibited a three-fold increase
in the PC antigen. All patients with squamous cell carcinoma exhibited elevated PC antigen levels. It should be noted that the 1
patient with asymptomatic myeloma had only 1 bone marrow cell which
had the PC antigen, whereas all patients with symptomatic myeloma
had higher PC antigen levels in both their bone marrow and peripheral blood.

TABLE I

Percentage of Surface Immunoglobulins, Plasma Cell
(PC) Antigen and E-Rosettes on Peripheral
Blood Lymphocytes from 120 Normal Individuals

Surface Marker	Mean (%)	Standard Deviation	Standard Error	Range
IgG	8.68	3.16	0.29	0-16
IgA	3.03	3.15	0.29	0-10
IgM	5.09	2.69	0.24	0-12
PC	11.89	2.43	0.21	6-17
E-rosettes	52.62	10.12	1.09	28-80

TABLE II

Percentage of Surface Immunoglobulins, Plasma Cell (PC)
Antigen and E-Rosettes on Peripheral Blood
Lymphocytes from Patients with Primary Immunodeficiencies

Type of Immunodeficiency	Patient	Mean (%)				
		IgG	IgA	IgM	PC	E-Rosettes
Severe	C.B.	22	10	20	43	0
Combined	M.M	54	41	12	42	0
	D.V.	6	5	3	50	6
	J.C.	4	0	0	2	48
	I.G.	4	0	1	1	35
Bruton's X-Linked	E.G.	1	3	6	6	ND
Agammaglobulinemia	J.B.	4	1	0	2	73
	G.H.	4	1	3	0	62
	C.H.	2	0	0	1	54
	W.T.	10	6	16	45	ND
X-Linked	J.T.	4	1	7	34	46
Dysgammaglobulinemia	M.S.	31	5	4	20	40
	S.S.	14	5	7	14	40
	G.A.	3	1	14	19	50
	R.T.	4	0	2	17	ND
Non X-Linked	C.V.	6	2	4	37	ND
Dysgammaglobulinemia	S.N.	7	1	6	15	43
	M.R.	9	1	2	18	67

TABLE III

Percentage of Surface Immunoglobulin, Plasma Cell (PC)
Antigen and E-Rosettes on Peripheral Blood Lymphocytes
from Patients with Severe Thermal Burns

Number of Patients	Time Postburn	Mean (%)				
		IgG	IgA	IgM	PC	E-Rosettes
5	0-24 hr	12	4	7	7	19
4	3 days	20	2	5	24	50
7	1 week	17	4	7	21	24
5	2 weeks	23	3	9	25	20
5	3 weeks	15	7	10	20	36
5	4 weeks	27	7	6	27	32
6	5 weeks	8	4	6	19	44
4	6 weeks	13	5	14	16	47

Burns listed here are all greater than 40% body surface area.

TABLE IV

Percentage of Surface Immunoglobulin, Plasma Cell (PC) Antigen and
E-Rosettes on Peripheral Blood (PB) and Bone Marrow (BM)
Lymphocytes from Patients with Different Types of Tumors

Tumor Type	Patient	Mean (%)					Lymphocyte Source
		IgG	IgA	IgM	PC	E-Rosettes	
Melanoma	F.S.	3	0	4	29	ND	PB
	N.S.	18	4	7	30	60	PB
	E.H.	10	3	7	12	47	PB
	K.H.	6	4	4	10	62	PB
	H.W.	17	13	7	11	61	PB
	P.F.	2	0	4	16	35	PB
Squamous Cell Carcinoma	W.P.	5	3	2	28	45	PB
	M.R.	6	0	0	26	41	PB
	S.C.	33	0	0	19	22	PB
	L.H.	6	2	6	23	45	PB
	K.K.	2	2	9	19	51	PB
	H.C.	3	0	0	18	33	PB
Asymptomatic Myeloma	J.C.	0	0	0	1	59	BM
Symptomatic Myeloma	C.O.	3	2	3	43	6	BM
		3	0	1	39	50	PB
	J.S.	2	1	0	29	3	BM
		13	1	3	33	26	PB
	B.H.	0	0	0	47	ND	BM
		0	0	0	41	57	PB
	H.B.	0	0	0	28	35	BM
		14	0	0	18	66	PB

DISCUSSION

The use of B-cell antigenic markers as a diagnostic tool has rapidly gained increasing popularity due to the availability of commercially produced, class-specific anti-immunoglobulin antisera. However, results produced with these antisera seem to be varied from patient to patient which reportedly have the same clinical disease. Not only do patients differ in their surface Ig markers, but also there is a widely divergent pattern in the surface-Ig markers among supposedly "normal" individuals. This divergent pattern may exist for many reasons, but it is reasonable to believe that the slightest bacterial infection may cause fluctuations in the surface-IG levels.

This new antigenic marker called the plasma cell or PC antigen seems to be relatively consistent. In 120 individuals who were selected as normal, the average occurrence of this marker remained near 12%. Further evidence for the stability of this new surface marker was given by the fact that persons ranging in age from 67 years to newborn (cord blood) have peripheral blood lymphocytes with the same level of PC antigen (unpublished observations).

The patients presented in this study showed varied PC antigen levels, but as each disease was investigated separately, some correlations appeared. Patients with primary immunodeficiencies had a variable increase of PC antigen in both the X-linked and non X-linked dsygammaglobulinemias, but in the severe combined and Bruton's X-linked immunodeficiencies a definite pattern existed. Severe combined immunodeficients did not have T-cells as reflected by their low E-rosettes and did not produce antibody; however, they did have a four-fold increase in the number of cells with the PC marker. The Bruton's X-linked patients, on the other hand, had normal levels of E-rosettes, but did not have even near normal levels of the PC antigen. This lack of PC antigen-bearing cells may reflect the absence of stem cells, or maybe the lack of ability to fully differentiate B-cells.

Patients with acute thermal burns showed a decrease in PC antigen levels to half-normal in the first 24 hr postburn. These values then increased to two-fold higher than normal and remained at that level until the sixth week postburn. These findings parallel those of Daniels, et al. (2) who showed by electrophoresis an initial fall in serum Ig with an increase to above normal values at 1 week postburn and continuing at that level for 8 weeks postburn. Since it is felt that infection plays a prominent role in mortality from severe burns (6), we feel that it is important to monitor the ability of burned individuals to produce immunoglobulins. The noted fall in PC antigen level and its recovery both precede the fall and recovery of serum Ig and thus may be a more sensitive assay for the immune status in severely burned individuals.

Patients with symptomatic and asymptomatic myeloma all showed a two to three-fold rise in PC antigen above normal levels on lymphocytes in their peripheral circulation. In their bone marrow, however, there was a striking difference in the PC antigen levels with the asymptomatic myeloma being quite low. Although it is difficult to make predictions based on a single observation, it may be possible to differentiate asymptomatic from symptomatic myeloma at an early stage, but more patients must be tested before drawing any conclusions. The higher number of E-rosettes seen in the bone marrow of this asymptomatic myeloma patient may also aid in early diagnosis of the disease.

Squamous cell carcinoma patients as well as 1/3 of the malignant melanoma patients showed a two-fold or greater increase in PC antigenic marker. Although we do not have a model to monitor blocking antibody levels in man, studies by other investigators (4) have demonstrated blocking antibody to exist in human tumor systems. It may be possible that the rise in PC antigen seen in these tumor patients may be an indicator of blocking antibody production, but again, no firm evidence exists for this hypothesis.

Since earlier data have shown that the PC antigen does not exist on cells which bear Ig receptors (3), and since it does exist on cells which increase in number preceding increased antibody production in severe thermal injury, we might speculate that this cell bearing the PC antigen is near the end of the sequence which leads to antibody production. This hypothesis may be confusing in view of the data associated with the different primary immunodeficiency diseases. It is possible, however, that several defects may influence antibody production. In patients with Bruton's X-linked agammaglobulinemia, a stem cell defect seems most probable. In those patients with severe combined immunodeficiencies and X-linked agammaglobulinemia, another type of defect is suggested. Since their cells differentiate to those lymphocytes with PC antigen, it would seem that the deficiency may be in the production and release of immunoglobulins. Additional studies are necessary to determine if this defect may be bypassed or corrected.

In summary, our studies present data on a new lymphoid cell surface antigen which we have called the plasma cell antigen (PC). This surface marker is highly consistent within different patient populations. This consistency suggests that it may be a useful tool in the diagnosis and possible treatment of primary immunodeficiency diseases as well as secondary immunodeficiency states such as carcinogenesis and acute thermal burns.

ACKNOWLEDGEMENT

This work was supported in part by the Shriners Burns Institute, Galveston Unit and NIH Grant Nos. DHEW 1 RO1 CA 15278-01 and DHEW 1 PO1 CA 16964-01.

REFERENCES

1. Cooper, M.D., Lawton, A.R. and Bockman, D.E., Lancet, 2 (1971) 791.
2. Daniels, J.C., Larson, D.C., Abston, S. and Ritzmann, S.E., J. Trauma, 14 (1974) 137.
3. Harris, N.S., Nature, 250 (1974) 507.
4. Hellstrom, K.E., Hellstrom, I., Sjogren, H.O. and Warner, G., Progress in Immunology (Ed. B. Amos), Academic Press, New York, (1971) 939.
5. Jondal, M., Holm, G. and Wigzell, H., J. Exp. Med., 136 (1972) 207.
6. Kefalides, N.A., Arana, J.A. and Bazan, A., N. Eng. J. Med., 267 (1962) 317.
7. Marchalonis, J.J. and Cone, R.E., Transplant Rev., 14 (1973) 3.
8. Minowada, J., Ohnuma, T. and Moore, G.E., J. Nat. Cancer Inst., 49 (1972) 891.
9. Warner, N.L., Advances in Immunology (Ed. F. Dixon and H. Kunkel), Academic Press, New York,(1974) 145.
10. Wybran, J.H., Fudenberg, H.H. and Sleisinger, M.H., Clin. Res., 19 (1971) 568.

SPONTANEOUS AUTOIMMUNE THYROIDITIS IN THE BUF RAT

N. R. ROSE[1], P. E. BIGAZZI*[2] and B. NOBLE**[2]

Wayne State University School of Medicine[1], Detroit,
Michigan (USA) and State University of New York at
Buffalo, School of Medicine[2], Buffalo, New York (USA)

The immune system has remarkable ability to distinguish self
from non-self. Autoimmune diseases represent exceptions to the
general rule of self-recognition. Their study provides clues about
the basis of the rule. In most human autoimmune diseases of solid
tissues, one sees infiltration by macrophages, plasma cells and
eosinophils in addition to large numbers of lymphocytes. Autoimmune lesions may be the result of damage initiated by autoantibodies, by lymphocytes sensitized to self-antigen, by activated
macrophages, or by the two or more of these factors combined.

A fruitful experimental approach to the study of autoimmune
disease has been the artificial stimulation of specific autoimmune
responses. For example, chronic thyroiditis, one of the best
characterized of the human autoimmune diseases, can be stimulated
by experimental immunization of rabbits and other animals. Administration of heterologous, cross-reacting thyroglobulin, chemically altered thyroglobulin, or homologous thyroglobulin in combination with adjuvants elicits circulating autoantibodies and the
delayed type of hypersensitivity to thyroglobulin as well as mononuclear and polymorphonuclear infiltration of the thyroid. Extensive research has been carried out to assess the relative roles of
humoral and cellular immunity in the pathogenesis of this
disease (18).

* Present address: University of Connecticut Health Center,
Farmington, Connecticut (USA)
** Present address: Royal College of Surgeons, Lincolns Inn
Fields, London (England)

Pathogenetic Mechanisms

Present evidence suggests that the immunological mechanisms producing experimental autoimmune thyroiditis are complex and may differ from species to species. Apparent discrepancies among experimental models may also reflect differences in the conditions used to elicit the immunological response to thyroid antigens. The humoral and cellular balance in experimental thyroiditis is altered by the nature of the antigen and adjuvants used, the immunization schedule and the innate responsiveness of the host. In guinea pigs, antibody titers are poorly correlated with thyroiditis but delayed hypersensitivity to thyroglobulin closely reflects the degree of infiltration of the thyroid (8). Furthermore, lymph node and spleen cells from actively immunized donors can effectively transfer the disease to recipients (9). Serum transfer has not yet been reported. Successful transfer with lymphocytes has been effected in rats and rabbits as well as guinea pigs (28,10). On the other hand, thyroid lesions have been induced in rabbits, mice and monkeys by administration of immune sera (11,17,29).

Ringertz and his colleagues (14) showed that a prominent feature of experimental thyroiditis in the rat is production of antibody-dependent cytotoxic lymphocytes. Clagett, Wilson and Weigle (3) suggested that thyroiditis in the mouse is initiated by disposition of antigen-antibody complexes along the thyroid follicular basement membrane. In our own investigations on passively induced thyroiditis in the mouse, there was an initial transient polymorphonuclear lesion suggestive of an immune complex or Arthus reaction followed by a second phase marked by lymphocytic infiltration which may represent secondary accumulation or even active immunization (26).

Caution must be exercised in extrapolating conclusions on experimental thyroiditis to its human counterpart. There are several obvious ways in which experimental and spontaneous disease differ. Most important, experimental induction of thyroiditis depends upon use of adjuvants or cross-reactive antigens. For this reason it is valuable to study the examples of spontaneously occurring autoimmune thyroiditis as a bridge between experimentally induced disease and the human disorder. Several examples of spontaneous autoimmune thyroiditis are available for study (1).

Spontaneous Autoimmune Thyroiditis

Obese strain (OS) chickens are derived from a closed flock of white Leghorn chickens raised by R.K. Cole at Cornell University, Ithaca, New York. They exhibit a severe hypothyroidism associated with extensive inflammation of the thyroid (32). The disease is

detectable soon after hatching.   Infiltration of the thyroid gland
consists of lymphocytes, plasma cells and macrophages.   Lymphoid
follicles or germinal centers are frequently seen.   Antibodies to
chicken thyroglobulin are present (33).   Positive wattle tests for
delayed hypersensitivity to thyroid antigens are also found (30).
Thymectomy increases and bursectomy decreases the incidence of
severity of the disorder, suggesting that bursa-dependent, B lymphocytes play a crucial role in the pathogenesis of this autoimmune
syndrome while the thymus (or T lymphocytes) may play a regulatory
or suppressor role (31).

TABLE I

Incidence of Thyroiditis in Untreated, Methylcholanthrene-
treated and Neonatally Thymectomized BUF Rats

Age	Untreated	Methylcholanthrene	Thymectomized
4-8 weeks	0	*10%	20%
9-12 weeks	0	29%	52%
13-20 weeks	13%	42%	26%

*thyroiditis over total

Spontaneous autoimmune thyroiditis has also been found in the
inbred BUF strain of rats.   BUF (previously Buffalo) rats are
descended from a line established in 1931 at Roswell Park Memorial
Institute, Buffalo, New York.   They have now been inbred for over
70 generations.   Several laboratories report high incidences of
spontaneous tumors in older animals of the strain (24,34).   Tumors
are seen primarily in the endocrine organs such as the pituitary,
adrenal and thymus glands.   Thyroid lesions in BUF rats, described
by Glover and Reuber (4) were first attributed to carbon tetra-
chloride with which the animals had been treated.   BUF thyroid
glands are also affected by treatment with a number of pharmacologically active agents, including methylcholanthrene, trypan blue
and dimethylbenzanthracine (5,6).   Spontaneous thyroiditis of BUF
rats was first recognized by Hajdu and Rona (7) and subsequently
confirmed by Reuber (13).   The autoimmune basis and incidence of
disease as a function of age and sex were described by Silverman
and Rose (20,23).   No spontaneous disease was found in animals
younger than three months, although 29% of methylcholanthrene-treated

BUF rats were positive by 12 weeks (Table I). The incidence rose
with increasing age, to a maximum at 20 weeks. Circulating antibodies were detected by indirect immunofluorescence in close association with the diffuse mononuclear infiltration of the thyroid.
Staining of colloid was observed, making it likely that the antibodies are directed to thyroglobulin. Neonatal thymectomy has a
dramatic effect on BUF rat thyroiditis (Table I). The incidence
of disease increases and thyroid lesions evolve at an earlier age
in thymectomized animals than in controls (21,22).

Autoantibodies in BUF Rats

     Assessment of circulating antibodies to thyroglobulin by means
of hemagglutination and to thyroid colloid by indirect immunofluorescence (IIF) provide valuable indices to thyroid disease in
BUF rats. The immunofluorescent pattern of colloid staining by
BUF thyroiditis sera is identical to that seen in human thyroiditis. On the other hand, cytoplasmic staining of the follicular
epithelium, attributed in the human to antibodies to microsomal
antigen, does not occur. The close correlation of IIF antibodies
with thyroid pathology makes it possible to identify various stages
of the disease in animals. In untreated BUF rats antibody first
appears at approximately 12 weeks of age. In neonatally thymectomized animals evidence of antibody formation can be found as
early as four weeks of age. Furthermore, it seems likely that at
least in thymectomized animals some regression of disease is
evident in older rats (see Table I). Animals having high titers
of antibody at earlier ages are sometimes negative for both
antibody and pathology when examined at 20 weeks or more.

     There is an unexpected relationship between the antibody titer
and degree of disease. Analysis indicates that the most severe
disease is associated with intermediate titers as measured by both
hemagglutination and IIF. Animals with severe disease frequently
have relatively lower titers as do animals with mild disease. The
simplest explanation for this finding is that animals with severe
disease have less available antigen to maintain a high level of
antibody production. It is also possible, however, that the lower
levels of detectable circulating antibody result from formation of
immune complexes.

Histology of BUF Thyroiditis

     Small thyroid follicles bound by cuboidal epithelium and
perivascular cuffing by lymphocytes are the first charges in spontaneous thyroiditis of the BUF rat. Later, dense collections of
small and medium sized lymphocytes with occasional germinal centers

replace the normal thyroidal architecture. No Hürthle cells are evident.

      In the mature lesions in the BUF rat, macrophages are another prominent component of the infiltrate making up about 12% of cells teased from the diseased thyroid. Medium sized lympho-cytes predominate over smaller ones and plasma cells are sometimes conspicuous. Both B and T lymphocytes can be identified in the thyroid infiltrates, the former being most numerous in early lesions and the latter comprising most of the cells found in advanced stages of disease.

      The histology of the thyroid gland in rats with spontaneous thyroiditis differs from that of animals with experimentally induced disease mainly by the presence of germinal centers in the sponta-neous disorder. Germinal centers are also found in chronic thyroid-itis of humans and in OS chickens.

                           TABLE II

          Comparison of Spontaneous and Experimentally-induced
                Autoimmune Thyroiditis in the BUF Rat

| | Antibody | | | | |
Type of Disease	IIF	CCH	TCH	DH	Path
Spontaneous	+	+	−	−	+
Experimental Immuniza-tion	+	+	+	+	+

IIF – indirect immunofluorescence
CCH – chromic chloride hemagglutination
TCH – tanned cell hemagglutination
DH – delayed hypersensitivity skin test
Path– pathological changes in thyroid

      Experimental and Spontaneous Thyroiditis in the Buf Rat

      There are several other important differences between experi-mentally induced thyroiditis and spontaneously appearing disease in the BUF rat (Table II). Animals that are immunized with rat thyroglobulin plus complete Freund's adjuvant develop high titers of antibody demonstrable by the tanned cell hemagglutination (TCH) test as well as by IIF and chronic chloride hemagglutination (CCH).

BUF rats with spontaneous thyroiditis may have moderate to high
titers of antibodies by IIF and CCF, but are generally negative in
TCH. This difference may represent a qualitative difference in the
antibody evoked by two entirely different methods of autoimmuniza-
tion or even suggest that different antigenic determinants are
involved in the two disease processes, spontaneous and induced.
Another important difference between spontaneously occurring
thryoiditis and induced disease in BUF rats is seen when the skin
test is used as a measure of delayed type hypersensitivity to
thryoid antigen. Experimentally immunized rats are strongly
positive in interdermal skin tests with rat thryoid extract or rat
thyroglobulin while animals with spontaneously developing disease
are negative to both antigens.

Neonatal Thymectomy

Bucsi and Staussen (2) found that neonatal thymectomy of rats
prevented induction of the experimental thyroiditis although the
thyroid antibodies were high both in thymectomized and non-thymec-
tomized animals. On the other hand, thymectomy at five weeks of
age followed by sublethal radiation provoked thyroiditis in 60 percent of Wistar rats (12). The majority of rats that developed severe thyroiditis were subsequently found to be leukopenic and to
have negligible mitogenic response to phytohemagglutin, indicative
of impaired T cell function. IN BUF rats, neonatal thymectomy enhanced the disease process. Thymectomy at the age of four weeks
had no discernible effect. This observation suggests that neonatal thymectomy frees B cells from suppressive regulation and permits
proliferation of autoreactive antibody-forming precursors (15).

At first, this concept of antibody-induced suppression is difficult to reconcile with the relative specificity of the effects of
neonatal thymectomy in BUF rats. The animals developed relatively
little or no evidence of other autoimmune responses. It is, therefore, likely that these animals have a genetically determined pre-
disposition to respond to thyroid antigen vigorously. Other investigations have indicated that response of rats to rat thyroglobulin
is genetically regulated. It is likely that the major genetic determinant of lymphocytic infiltration of the thyroid is linked to the
major histocompatibility (Ag-B) locus (16). In this respect, rats
are similar to mice in which an H-2 linked immune response gene has
been defined (27). One might suppose, therefore, that two genetic
traits are necessary for the spontaneous occurrence of autoimmune
thyroiditis in BUF rats. The first is the loss of thymic suppressive regulation and the second, the inheritance of an unusually
vigorous response to thyroglobulin, particularly in terms of lymphocytic infiltration of the thyroid.

Circulating Lymphocytes

Normal BUF rats have abnormally low numbers of lymphocytes in the peripheral blood. Preliminary results indicate that a further depression of the number of circulating lymphocytes is characteristic of BUF rats with thyroiditis when compared with those that have not yet developed the disease. The BUF lymphopenia resembles the effect of neonatal thymectomy and suggests that a functional T cell deficiency is present, particularly in aging animals. In this respect, BUF rats resemble the aging NZB mouse (25).

One may speculate that lymphopenia of adult BUF rats reflects a premature involution of thymic function permitting the expression of immunity by self-responsive B cells which are thymus-independent after induction. Such autoimmune B cell responses are manifested by high titers of autoantibody and infiltration of the thyroid by plasma cells and B cells. T cells, macrophages and other inflammatory cells are attracted secondarily. Neonatal thymectomy appears to mimic the age-related loss of thymic-dependent suppression.

SUMMARY

The inbred BUF rat develops autoimmune thyroiditis spontaneously. The incidence is related to the age of the animal and is increased by neonatal thymectomy and by treatment with methylcholanthrene. Autoantibodies to thyroglobulin can be demonstrated by indirect immunofluorescence and hemagglutination, but skin tests are negative. The "spontaneous" autoimmune disease may be due to the conjunction of an unusually vigorous immunological response to thyroglobulin and the loss of thymic suppressor function.

ACKNOWLEDGMENT

This work was supported in part by NIH research grants CA 02357, CA 16426 and contract HD 2841.

REFERENCES

1. Bigazzi, P.E. and Rose, N.R., Prog. in Allergy, (1975) 245.
2. Bucsi, R.A. and Straussen, H.R., Experientia, 28 (1972) 194.
3. Clagett, J., Wilson, C. and Weigle, W.O., J. Exp. Med., 140 (1974) 1439.
4. Glover, E.L. and Reuber, M.D., Endocrinology, 80 (1967) 361.
5. Glover, E.L. and Reuber, M.D., Arch. Path., 86 (1968) 542.
6. Glover, E.L., Reuber, M.D. and Godfrey, E.F., Arch. Environm. Hlth., 18 (1969) 901.

7. Hajdu, A. and Rona, G., Experientia, 25 (1969) 1325.
8. McMaster, P.R.B., Lerner, E.M. and Exum, E., J. Exp. Med.,
    113 (1961) 611.
9. McMaster, P.R.B. and Lerner, E.M., J. Immunol., 99 (1967) 208.
10. Nakamura, R. and Weigle, W.O., Int. Arch. Allergy, 32 (1967)
    506.
11. Nakamura, R.M. and Weigle, W.O., J. Exp. Med., 130 (1969) 263.
12. Penhale, W.J., Farmer, A., McKenna, R.P. and Irvine, W., Clin.
    Exp. Med., 15 (1973) 225.
13. Reuber, M.D., Arch. Environmn. Hlth., 21 (1970) 734.
14. Ringertz, B., Wasserman, J., Packalen, Th. and Perlmann, P.,
    Int. Arch. Allergy, 40 (1971) 917.
15. Rose, N.R., Ann. N.Y. Acad. Sci., 249 (1975a) 116.
16. Rose, N.R., Cellular Immunol., 18 (1975b) 360.
17. Rose, N.R. and Kite, J.H., In: Int. Convoc. on Immunol. (Eds.
    Rose and Milgrom) S. Karger, Basel, Switzerland, (1969) 247.
18. Rose, N.R. Seminars in Thrombosis & Hemostasis 1 (1975) 319.
19. Rose, N.R., In: Textbook of Immunopathology, (Eds. Miescher
    and Muller-Eberhard) Grune and Stratton, Inc., New York, New
    York (1976)
20. Silverman, D. and Rose, N.R., Proc. Soc. Exp. Biol. Med., 138
    (1971) 579.
21. Silverman, D. and Rose, N.R., Science, 184 (1974a) 162.
22. Silverman, D. and Rose, N.R., J. Nat. Cancer Inst., 53 (1974b)
    1721.
23. Silverman, D. and Rose, N.R., J. Immunol., 114 (1975) 145.
24. Snell, K.C., In: The Pathology of Laboratory Animals, (Eds.
    Ribelin and McCoy) Charles C. Thomas, Springfield, Illinois
    (1971) 241.
25. Talal, N., Dauphinee, M., Pillafisetty, R. and Goldblum, R.,
    Ann. N.Y. Acad. Sci., 249 (1975) 438.
26. Tomazic, V. and Rose, N.R., Clin. Immunol. and Immunopathol., 4
    (1975) 511
27. Tomazic, V., Rose, N.R. and Shreffler, D.C., J. Immunol., 112
    (1974) 965.
28. Twarog, F.J. and Rose, N.R., J. Immunol., 104 (1970) 1467.
29. Vladutiu, A.O. and Rose, N.R., J. Immunol., 106 (1971) 1139.
30. Welch, P. and Kite, J.H., Fed. Proc., 30 (1971) 306.
31. Wick, G., Kite, J.H., Cole, R.K. and Witebsky, E., J. Immunol.,
    104 (1970) 45.
32. Wick, G., Sundick, R.S. and Albini, B., Clin. Immunol. Immuno-
    path., 3 (1974) 272.
33. Witebsky, E., Kite, J., Wick, G. and Cole, R.K., J. Immunol.,
    103 (1969) 708.
34. Yamada, S., Masuko, K., Ito, M. and Nagayo, T., Gann. 64 (1973)
    287.

ANTIBODY-DEPENDENT CELL-MEDIATED CYTOTOXICITY TO TARGET CELLS

INFECTED WITH HERPES SIMPLEX VIRUSES

S. L. SHORE, F. M. MELEWICZ, H. MILGROM and A. J.
NAHMIAS
Center for Disease Control and Emory University School
of Medicine
Atlanta, Georgia (USA)

Two major mechanisms have been described by which lymphoid cells damage or destroy target cells bearing foreign antigens on their surface. The first of these mechanisms, cell-mediated cytotoxicity (CMC), is important in allograft rejection and tumor immunity (4) and is carried out by sensitized T lymphocytes from specifically immunized hosts. The second mechanism, antibody-dependent cell-mediated cytotoxicity (ADCC), has as yet an undefined role in vivo and is conveyed by the action of mononuclear effector cells on target cells sensitized with antibody to surface membrane antigens (7,15). Unlike CMC, the effector cell in ADCC is found in nonimmunized, as well as immunized hosts. Clearly not a T lymphocyte, in some ADCC systems it appears to be a K cell (3,22), a nonphagocytic lymphoid cell with an Fc receptor, but no readily detectable surface immunoglobulin. A variety of target cells have been employed in vitro to demonstrate ADCC; these have included xenogeneic (8,14), allogeneic (9,10), and various tumor cells (16), as well as cells passively coated with antigens (2,21). We recently demonstrated for the first time ADCC using viral-infected target cells (17,19).

Our system employed cells acutely infected with either herpes simplex type 1 (HSV-1) or herpes simplex type 2 (HSV-2), human serum antibody arising from natural infection, and human peripheral blood mononuclear cells. We report here the characteristics of HSV-ADCC reaction and the nature of the mediating antibody; in addition, we present preliminary data on the identity of the human effector cell.

MATERIALS AND METHODS

The methods employed for the HSV-ADCC system have been detailed elsewhere (17,19). In brief, Chang liver (CL) cells, infected
in monolayer culture 16-24 hr previously with either HSV-1 or HSV-2
(1-2 PFU/cell), were harvested either by gentle shaking with glass
beads or by adding 0.25% trypsin - 0.02% EDTA. Both methods of
harvesting gave similar results in the ADCC assay, although trypsin-
EDTA treatment yielded higher cell viability. The cells were then
labelled with ^{51}Cr, washed, and adjusted to a concentration of 5 x
10^4/ml in Eagles minimum essential medium (MEM) containing 10% heat-
inactivated fetal calf serum and antibiotics. The mononuclear cell
(MC) population was isolated from heparinized human blood by the
Ficoll-Hypaque centrifugation technique (6); the interface cells
were washed 4 times in Hanks BSS and adjusted to an appropriate
concentration in MEM. The human sera to be tested or used as a
source of antibody were heat-inactivated at 56 C for 30 min. Human
donors of serum and/or MC were classified as "immune" or "nonimmune"
based on whether their serum contained detectable neutralizing antibodies and whether their lymphocytes transformed in culture in the
presence of HSV antigens (20).

The ADCC assay was carried out in 12 x 75 mm plastic tissue
culture tubes in a final volume of 1.0 ml. Each culture consisted
of 2 x 10^4 target cells (0.4 ml), mixed with the appropriate number of MC (0.4 ml) and serum (0.2 ml). The cultures were placed in
a CO_2 incubator at 37 C for 16-18 hr. The cultures were then agi-
tated and centrifuged, and the ^{51}Cr contents of the supernatant and
cell pellet were determined. Cytotoxicity was defined as the percent isotope released by MC in the presence of immune serum less
the percent isotope released by MC alone (when lymphocyte-dependent
antibody was being measured), or less the percent isotope released
by MC in the presence of nonimmune serum (when effector cell activity was being measured).

The background release of isotope was about 15% over 16-18 hr
for HSV-1-infected CL cells, and about 25% for HSV-2-infected and
control CL cells. Three freeze-thaw cycles released about 85% of
the total cell-associated isotope.

RESULTS AND DISCUSSION

In the presence of pooled serum positive for HSV antibodies
(immune serum), MC cytotoxicity to HSV-1-infected target cells was
induced (Table I). Activity was detected in the serum up to a dilution of 1:100,000. Pooled negative control serum (nonimmune) had
no such activity. It is unlikely that the cytotoxic reaction involved the classical complement cascade since the antibody-contain-

ing serum was heat-inactivated; in addition, as shown in Table I, cytotoxic activity was detectable at a serum dilution at least 100-fold in excess of that at which complement-dependent cytolytic antibody is usually detectable.

TABLE I

Induction of ADCC to HSV (VR$_3$ Strain)-infected Target Chang Liver (CL) Cells by HSV Antibody Positive Serum

Serum*	Dilution	% Cytotoxicity**	
		With MC***	With Complement****
Immune (5)	1:100	41.6	33.8
	1:1000	40.9	6.7
	1:10,000	34.8	0.4
	1:100,000	13.9	1.4
	1:1,000,000	1.2	2.3
Nonimmune (5)	1:100	0.8	0.6

*Figure in parentheses represents the number of donors whose sera were pooled.
**No cytotoxic effect on uninfected CL target cells was noted.
***MC from immune donor; effector:target ratio 30:1.
****Fresh human HSV-antibody negative serum at final dilution of 1:10

Tissue culture cells infected with either type of HSV were suitable targets for the ADCC process (Table II). When tested at a 1:100 dilution, serum specimens showing an HSV-1 or HSV-2 antibody profile were equally reactive against targets infected with the homologous or heterologous virus types. This demonstrated the presence of antigens common to both HSV-1 and HSV-2 on the surface of infected cells, a fact that is well recognized for HSV (12). However, these data do not rule out the presence of type-specific antigens.

The ADCC reaction with HSV-infected cells was not peculiar to only 1 tissue culture line. The reaction occurred with HSV infection in at least 4 cell lines, 3 of which were allogeneic -Chang liver, human amnion Edmonton and HEp-2- and 1 xenogeneic line -BHK 21 (17,19). In addition, at least 2 strains of HSV-1, HEA (a recent isolate from a human neonate) and VR3 (a laboratory strain) were used to infect targets in the ADCC assay.

TABLE II

Cytotoxic Effect of Human Sera* on
HSV-infected and Control CL Target Cells

| Serum Specimens Tested | Antibody Type** | % Cytotoxicity for Target Cells | | |
		HSV-1 (HE Strain)	HSV-2 (FLO Strain)	Uninfected
3	HSV-1	28.2+1.9***	33.5+4.3	0.4+1.2
3	HSV-2	27.5+8.2	29.0+5.8	1.8+0.4
3	Seronegative	2.2+1.0	-0.4+0.4	-0.3+0.2

*All sera were tested at a 1:100 dilution in the presence of MC from an immune donor at an effector:target cell ratio of 25:1.
**Serum specimens were typed as containing either HSV-1 or HSV-2 neutralizing antibodies, or seronegative as previously described (11).
***Mean % cytotoxicity +1 S.D.

We have previously reported on the kinetics of the HSV ADCC reaction (17,19). A small but reproducible cytotoxic effect could be measured at 1 hr post-incubation. The reaction was complete between 4 and 8 hr post-incubation. If effector cells and target cells were brought into contact in a more rapid fashion, such as by centrifugation, a small cytotoxic effect could be seen within 15 min (unpublished observations).

For a fixed number of target cells (2×10^4 per culture), as little as 1 to 3 effector cells per target cell (E:T) produced a definite cytotoxic response (Table III). For most human donors, the cytotoxicity was directly proportional to the $\log_{10}$ of the E:T between E:T of 3:1 and 30:1, and then reached a plateau at higher E:T values. This plateau was not observed with MC from some donors, even up to an E:T of 100:1, presumably reflecting a lower number of circulating effector cells.

TABLE III

Relationship Between Effector:Target
Cell Ratio and Cytotoxicity

E:T	% Cytotoxicity
100:1	48.9
30:1	69.3
10:1	47.1
3:1	23.7
1:1	9.2
0.3:1	3.7

The MC were from an immune donor and were treated in the presence of a 1:100 dilution of antibody-positive serum. The target cells were HSV-1 infected CL cells. The decreased cytotoxicity noted at E:T of 100:1 was due to an increase in ^{51}Cr-release in the presence of MC alone. This reduced the calculated value for cytotoxicity at this E:T even though total ^{51}Cr-release in the presence of antibody was equivalent to that observed at an E:T of 30:1. The nature of this MC-induced cytotoxicity, usually seen only at high E:T ratios, is under investigation.

    In other ADCC systems, intimate contact between the effector cell and the antibody-coated target cell was required and no soluble cytotoxic factor was found in the supernate (15). One way of demonstrating the requirement for intimate contact between effector and target cells and the lack of secretion of a soluble non-specific cytotoxin is to introduce "bystander" cells into cultures in which ADCC is proceeding to determine if these cells are also injured. As shown in Table IV, ^{51}Cr-labelled but unsensitized HSV-1 infected CL target cells were added to a culture consisting of unlabelled, but antibody-coated HSV-1 target cells, and human MC. The unsensitized bystander cells were not injured, whereas a control culture of ^{51}Cr-labelled sensitized target cells demonstrated a cytotoxic reaction. The results indicated that the MC attack was specific for antibody-coated target cells, and that physical contact between effector and target cell was necessary.

TABLE IV

Lack of Injury of Bystander Target Cells on Herpes ADCC*

Target Cells	Bystander Cells	% Specific ^{51}Cr-Release
^{51}Cr-labelled, sensitized (2×10^4)	None	60.2 ± 9.1
^{51}Cr-labelled, sensitized (2×10^4)	unlabelled, unsensitized (2×10^4)	55.0 ± 4.3
None	^{51}Cr-labelled, unsensitized (2×10^4)	4.1 ± 0.4
Unlabelled, sensitized	^{51}Cr-labelled, unsensitized (2×10^4)	6.6 ± 2.0

*1×10^6 MC from a nonimmune donor were added to each culture.
Target cells were sensitized with a 1:10 dilution of pooled HSV
antibody-positive serum and washed 3 times.

We have previously shown that the factor in human immune serum
which acted synergistically with human MC to damage HSV-infected
target cells was an antibody with specificity for HSV membrane antigens (17,19). Evidence for this was the following: a) the factor
could be removed from immune serum by absorption with HSV-infected
cells, but not with uninfected cells; b) the factor was present in
the serum of immune adult donors, but not of nonimmune adult donors;
c) the factor showed a significant increase in titer in paired sera
taken from patients with primary herpetic gingivostomatitis; d) target cells were sensitized to subsequent attack by MC by prior incubation with immune serum even after thorough washing to remove
free unreacted factor; conversely similar attempts to sensitize or
"arm" MC with immune serum did not increase their cytotoxic capacity.

We have also shown that the antibody present in pooled human
serum belonged to the IgG class of immunoglobulins. Thus, chromatography of pooled antibody-positive serum on columns of either Sephadex G-200 or DEAE-Sephadex A-50 demonstrated that the antibody co-
eluted with serum IgG (17). The antibody also quantitatively crossed
the human placenta (18), a characteristic of IgG. Finally, as shown

in Table V, all but a small fraction of the serum antibody was removed after absorption with Protein A-containing S. aureus, but not with a S. aureus strain negative for Protein A. We have not ruled out the possibility that IgM antibody can also mediate ADCC, since the serum pool contained no IgM antibody to HSV as determined by an indirect immunofluorescence test (13).

TABLE V

Lymphocyte-dependent Antibody Absorption
with Protein A-containing S. aureus*

Serum Absorbed With	Serum Dilution Tested	% Cytotoxicity	% Reduction of Cytotoxicity
Protein A(+) Strain**	1:100	10.8	85
Protein A(+) Strain	1:1000	-1.6	100
Protein A(-) Strain***	1:100	73.6	0
Protein A(-) Strain	1:1000	70.1	5
Not absorbed	1:100	73.6	0
Not absorbed	1:1000	74.1	0

*0.6 ml of heat inactivated pooled antibody-positive serum was absorbed with an equal volume of pelleted bacteria for 30 min at room temperature to remove most of the IgG (1). The CL target cells were infected with HSV-1 and incubated with MC from a nonimmune donor at E:T 30:1.

Pepsin-digestion of IgG with high titers of HSV antibody demonstrated that the Fc fragment must be intact for cytotoxicity to occur (Table VI). Thus, the $F(ab')_2$ fragment was incapable of mediating both ADCC and complement-dependent cytolysis. It is known that the latter is dependent on an intact Fc fragment since binding of C1q is to this portion of the IgG molecule. Note, however, that the $F(ab')_2$ fragment was still fully capable of agglutinating HSV-sensitized sheep RBC's and neutralizing infectious virus, both of which are Fc-independent functions. We showed before (unpublished) that the $F(ab')_2$ also inhibited ADCC by binding to the target cell; this competitive inhibition could only be demonstrated when a suboptimal amount of IgG antibody was used to sensitize the target cell.

TABLE VI

Comparative Antibody Activities of IgG
and F(ab')$_2$* in ADCC Against HSV

Type of Antibody	Lymphocyte-dependent	Complement-dependent cytolytic	Indirect Hemagglutinin	Neutralizing
IgG	0.1**	30	39	63
F(ab')$_2$	>100	>1000	19	32

*IgG in Cohn Fraction II from pooled human serum and F(ab')$_2$ prepared by pepsin digestion (modification of Nisonoff's method).
**Lowest concentration in µg/ml showing activity.

We have also attempted to identify the dominant effector cell in the HSV ADCC reaction. That the effector cell was unlikely to be a sensitized T cell was supported by the following observations: a) effector cell activity was similar in either HSV immune or non-immune individuals (17,19); b) the effector cell was present in the cord blood of healthy newborns, who presumably lack sensitized T cells to HSV (18); c) there was enrichment rather than decrease of cytotoxicity when T cells were removed. For these experiments (Table VII) fractionation of blood MC on nylon wool columns (5) was followed by depletion of E-rosetting cells from the nonadherent cell population. The nonadherent cells, which represented about 50% of the MC applied to the column, were as cytotoxic as the unfractionated MC. The cytotoxic capacity of the nonadherent MC was evident, even though the phagocytic cell population had been reduced a mean of 29-fold and the B-cell population (as indicated by surface Ig-bearing cells) had been reduced 21-fold. Interestingly, passage through the column did not result in enrichment of cytotoxic effectors, despite a loss of 50% of the starting cell population. This indicated either that the effector cell was somewhat adherent or, alternatively, that the B-cell or monocyte populations lost on the column made some contribution to effector cell activity.

TABLE VII

Characterization of ADCC Effector Cell
for HSV-1 Infected Target CL Cells

| Treatment* | % Mean Cytotoxicity at E:T Ratio of: | | | |
	30:1	10:1	3:1	1:1
Unfractionated (8) (L-11.6;SIg-12.6; Fc-14.1; E-72.9)	44.3	20.4	8.3	---
Nylon wool nonadherent (8) (L-0.4; SIg-0.6; Fc-5.7; E-88.4)	42.9	20.3	9.6	4.1
Nylon wool nonadherent, E-rosette depleted (5) (L-1.8; SIg-4.1; Fc- 40.6; E-6.6)	---	48.6	29.2	13.9

*L-% of MC ingesting latex particles; SIg-% of latex-negative MC
with surface Ig; Fc-% of latex-negative MC with surface Fc receptor;
E-% of latex-negative MC with receptor for SRBC. Number of healthy
adult donors tested indicated in parentheses. E-rosette positive
cells were removed by incubating the nylon wool nonadherent cells
with SRBC's, followed by centrifugation on Ficoll-Hypaque and harvesting of the interface cells.

After depletion of E-rosetting cells, the nonadherent cell
population showed a striking enrichment of cytotoxic activity at
each E:T ratio tested (Table VII). This enrichment occurred despite a 14-fold reduction in the T-lymphocyte population, again
indicating that the sensitized T cell was not the effector cell
(3 of the 5 MC donors were HSV immune). There was also a striking
increase in the number of Fc-positive, surface Ig-negative cells
from 5%, before E-rosette depletion, to 36% after. Since the K
cell was defined as an Fc+, surface Ig-, E-rosette-, nonphagocytic
cell, it was reasonable to infer that the enrichment in cytotoxicity
noted was secondary to the enrichment of the K cell population.
The presence of normal HSV ADCC effector cells in 3 patients with
congenital sex-linked agammaglobulinemia (Bruton) with almost total absence of circulating Ig-bearing lymphocytes offered further

evidence that the effector is not a classical B cell.

Earlier (unpublished), we demonstrated the presence of an Fc receptor on the effector cell by showing inhibition of ADCC both with heat-aggregated human IgG and with third party cellular immune complexes (uninfected CL complexed with rabbit antibody). The effector cell could also be removed by adsorbing MC onto plastic dishes coated with soluble immune complexes (human serum albumin [HSA] - rabbit anti HSA).

ACKNOWLEDGEMENTS

This work was supported in part by Grant No. CP43393 within the Virus Cancer Program of the NCI, NIH, PHS and by Grant No. 5-R01-DE03924 from the NIDR, NIH, PHS.

REFERENCES

1. Ankerst, J., Christensen, P., Kjellen, P. and Kronvall, G., J. Infect. Dis., 130 (1974) 268.
2. Calder, E.A., Penhale, W.J., McLeman, D., Barnes, E.W. and Irvine, W.J., Clin. Exp. Immunol., 14 (1973) 153.
3. Calder, E.A., Urbaniak, S.J., Penhale, W.J. and Irvine, W.J., Clin. Exp. Immunol., 18 (1974) 579.
4. Cerottini, J.C. and Brunner, K.T., Adv. Immunol., 18 (1974) 67.
5. Greaves, M.F. and Brown, G., J. Immunol., 112 (1974) 420.
6. Harris, R. and Ukaejiofo, E.O., Lancet, ii (1969) 327.
7. MacLennan, I.C.M., Transplant. Rev., 13 (1972) 67.
8. MacLennan, I.C.M. and Loewi, G., Clin. Exp. Immunol., 3 (1968) 385.
9. MacLennan, I.C.M., Loewi, G. and Howard, A., Immunology, 17 (1969) 897.
10. McConnachie, P.R., Rachelefsky, G., Stiehm, E.R. and Terasaki, P.I., Pediatrics, 52 (1973) 795.
11. Nahmias, A.J., Josey, W.E., Naib, Z.M., Luce, C.F. and Duffey, A., Am. J. Epidemiol., 91 (1970) 539.
12. Nahmias, A.J. and Roizman, B., N. Eng. J. Med., 289 (1973) 667.
13. Nahmias, A.J., Shore, S.L. and Del Buono, I., Viral Immunodiagnosis (Ed. E. Kurstak) Academic Press, New York, New York, (1974) 157.
14. Perlmann, P. and Holm, G., Transplant. Rev., 11 (1969) 117.
15. Perlmann, P., Perlmann, H. and Wigzell, H., Transplant. Rev., 13 (1972) 91.
16. Pollack, S. and Nelson, K., J. Immunol., 110 (1973) 1440.
17. Shore, S.L., Black, C.M., Melewicz, F.M., Wood, P.A. and

Nahmias, A.J., J. Immunol., 116 (1976) 194.
18. Shore, S.L., Milgrom, H., Wood, P.A. and Nahmias, A.J., Pediat. Res., 9 (1975) 336.
19. Shore, S.L., Nahmias, A.J., Starr, S.E., Wood, P.A. and Mc Farlin, D.E., Nature, 251 (1974) 350.
20. Starr, S.E., Karatela, S.A., Shore, S.L., Duffey, A. and Nahmias, A.J., Infect. Immun., 11 (1975) 109.
21. Wasserman, J., Packalen, T., Perlmann, P. and Perlmann, H., Int. Arch. Allerg., 36 (1969) 115.
22. Wisløff, F., Frøland, S.S. and Michaelsen, T.E., Int. Arch. Allerg., 47 (1974) 139.

The RES in

Transplantation Immunology

IMMUNOGENETIC ASPECTS OF ALLOTRANSPLANTATION

E. J. YUNIS[1], B. DUPONT[2] and J. HANSEN[2]

University of Minnesota[1], Minneapolis, Minnesota (USA)
and Sloan-Kettering Institute[2], New York, New York (USA)

Histocompatibility antigens may be defined as those antigens present on various cells which evoke an immune response after transplantation of an organ or tissue from one individual into another.

The first suggestion that the allograft rejection was immuno-logically-based was made by Murphy in 1912 (8). But the proof that the mechanism of allograft rejection was immunologic was established by Medawar (88-90). He described the phenomenon of first set allograft rejection of skin which occurs following revascularization of the allograft approximately nine to 12 days after grafting when there is lymphocyte infiltration of the graft accompanied by ischemia and sloughing. Regrafting of tissues from the same donor results in a lack of vascularization of the allo-graft or an accelerated rejection (second set phenomenon). He and his collaborators also developed the concept that graft rejection was analogous to delayed hypersensitivity and that the actual attack of the graft was not by humoral antibodies, but by lymphoid cells from lymphatic tissues (18,95). The experiments that established the basic principles of the genetics of tissue trans-plantation were performed in inbred strains of mice by Little and Tyzzer (1916) and the early studies that followed were summarized by Little (79,80). Skin grafts within inbred strains are success-ful; grafts between inbred strains are not successful (allografts fail). While skin grafts from either inbred parent strain to the F_1 hybrid succeed, grafts in the reverse direction fail. Skin grafts from F_2 or subsequent generations also survive in all F_1 mice. In addition, Little showed that acceptance of a tumor allograft depended upon the graft tissue and host possessing a

number of susceptible factors in common or different, each determined by independent genes. If the combination of factors present
in the tumor graft did not match those possessed by the host the
tumor graft could not grow. Using these studies it was estimated
that more than 20 loci are involved in transplantation in mice.
The number of loci were calculated by the formula $(.75)^n$ (since
each locus examined in two pure strains of mice will be expressed
(codominant) in 75% of the F_2). The exponential indicates the
number of loci. Gorer established that alloantigens of the mouse
were implicated in allograft rejection using serological reagents,
and that H-2 was a chromosomal region with several closely-linked
genes controlling specific alloantigens (48-50). Gorer, Lyman and
Snell showed that the major histocompatibility which is linked to
a tail abnormality was identified serologically as H-2 (51). Snell
named the genes determining the fate of allografts histocompatibility or H genes and established the basis for the study of histocompatibility genes (120,121).

    The antigenic phenotype of individuals belonging to the
population may differ because of genetic polymorphism or alleles
in one or more loci. Therefore, the analysis of genetic histocompatibility was aimed at a definition of the number of loci that
control the codominant antigens and the number and frequencies of
their alleles. The H-2 system of the mouse has been of interest
to immunologists since it is an important model for the major
transplantation antigen systems of other mammals. The antigens of
this major system are more immunogeneic than the antigens of other
systems, and matching at H-2 results in a delay of skin graft
rejection. The H-2 complex is located in the IX linkage group of
chromosome 17 of the mouse, and is composed of closely-linked genes
that control many traits of immunological implication: histocompatibility, immune responses, serum protein variants (complement),
thymus leukemia cellular alloantigens, tumor virus susceptibility,
mixed lymphocyte reaction, graft-versus-host reaction, cell-mediated
lympholysis, hybrid resistance, the I region lymphocyte antigens,
testosterone levels, T cell-B cell interactions, and liver cyclic
adenosine monophosphate levels. The serologically detected cellular alloantigens are controlled by the two major markers of the
region: the H-2K and H-2D loci controlling many alleles (polymorphism) (119,121).

HLA Linkage Group

    The serology and inheritance of human leukocyte alloantigens
(HLA-A, B and C) have been the result of the studies of many
investigators during the last 18 years. This development has been
achieved by international collaboration in six histocompatibility
workshops. The first characterization of a human leukocyte allo-

antigen, Mac, was done by Dausset (20). Following this, van Rood
(10) described the first leukocyte group system with two alleles,
4a and 4b, controlled by one locus. In addition, he introduced
the use of 2 x 2 association analysis (computer) to identify
patterns of serological reactions. A second allelic group of
leukocyte antigens called LA was elucidated by Payne et al. (99).
Based on population studies (22), a single genetic system of
leukocyte alloantigens, Hu-1 was postulated (98). This designation
was later changed to HL-A (98). This assumption of control by a
single genetic locus was proven in family studies by Ceppellini,
Amos and Dausset (6,21,31). According to this concept, this single
genetic locus controls a large group of alleles which could identify
the segregation of paternal and maternal antigens by the serological
reactions transmitted to the children. Therefore, only six genotypes could be identified in a given family. The paternal genes
were usually designated as A and B (AB genotype), the maternal C
and D (CD genotype) with the possible presence of a maximum of
four genotypes in the children: AC, BC, AD and BD. The one gene
hypothesis was confirmed by the report that allogeneic stimulation
in mixed culture was correlated with the serological reactions
of the family members (12), i.e., children inheriting the same
lymphocyte antigens failed to stimulate in mixed lymphocyte
culture. It was also deduced that the genes controlling the
production of these leukocyte antigens were codominant since
they were expressed in each member of the family. The involvement
of cell surface antigens in transplant rejection has been reviewed
(9), and was demonstrated in recipients immunized by skin
allografts or leukocyte injections from donors characterized
serologically (23,112). Intrafamilial skin allografts in some
cases survived for over 20 days, suggesting the potential use of
sibling donors in transplantation (32). Prospective matching
between siblings for kidney and skin allografts were then reported
(8). These kidney grafts have been very successful (9). Using the
knowledge of genetic inheritance of HLA antigens, characterization
of haplotypes led to the finding of longer survival of skin grafts
among siblings HLA identical, than haplo-identical siblings and
shorter survival among siblings HLA different. There was a mean
skin graft survival of 24, 14 and 11 days, respectively (9).

The most direct proof that HLA antigens are important in
transplantation comes from the observation that allografts are
rejected in an accelerated manner when performed in the presence
of alloantibody specific to antigen(s) of the donor (crossmatch).
But the importance of the human HLA antigens in the allograft
reaction in the absence of immunization has been more difficult to
ascertain. The reports of longer allograft survival in better
matched unrelated donor-recipient combinations could be explained
by the assumption that the HLA-A and B antigens themselves elicit

the allograft immunity (first set phenomenon) or that there are
loci independent from but closely linked to the known HLA loci
that are responsible for allograft immunity and that they are in
genetic disequilibrium with HLA-A and HLA-B (10, 146).

Three Linked Loci

Based on population and family studies (24,31,83) it was
concluded that the HLA specificities are genetically related and
the products of at least two different genetic regions. These two
regions now known as HLA-A (LA or first locus) and HLA-B (FOUR or
second locus) are controlled by two sites closely linked on the
same chromosome. The concept of at least two loci has been proven
by the finding of crossing over between the two loci or two segregant series (67,141). From analysis of the haplotypes of
Scandinavians, it was estimated that the recombination frequency
between the HLA-A and HLA-B loci is 0.81% $\pm$ 0.24 (132).

The HLA alleles or the HLA antigens studied in families or in
populations can be grouped in two linked "segregant series" or loci.
Within the family the alleles cannot be transmitted with the same
gamete and most of the time the alleles of one locus cannot be
brought together by cross over. In most families they behave as
mutually exclusive allelic traits and are inherited as a genetic
unit changed only by cross over. The combination of the two factors or alleles transmitted together with the same chromosome is
called a haplotype and a genotype is formed by two haplotypes
(Table I).

Using computer programs (83) designed to study alleles or
segregating phenotypes from population and family studies it was
concluded that: 1) the HLA specificities tested for could be
grouped into two non-overlapping series corresponding to the two
HLA loci; 2) the combined frequencies were distributed according to Hardy-Weinberg equilibrium which assures mutually exclusive
alleles; 3) haplotype frequencies can be calculated from phenotypes
in the population by calculating the haplotype frequency in terms
of gene frequency and a correction factor (genetic association),
and 4) in families, the zygotic assortment of the four paternal
haplotype was such that one allele of one locus appeared only in
one haplotype.

The finding that some sera react with cells from some but not
all individuals phenotypically or genotypically identical suggested
the existence of a third locus now named HLA-C (103,122,140).
Direct evidence has been presented in studies of recombinant
families to suggest that this segregant series is controlled by a
factor placed between the HLA-A and HLA-B loci (133).

TABLE I

Inheritance of HLA Antigens

Antisera		♂ AB	♀ CD	1	2	3	4	5	Chromosome Marker
				Siblings					
Anti	HLA-A	+	–	+	–	+	–	+	A
	HLA-B	+	–	+	–	+	–	+	
Anti	HLA-A	+	–	–	+	–	+	–	B
	HLA-B	+	–	–	+	–	+	–	
Anti	HLA-A	–	+	+	–	–	+	+	C
	HLA-B	–	+	+	–	–	+	+	
Anti	HLA-A	–	+	–	+	+	–	–	D
	HLA-B	–	+	–	+	+	–	–	

Genotypes:

	♂ AB	♀ CD	1	2	3	4	5
Paternal Haplotypes	A(A)B A(B)B		A(A)B	A(B)B	A(A)B	A(B)B	A(A)B
Maternal Haplotypes		A(C)B A(D)B	A(C)B	A(D)B	A(D)B	A(C)B	A(C)B

Two important phenomena among many have complicated the
serological study of the HLA antigens: 1) cross reactivity and 2)
genetic linkage disequilibrium. Cross reactivity between several
alleles of each locus is probably related to the sharing of common
determinants between allelic products. Most HLA antibodies have
broad reactivity that includes reactions with several allelic
products of the same locus. The definition of new HLA specificities
are recognized by a committee of the W.H.O. only when a sufficient
number of antibodies of restricted specificity have been found.
When the antibodies that give similar reactivity are studied at a
workshop they are given a provisional number (HLA-W#) (Table II).
A list of the HLA specificities is given in Table II. Definition
of HLA specificities can be further complicated by linkage disequilibrium: certain genes of the HLA-A locus exist in strong
linkage disequilibrium with certain genes of the HLA-B locus. This
means that some alleles of these loci are associated with each other
at a much higher frequency than expected. The degree of association is measured as a constant (Delta: Δ) from the difference between
the observed haplotype frequency and that expected from the product
of the gene frequencies (30). The deltas vary from population to
population, i.e., in European caucasoids the strongest delta values
are found for the haplotypes 1-8, 2-12 and 3-7. The reason for
these associations is unknown. Nevertheless, two important basic
mechanisms may be involved: 1) selection pressure from disease,
and 2) genetic control of recombination frequency of the major
transplantation linkage group by the T alleles (9).

Genetic Control of Mixed Lymphocyte Culture Reaction in Man

When lymphocytes from two different individuals are mixed
in vitro this usually results in a proliferative response of the
lymphocytes from both individuals (16). A unidirectional mixed
lymphocyte reaction (MLR) can be obtained by pretreatment of one
of the donor cells with Mitomycin-C (14), or X-irradiation (65).
These agents block the lymphocyte proliferation by inhibition of
DNA synthesis, and the cells will function in the in vitro
culture as stimulating cells. Dutton (29) demonstrated that
mouse strains differing at H-2 showed positive MLR while H-2
identical strains did not result in positive MLR. Bach and Amos
(7,12) proved that the MLR in man was genetically controlled
by genes within the major histocompatibility complex (HLA). The
existence of a separate MLR-stimulating locus (now designated HLA-D
locus) was obtained in studies of HLA-A and HLA-B recombinants and
in a family study reported by Yunis et al., 1971 (145,146). In
this family two different pairs of HLA sero-identical siblings
were studied. One pair had the HLA sero-haplotypes A/C and the
other pair the haplotypes A/D. The A/C x A/C combinations were
mutually MLR identical but the A/D x A/D combinations were

TABLE II

Specificities Determined by Four HLA Loci

HLA-A		HLA-B**		HLA-C		HLA-D	
1975	Previously	1975	Previously	1975	Previously	1975	Previously
HLA-A1	HL-A1	HLA-B5	HL-A5	HLA-CW1	T1	HLA-DW1	LD101(W5a)
HLA-A2	HL-A1	HLA-B7	HL-A7	HLA-CW2	T2	HLA-DW2	LD102(7a)
HLA-A3	HL-A3	HLA-B8	HL-A8	HLA-CW3	T3	HLA-DW3	LD103(8a)
HLA-9	HL-A9	HLA-B12	HL-A12	HLA-CW4	T4	HLA-DW4	LD104(12a)
HLA-AW23	W23	HLA-B13	HL-A13	HLA-CW5	T5	HLA-DW5	LD105(16a)
HLA-AW24	W24	HLA-B14	W14			HLA-DW6	LD106(15a)
HLA-A10	HL-A10	HLA-B18	W18				
HLA-AW25	W25	HLA-B27	W27				
HLA-AW26	W26	HLA-BW15	W15				
HLA-A11	HL-A11	HLA-BW16	W16				
HLA-A28	W28	HLA-BW38	W16.1				
HLA-A29	W29	HLA-BW39	W16.2				
HLA-AW30	W30	HLA-BW17	W17				
HLA-AW31	W31	HLA-BW21	W21				
HLA-AW32	W32	HLA-BW22	W22				
HLA-AW33	W10.6	HLA-BW35	W5				
HLA-AW34	Malay 2	HLA-BW37	TY				
HLA-AW36*	Mo*	HLA-BW40	W10				
HLA-AW43*	BK	HLA-BW41	Sabell				
		HLA-BW42*	MWA				

(HLA-A29 through HLA-AW33 bracketed as HLA-AW19)

*Specificities most frequently found in certain African groups.
**W4 (4a) and W6 (4b) remain unchanged.

mutually storngly MLR responsive. However, one of the children
with the A/D HLA genotypes did not stimulate or respond against
the cells of the two A/C children. This family study demonstrated
that the MLR is controlled by a separate genetic factor different
from the serologically identified HLA determinants. A recombinational event was postulated to have occurred between the two
maternal HLA haplotypes C and D and this recombinant haplotype (D)
made one A/D child identical for the MLR stimulating determinants
with the two A/C children. Five additional families with two sets
of HLA identical sibling pairs demonstrating similar MLR response
patterns have subsequently been described (36,67,115,135). These
family studies constitute today the best indirect proof for the
existence of a separate genetic factor controlling MLR stimulation
in man. Family studies involving families with recombination
between the HLA-A locus and the HLA-B locus have shown that the
HLA-D determinant segregates together with the HLA-B specificity
(25,36,47,75,93,106,146-148). Independent genetic control of the
MLR in man by a separate genetic locus closely linked to the genes
coding for the serologically defined HLA antigens is supported by
the following observations: 1) identification of other examples
of MLR positive HLA sero-identical sibling combinations (7,12,102),
and MLR non-responsive sibling combinations differing by one HLA
haplotype (117); 2) MLR non-responsive parent-child combinations
(2,7,10,12); 3) most HLA sero-identical unrelated combinations are
MLR positive, but only a small proportion of them showed MLR
identity or weak MLR (69,109, 127).

   Estimation of the recombination fraction between the HLA-B
locus and the HLA-D locus has been made based on family MLC studies
(66). Thirty-nine families with 167 children were studied and 2
children were identified by indirect proof as HLA-B/HLA-D recom-
binations, for a calculated recombination fraction of 0.0074. However, the direct proof of a recombinational event can only be made
with the positive identification of the 2 HLA-D and HLA-B gene
products between which the recombination occurred, and the identification of these determinants on the recombinant haplotype.

   The genetic control of the MLR in the mouse has been shown
to depend primarily on genetic determinants within the H-2 complex
(119). Differences at the K-end of the H-2 complex in general
provoke much stronger MLR than H-2D differences (113,144). Strong
MLR in the mouse is primarily controlled by genetic differences
within the I-region (15,94). Other genes within the H-2 complex
are also capable of generating MLR; however, they usually induce
weak MLR as compared to the I-region differences (1,94,104,142).
One additional system, outside of H-2, can also induce MLR (M-locus)
(41). By modification of the MLC culture technique with decreasing serum concentration in the cultures the weak MLR generated by
genes different from the MLR stimulating genes within the I region

can produce strong responses (100,101).

The complex genetic control of MLR in the mouse and the considerable homology between H-2 and HLA indicates that a single gene hypothesis for MLC stimulation in man is an oversimplification. A few family studies have been reported suggesting that another gene controlling weak stimulation in MLR is located between the HLA-A locus and the HLA-B locus (27,36,93,136). It is not known whether or not these weak stimulations are induced by the HLA-A or the HLA-B antigens or by separate genetic factors. Proof for the existence of these genetic determinants has, however, not been presented.

Identification of MLR Stimulating Specificities by HLA-D Homozygous Typing Cells (HLA-D Locus Typing)

Two different principles have been used to identify the specificity of the HLA-D determinants: 1) HLA-D locus typing with HLA-D homozygous typing cells and 2) serological identification of lymphocyte membrane components which are shared by individuals identical for the HLA-D determinants. The development of HLA-D typing with D-homozygous cells is based on the concept that 1) there is a qualitative difference between strong and weak MLR; 2) the strong MLR is controlled by a separate genetic determinant closely linked to the HLA-B determinant; 3) the weak MLR obtained in MLC testing between 10 percent of HLA sero-identical unrelated individuals depends on strong genetic linkage disequilibrium between the HLA-B determinants and the HLA-D determinants. This last assumption was strongly supported by the observation of nonresponsive parent-child combinations in some situations where the parents both carried one HLA sero-identical haplotype (26,73,91,92,115,117,135). Even parent-child combinations with identical HLA-B determinants were observed frequently to be non-responsive (26). Also, some unrelated HLA-B identical combinations were observed to give only weak MLR (60,76, 93,147). These "atypical" or unexpected weak MLR were most frequently observed when the HLA-B antigens involved were the common B7, B8, BW35, B12.

The HLA-D homozygous typing cell (TC) is defined as a cell homozygous for one D specificity. The typing reaction is demonstrated by the lack of MLR or only weak MLR induced in a responder cell when stimulated by the TC. Because of this weak MLR, the typing reaction is called a Typing Response (for a detailed report on statistical considerations and criteria for Typing Responses consult Joint Report from the V[1] International Histocompatibility Workshop 1975, Histocompatibility Testing 1975) (61). A TC is defined operationally as a D-homozygous TC if the MLR in the population shows discrimination between weak responses (Typing Responses) and strong responses. D-homozygous TC are primarily

obtained from the outbred population (random matings) by selecting
families where the parents share one HLA serophalotype and an HLA-
homozygous child is identified (26,91,92). TC may also be obtained
from inbred families – primarily first cousin marriages, where the
parents share one HLA haplotype and the HLA homozygous child is
HLA homozygous by descent (63,139). The D-locus homozygous child
should elicit Typing Responses from both parents.

Six distinct HLA-D specificities (DW1-DW6) and two less well-
defined specificities, LD-107 and LD-108, have been defined during
the VI International Histocompatibility Workshop 1975. Thirty eight
of the 62 Workshop TC were found to belong to 1 of these 8 groups.
A clear positive association between some HLA-B antigens and the
DW specificities were observed: DW1 to BW35, DW2 to B7, DW3 to B8,
DW4 to BW15 and CW3 and LD-107 to B12. Each of the DW specificities
were defined by two-to-six TC. In the Workshop data the phenotype
distribution fit the Hardy-Weinberg distribution suggesting that
the DW specificities belong to one genetic system.

In 48 MLC combinations between individuals carrying the same
two DW specificities, only 66 percent showed low MLR (153). In
another study, 12 combinations sharing the same two DW specificities showed varying degrees of low to intermediate MLR (64). These
observations together with the finding of mutual weak to moderate
MLR between TC belonging to the same DW specificity group or strong
MLR between TC in the same group indicates that the DW specificities as defined today may be composed of either several cross-
reacting subspecificities of one allelic system (57,105), or the
products of genetic subloci (28).

MLC Activated Cells or Primed Cells: Killer and Memory Cells

The triggering of allogeneic response (in vitro) and the resulting cellular differentiation might enable us to study the
cellular events leading to a rejection of a graft and possibly to
understand the basis of graft-versus-host reactions. Mixture of
allogeneic leukocytes different at HLA-D results in proliferation
of lymphocytes measured in the mixed lymphocyte reaction (MLC and
the generation of killer cells are measured by the cell-mediated
lysis (CML) (12,13,14,15,16,33,59,78,79,123). Both MLC and CML
have immunologic specificity as demonstrated by the fact that the
primed lymphocytes produced display allogeneic memory (secondary
MLC and CML) (11,33,34,42,59,86,88,118). The proliferative phase
is initiated by differences at determinants genetically controlled
by genes linked to HLA-B now named HLA-D (5,19,36,37,62,87,87a,116,
138). But the determinants initiating the production of killer
cells as well as those involved in the lysis of the target are
probably genetically controlled by genes linked to HLA-A and HLA-B

and different from HLA-D (HDR, CML-S) (36-39,40,55-57,59,60,82,86,
87a,123,127,128,138,146). It appears that HLA-A, B and C are
important primarily in transplantation, especially in an immunized
individual. HLA-D is important in prevention of the graft-versus-
host reaction in bone marrow transplants and the hypothetical locus
HDR (hypersensitivity delayed reaction) or CML-S may be important
in the generation of killer cells and the first-set phenomenon of
the allograft rejection (10,146). The cells involved in the
triggering of allogeneic responses are lymphocytes and macrophages.
T cells are primarily the responding cells, and are also primarily
the effectors in CML whether the stimulation is produced by T cells,
B cells or null cells. The specificity of the effector cells
generated by either T, B or null cells is not directed toward a
determinant restricted to one type of cell. The stimulating cells
appear to be primarily the B cells, but may also be the T and null
cells and the macrophages. The macrophages are necessary for cell
cooperation to produce a T cell response, but they do not appear
to be necessary in the stimulation phase (35,107,108,125,126). The
genes controlling the response of the T cells by allogeneic (HLA-D)
cells may be similar to Ir genes and have been tentatively mapped
together with HLA-B. A determinant controlling the level of response
in MLC, MLR-R has been postulated. This determinant may explain
some examples of asymmetric responses in some family studies (28,
146).

Serological Identification of HLA-D Association Alloantigens
            with Restricted Tissue Distribution

    A number of I region associated alloantigens (Ia antigens)
have recently been defined in the mouse (119). A presumably similar
alloantigen system is presently being identified in man. These
alloantigens occur primarily on B lymphocytes (143,144,182) and
can be identified by indirect immunofluorescence (110,143,144) or
by complement-dependent, antibody-mediated cytotoxicity assays
performed on B-cell enriched or purified B lymphocyte suspensions
(82,110,144). Preliminary studies indicate that the genetic control
of some of these B alloantigens is closely linked to the HLA-D
locus (110,144). Other specificities may, however, be controlled
by genes in other segments of the HLA complex (82). Some B cell
alloantigens seem to be coded for by genes transmitted independently of the HLA complex (74,144). Some of the antisera directed
against B alloantigens demonstrate specific blocking of MLC stimulation (52).

Identification of Other Genes in Linkage with the HLA-Complex

    The HLA-complex was assumed to be located on chromosome number

6 by somatic cell hybridization studies between Man-Chinese hamster
cells (62,124).   In addition, in one family chromosome number 6 with
a pericentric inversion segregated in the family with one HLA haplotype (72).   The gene coding for the different types of phosphoglucomu-
tase 3 (PGM_3) have been shown (71,72) to be in linkage with HLA,
and was also found in the cell hybrids containing human chromosome
6.   The sequence on the chromosome is likely to be PGM_3 - HLA-D -
HLA-B - HLA-C - HLA-A.

     Recently, several components of the serum complement system
have been shown to be in linkage with the HLA complex.   Linkage
of HLA to the glycine-rich B-2 glycoprotein (GBG, Bf or factor B)
have been established (3).   The genes involved in the synthesis of
the second component of complement (C2) are in linkage with the
serologically defined HLA antigens (45,46).   It has subsequently
been shown that the gene for C2 synthesis is strongly associated
with the HLA-DW2 specificity, since most individuals with half
levels of C2 are HLA-DW2 heterozygous and individuals with complete
lack of C2 are HLA-DW^2 homozygous.   The DWs specificity segregates
in these families with the C2 deficiency gene (44,46).   Recently it
has been demonstrated that the gene controlling synthesis of the
fourth component of complement (C4) segregated in one family
together with the HLA haplotypes and that the C4 deficient child
was HLA homozygous (106).   Mapping of the Bf locus in relation
to HLA has been attempted (134).   The data, however, are conflicting and mapping of the Bf gene in relation to the different loci
of HLA cannot be made at present.   The C2 deficiency gene seems to
be located close to the HLA-B locus outside the HLA-A HLA-B segment
of HLA.   It cannot at present be determined if the C2 deficiency
gene is to be mapped between HLA-B and HLA-D or outside the HLA-D
locus (44,46).   However, in rhesus monkeys, Bf was found to be in
close linkage with the Rh L-A complex probably between MLR and Ir-
GA, outside the loci controlling serologically detectable alleles
(149).

Genetic Control of Immune Responsiveness and HLA-Disease
Associations

     In the mouse the I and S regions are mapped between the K end
and the D end of the H-2 complex.   The S region controls variations
in serum proteins (complement) which appear to be functionally
distinct.   The I region traits include the Ir genes, the MLC, Ia
antigens and also GVHR.   Since it is believed that the H-2 complex
of the mouse is the homolog of HLA of man, it would be expected
that several functions genetically controlled by genes near the K
end will be controlled by genes near the HLA-B locus (disease
susceptibility, Ir genes, complement components, lymphocyte activation and GVHR) (119) (Fig. 1).

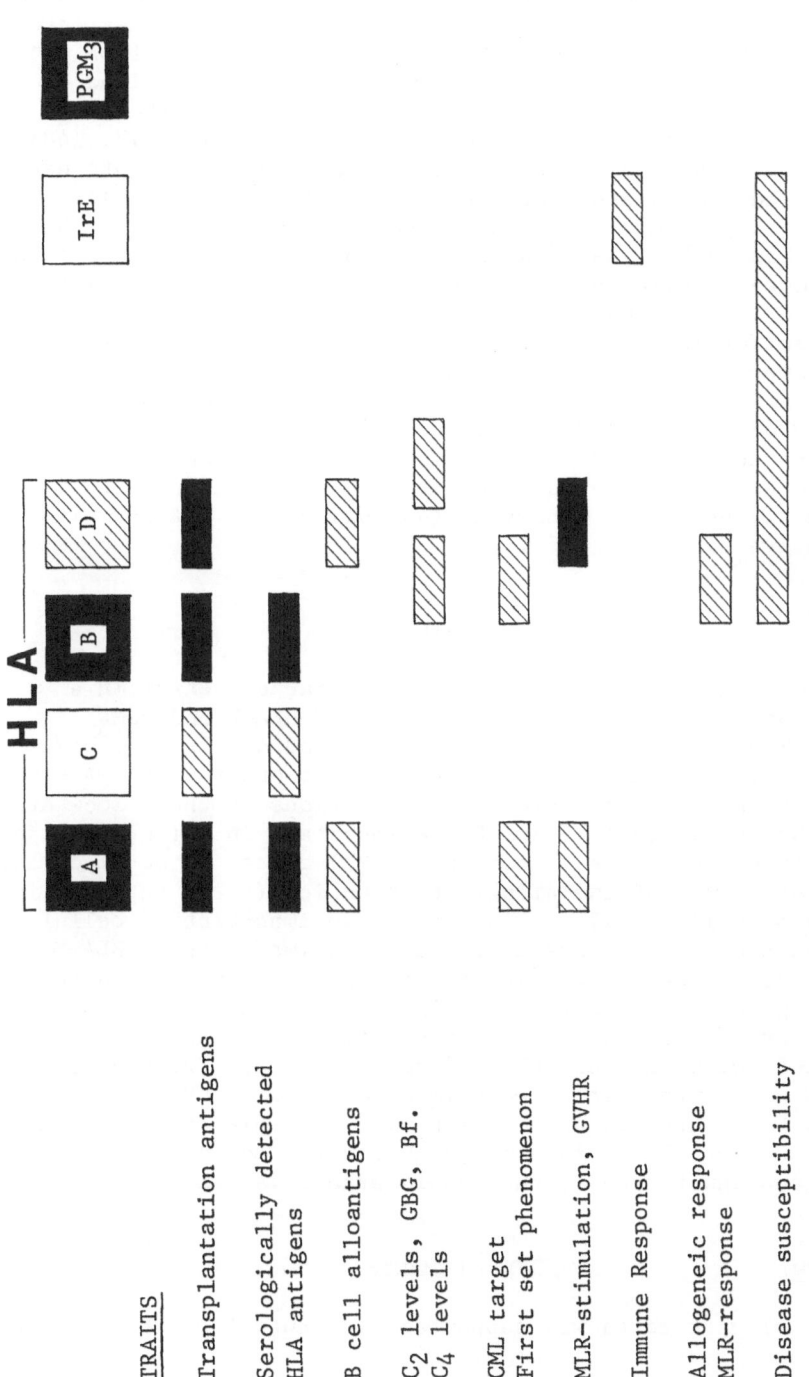

Fig. 1. Map of the LHA-Ir-PGM3 linkage group showing probable locations of genes controlling variant traits associated with the complex. The top shows the genes; the black indicates definitive loci; the hatched HLA-D probable and the empty boxes possible locations. The lower part of the drawing shows the traits associated with those loci. Hatched areas indicate uncertainty about localization to that region.

Attempts to study and map Ir genes in man are difficult mainly because of ethical considerations of human experimentation, the lack of controlled experimental conditions to study immune response in man, and more specifically, the lack of antigens of restricted antigenicity that can be used. So far, the only information available about tentative mapping of Ir genes in man was obtained studying families with ragweed allergy. In such families, the HLA-D locus was tentatively placed between HLA-B and the IrE (17, 148). Determination of HLA-Ir genotypes will be of particular interest in attempts to determine associations with disease susceptibility. There are 2 main groups of diseases associated with HLA: the HLA-B27 associated arthritides (ankylosing spondilitis, reactive arthritis, and acute anterior uveitis group) and the HLA-B8 associated immune based disease: Myasthenia gravis, coeliac disease, dermatitis herpetiformis, Addison's disease, Graves disease, juvenile onset diabetes and chronic active hepatitis. Although only 6 HLA-D determinants have been characterized, 3 of them have already served to classify diseases according to the HLA-D typing. HLA-DW2 was found increased in multiple sclerosis and C2 deficiency, HLA-DW3 in juvenile onset diabetes and Addison's disease and HLA-DW4 in adult type of rheumatoid arthritis (44,46,129,133) but not with chronic active hepatitis (98).

CONCLUSIONS

It now appears unequivocal that three markers exist in a linkage group in chromosome 6 of man: HLA-A, HLA-B and PGM_3 (Fig. 1.) Tentatively, two other HLA loci and one Ir gene have been mapped close to HLA-B. The probable map order is HLA-A - HLA-C - HLA-B - HLA-D - Ir. The biological functions of these loci are unknown. However, HLA-A, B and C are important in allograft rejection. Other closely linked loci (HDR, CML) appear to be important in the first events of the allograft rejection (first set) and in generation of killer cells. HLA-D might be important in cellular recognition and graft-versus-host reactions (matching at HLA-D decreases the incidence and severity of graft-versus-host disease), and the Ir genes in the defense against infections. HLA-B and HLA-D loci are important markers in studies of disease susceptibility. HLA-B locus antigens HLA-B27 and HLA-B8 are frequently associated with arthritic or autoimmune disorders. HLA-D determinants have been found in association with multiple sclerosis and C2 deficiency (HLA-DW2); juvenile diabetes and Addison's disease (HLA-DW3) and adult type of rheumatoid arthritis (HLA-DW4).

ACKNOWLEDGMENTS

Original work cited was supported by grants from the American

Cancer Society, U.S. Public Health Grants HL-06314, 1-RO1 HD 08145, NO1-CB-43853 and CA-08748-0851, and from N.C.I. Program Project Grant NCI-CA 17404-01.

REFERENCES

1. Abbasi, K., Demant, P., Festenstein, H., Holmes, J., Huber, B., Rychlikova, M., Transplant. Proc. 5 (1973) 1329.
2. Albertini, R. and Bach, F.H., J. Exp. Med., 128 (1968) 639.
3. Allen, F.H., Jr., Vox Sang. 27 (1974) 383.
4. Alter, B.J., Bach, F.H., Cell Immunol., 1 (1970) 207.
5. Alter, B.J., Bach, F.H., J. Exp. Med., 140 (1974) 141.
6. Amos, D.B., Transplant. 5 (1967) 1015.
7. Amos, D.B. and Bach, F.H., J. Exp. Med., 128 (1968) 623.
8. Amos, D.B., Hattler, B.G., Hutchin, P., McCloskey, R. and Zmijewski, C.M., Lancet 1 (1966) 300.
9. Amos, D.B. and Ward, F., Physiol. Reviews 55 (1975) 206.
10. Amos, D.B. and Yunis, E.J., Cell Immunol., 2 (1971) 517.
11. Anderson, L.C., Hayry, P., Europ. J. Immunol., 3 (1973) 595.
12. Bach, F.H. and Amos, D.B., Science, 156 (1967) 1506.
13. Bach, F.H. and Hirschhorn, K., Science, 143 (1964) 813.
14. Bach, F.H. and Voynow, N.K., Science, 153 (1966) 545.
15. Bach, F.H., Widmer, M.B., Bach, M.L. and Klein, J., J. Exp. Med., 136 (1972) 1430.
16. Bain, B. and Lowenstein, L., Science, 145 (1964) 1315.
17. Billingham, R.E., Brent, L., and Medawar, P.B., Proc. Roy. Soc. (Biol) 143 (1954) 58.
18. Blumenthal, M.N., Amos, D.B., Noreen, H., Mendell, N.Y. and Yunis, E.J., Science, 184 (1974) 1301.
19. Bonnard, G.D., Chaplis, M., Glasuer, A., Mempel, W., Baumann, P., Grosse-Wilde, H. and Albert, E.D., Transplant Proc. 5 (1973) 1679.
20. Ceppellini, R., In: Progress in Immunology (Ed. D. B. Amos) Academic Press, New York, New York (1971) 973.
21. Ceppellini, R., Curtoni, E.S., Mattiuz, P.L., Miggiano, V.C., Scudeller, G. and Serra, A., In: Histocompat. Testing 1967 (Ed. E.S. Curtoni, P.L. Mattiuz and R.M. Tosi) Munksgaard, Copenhagen, Denmark (1967) 149.
22. Ceppellini, R., Curtoni, E.S., Leicheb, G., Mattiuz, P.L., Miggiano, V.C. and Visetti, M., In: Histocompat. Testing 1965 (Ed. H. Balner, F.J. Cleton and J.G. Eernisse) Munksgaard, Copenhagen, Denmark (1965) 13.
23. Cerottini, J.C., Engers, H.D., MacDonald, H.R. and Brunner, K. T., J. Exp. Med. 140 (1974) 703.
24. Charmot, D., Mawas, C., Sasportes, M., Legrand, L. and Dausset, J., Histocompat. Testing 1975, Munksgaard (in press)
25. Chess, L., McDermott, R.P., Schlossman, S.F., J. Immunol., 113 (1974) 1122.

26. Dausset, J., Acta Haematol., 20 (1958) 156.
27. Dausset, J., Colombani, J., Legrand, L. and Fellows, M., In:
    Histocompat. Testing 1970 (Ed. P.I. Terasaki) Munksgaard,
    Copenhagen, Denmark, (1970) 53.
28. Dausset, J., Ivanyi, P., Colombani, J., Feingold, N. and
    Legrand, L., In: Histocompat. Testing 1967 (Ed. E.S. Curtoni,
    P.L. Mattiuz and R.M. Tosi), Munksgaard, Copenhagen (1967) 188.
29. Dausset, J., Ivanyi, P. and Ivanyi, D., In: Histocompat.
    Testing 1965 (Ed. H. Balner, F.J. Cleton and J.G. Eernisse)
    Munksgaard, Copenhagen, Denmark (1965) 51.
30. Dausset, J., Rappaport, F.T., Ivanyi, P. and Colombani, J.,
    In: Histocompat. Testing 1965 (Ed. H. Balner, F.J. Cleton,
    and J.G. Eernisse) Munksgaard, Copenhagen, Denmar, (1965) 63.
31. Dupont, B., Good, R.A., Hansen, G.S., Jersild, C., Nillsen,
    L.S., Park, B.H., Svejgaard, A., Thomsen, M. and Ynis, E.J.,
    Proc. Nat. Acad. Sci., 71 (1974) 52.
32. Dupont, B., Jersild, C., Hansen, G.S., Staub-Nielsen, L.,
    Thomsen, M. and Svejgaard, A., Transplant. Proc. 5 (1973) 1543.
33. Dupont, B., Nielsen, L.S., Svejgaard, A., Lancet, 2 (1971) 1336.
34. Dupont, B., Yunis, E.J., Hansen, J.A., Reinsmoen, N., Suciu-
    Foca, N., Mickelson, E. and Amos, D.B., In: 6th International
    Histocompat. Workshop, (1975) in press.
35. Dutton, R.W., J. Exp. Med., 123 (1966) 655.
36. Eijsvoogel, V.P., Du Bois, M.J.G.J., Melief, C.J.M., de Groot-
    Kooy, M.L., Koning, C., van Rood, J.J., van Leeuwen, A., Du
    Toit, E. and Schellekens, P.Th.A., Histocompat. Testing,
    Munksgaard, (1972) 501.
37. Eijsvoogel, V.P., du Bois, R., Melief, C.J.M., Zeylemaker, W.P.,
    Raat-Koning, L. and de Groot-Kooy, L., Transplant. Proc. 5
    (1973) 415.
38. Eijsvoogel, V.P., du Bois, M.J.G.J., Meinesz, A., Bierhorsh-
    Eijlander, A., Zeylemaker, W.P. and Schelleken, P.Th.A.,
    Transplant. Proc. 5 (1973) 1675.
39. Eijsvoogel, V.P., Sem. Hemat. 11 (1974) 305.
40. Festenstein, H., Demant, P., and Rogers, A., Histocompat.
    Testing, Munksgaard (1975) in press.
41. Festenstein, H. and Demant, P., Transplant. Proc. 5 (1973) 1321.
42. Fradelizi, D. and Dausset, J., Mixed lymphocyte reactivity of
    human lymphocytes primed in vitro, Eur. J. Immunol., 5 (1975)
    295.
43. Fradelizi,D., Mawas, C., Charmot, D. and Sasportes, M., Histo-
    compat.Testing, Munksgaard (1975) in press.
44. Friend, P.S., Handweger, B.S., Youngki, K., Michael, A.F. and
    Yunis, E.J., Immunogenetics (in press)
45. Fu, S.M., Kunkel, H.G., Brusman, H.P., Allen, F.H., Jr. and
    Fotino, M., J. Exp. Med., 140 (1975) 1108.
46. Fu, S.M., Stern, R., Kunkel, H.G., Dupont, B., Hansen, J.A.,
    Day, N.K., Good, R.A., Jersild, C. and Fotino, M., J. Exp. Med.
    142 (1975) 495.

47. Gatti, R.A., Meuwissen, H.J., Terasaki, P.I., and Good, R.A., Tissue Antigens, 1 (1971) 239.
48. Gorer, P.A., J. Pathol. Bacteriol., 44 (1937) 691.
49. Gorer, P.A., J. Pathol. Bacteriol. 47 (1938) 231.
50. Gorer, P.A., In: Biological Problems of Grafting, (Eds. F. Albert and P.B. Medawar) Blackwell Scientific Publications, Ltd. Oxford, England (1959)
51. Gorer, A., Lyman, S. and Snell, G.D., Proc. Roy. Soc. B-135 (1948) 499.
52. Greenberg, L.J., Reinsmoen, N., Yunis, E.J., Transplant. 16 (1973) 520.
53. Grosse-Wilde, H., Netzel, B., Mempel, W., Ruppelt, W., Brehm, G., Bertrams, J., Ewald, R., Lenhard, V., Rittner, Ch., Scholz, S. and Albert, E.D., Histocompat. Testing, Munsgaard (1975) in press.
54. Grunnet, N. and Kristensen, T., Histocompat. Testing, Munksgaard (1975) in press.
55. Grunnet, N., Kristensen, T., Jorgensen, F. and Kissmeyer-Nielsen, F., Tissue Antigens, 4 (1974) 218.
56. Grunnet, N. and Kristensen, T., Histocompat. Testing, Munksgaard (1975) in press.
57. Hansen, J.A., Dupont, B., Rubinstein, P., Suciu-Foca, N., Fu, S.M., Mickelson, E., Whitsett, C., Jersild, C., Kunkel, H.G., Day, N.K., Good, R.A., Thomas, E.D., Reempsma, K., Allen, F.H., Jr. and Fotino, M., Histocompat. Testing, Munksgaard (1975) in press.
58. Hayry, P. and Defendi, V., Science, 10 (1970) 133.
59. Hodes, R. and Svedmyr, E.A.J., Transplant. 9 (1966) 478.
60. Horwitz, S.D., Groshony, T., Bach, F.H., Hong, R. and Yunis, E.J., Lancet, (1975)
61. Joint Report from the VI International Histocompat. Workshop 1975, Histocompatibility Testing (1975)
62. Jongsma, A.H., van Someren, H., Westerveld, A., Hegemeijer, A. and Pearson, P., Humangenetik, 20 (1973) 195.
63. Jorgensen, F., Lamm, L.U. and Kissmeyer-Nielsen, F., Tissue Antigens, 3 (1973) 323.
64. Jorgensen, F., Lamm, L.U., Ferrara, G.B., Ipsen, S., Jørgensen, H. and Kissmeyer-Neilsen, F., In: Histocompat. Testing, Munksgaard (1975) in press.
65. Kasakura, S. and Lowenstein, L., Nature, 208 (1965) 974.
66. Keuning, J.J., Tweel, J.G. van den, Gabb, B.W., Termijtelen, E., Goulny, E., Blokland, E., Elferink, B.G., and Rood, J.J. van, Tissue Antigens (1975) in press
67. Kissmeyer-Nielsen, F., Svejgaard, A., Ahrons, S. and Staub-Nielsen, L., Nature 224 (1969) 75.
68. Kissmeyer-Nielsen, F., Svejgaard, A., Sørensen, S.F. and Staub-Nielsen, L. and Thorsby, E., Nature, 228 (1970) 63.
69. Kristensen, T. and Grunnet, N., Histocompat. Testing, Munksgaard (1975) in press

70. Kristensen, T. and Grunnet, N., Histocompat. Testing,
    Munksgaard (1975) in press
71. Lamm, L.U., Svejgaard, A. and Kissmeyer-Nielsen, F., In:
    Intern. Congr. Human Genetics, Paris, Excerpta Med. Found.
    Intern. Congr. Ser. No. 233 (1971) 107.
72. Lamm, L.U., Friedrich, U., Petersen, G.B., Jørgensen, J.,
    Nielsen, J., Therkelsen, A.J. and Kissmeyer-Nielsen, F.,
    Human Heredity, 24 (1974) 273.
73. Lebrun, A., Sasportes, M., Dausset, J., Transplant. Proc., 5
    (1973) 363.
74. Legrand, L., Dausset, J., Proc. Sixth Internat. Histocompat.
    Workshop, (1975) in press
75. Lebrun, A., Sasportes, M., Lebrun, D., Dausset, J., C.R. Acad.
    Sci. 273 (1971) 2130
76. L'Esperance, Hansen, J.A., Jersild, C., O'Reilly, R., Good, R.A.,
    Thomsen, M., Nielsen, L.S., Svejgaard, A. and Dupont, B.,
    Transplant. Proc. 7 (1975) 823.
77. Lightbody, J.J. and Bach, F.H., Transplant. Proc. 4 (1972) 307.
78. Lightbody, J., Bemoco, D., Miggiano, V.C. and Ceppellini, R.,
    G. Batt. Virol. Immunol., 64 (1971) 243.
79. Little, C.C. and Tyzzer, E.E., J. Med. Res., 33 (1916) 393.
80. Little, C.C., In: Biology of Laboratory Mouse, (Ed. G.D. Snell)
    Blakiston, Philadelphia, Pennsylvania (1941)
81. Long, M.A., Handwerger, B., Yunis, E.J., Histocompat. Testing,
    Munksgaard (1975) in press
82. Mann, D.L., Abelson, L., Henkart, P., Harris, S. and Amos, D.B.,
    Proc. Sixth Intern. Histocomp. Workshop (1975) in press
83. Mattiuz, P.L., Ihde, D., Piazza, A., Ceppellini, R. and Bodmer,
    W.F., In: Histocompat. Testing (1970) 193.
84. Mawas, C.E., Charmot, D., Sasportes, M., Histocompat. Testing,
    Munksgaard, (1975) in press
85. Mawas, C., Christen, Y., Legrand, L., Dausset, J., Transplant.
    Proc., 5 (1973) 1691.
86. Miggiano, V.C., Bemoco, D., Lightbody, J., Trinchieri, G. and
    Ceppellini, R., Transplant. Proc. 4 (1972) 231.
87. Mayr, W.R., Bemoco, D., deMarchi, M. and Ceppellini, R.,
    Transplant. Proc., 5 (1973) 1581.
88. Medawar, P.B., J. Anat., 78 (1944) 176.
89. Medawar, P.B., J. Anat., 79 (1945) 157.
90. Medawar, P.B., Brit. J. Exp. Path., 27 (1946) 15.
91. Mempel, W., Grosse-Wilde, H., Albert, E., Thierfelder, S.,
    Transplant. Proc. 5 (1973) 401.
92. Mempel, W., Grosse-Wilde, H., Baumann, P., Netzel, B., Stein-
    bauer, R., Rosenthal, I., Scholz, S., Bertrans, J. and Albert,
    E.D., Transplant. Proc., 5 (1973) 1529.
93. Mempel, W., Albert, E.D., Burger, A., Tissue Antigens, 2 (1972)
    250.
94. Meo, T., Vives, J., Miggiano, V., Shreffler, D., Transplant.
    Proc., 5 (1973) 377.

95. Mitchison, N.A., Proc. Roy. Soc. (Biol) 161 (1964) 275.
96. Murphy, J.B., J.A.M.A., 59 (1912) 874.
97. Nomenclature Committee, Nomenclature for factors of the HL-A
    System, Bull. World Health Organ. 39 (1969) 483.
98. Page, A.R., Sharp, H.L., Greenberg, L.J. and Yunis, F.J.,
    J. Clin. Invest. 56 (1975) 530.
99. Payne, R., Tripp, M., Weigle, J., Bodmer, W. and Bodmer, J.,
    Cold Spring Harbor Symp.
100.Peck , A.B. and Click, R.E., Transplant. 16 (1973) 331.
101.Peck , A.B., Katz-Heber, E., Click, R.E., Eur. J. Immunol., 3
    (1973) 516.
102.Plate, J.M., Ward, F.E., Amos, D.B., In: Histocompat. Testing1970
    (Ed. P.I. Terasaki) Munksgaard (1970) 153.
103.Pierres, M., Fradelizi, D., Neanport-Santes, C. and Dausset,
    J., Tissue Antigens, 4 (1975) 266.
104.Plate, J.M., Transplant. Proc. 5 (1973) 1351.
105.Reinsmoen, N., Stewart, M., Emme, L., Hanrahan, L.A., Dupont,
    B., Hansen, J.A., Friend, P., Amos, D.B. and Yunis, E.J.,
    Histocompat. Testing (1975) in press
106.Rittner, C.H., Hauptmann, G., Grosshans, E. and Mayer, S.,
    VI Histocompat. Workshop Conf., Munksgaard, (1975) in press
107.Rocklin, R., Chess, L., MacDermott, R.P., Schlossman, S.F.,
    David, G., J. Exp. Med., 140 (1974) 1303.
108.Rode, H.N., Gordon, J., J. Cell Immunol., 13 (1974) 87.
109.Rood, J.F. van and Eijsvoogel, V.P., Lancet, 1 (1970) 698.
110.Rood, J.J. van, Leeuwen, A. van, Keuning, J.J. and Blusse,
    van oud Alblas, A., Tissue Antigens, 5 (1975) 73.
111.Rood, J.J. van and Leeuwen, A. van, J. Clin. Invest., 42 (1963)
    1382.
112.Rood, J.J. van, Leeuwen, A. van, Eernisse, J. G., Fredericks,
    E. and Bosch, L.J., Ann. N.Y. Acad. Sci., 120 (1964) 285.
113.Rychkikova, M., Demant, P., Ivanyi, P., Nature (New Biol),
    230 (1971) 271.
114.Rychlikova, M., Demant, P., Ivanyi, P., Folia Biol., 16
    (1970) 218.
115.Sasportes, M., Lebrun, A., Rapaport, F.T., Dausset, J., Trans-
    plant. Proc., 4 (1972) 209.
116.Schendel, D.J. and Bach, F.H., J. Exp. Med., 140 (1974) 1534.
117.Sigler, H.F., Ward, F.E., Amos, D.B., Phaup, M.B. and Stickel,
    D.L., JEM 133 (1971) 411.
118.Sheehy, M.J., Sondel, P.M., Bach, M.L., Wank, R. and Bach, F.H.,
    Science, 188 (1975) 1308.
119.Shreffler, D. and David, C.S., Adv. in Immunol., 20 (1975) 125.
120.Snell , G.D., J. Genet., 49 (1948) 87.
121.Snell , G.D. and Stimpfling, J.H., In: Biology of the Lab. Mouse,
    Blakiston , Philadelphia, Pennsylvania (1966) 457.
122.Solheim, B.G., Bartlie, A., Sandberg, L., Staub-Nielsen, L.
    and Thorsby, E., Tissue Antigens, 3 (1973) 439.
123.Soliday, S., Bach, F.H., Science, 170 (1970) 1406.

124. Someren, H. van, Westerveld, A., Hagemeijer, A., Mees, J.R., Meers, K.P. and Zaalberg, O.B., Proc. Nat. Acad. Sci., 71 (1974) 961.
125. Sondel, P.M., Chess, L., MacDermott, R.P. and Schlossman, S.F., J. Immunol., 114 (1975) 982.
126. Sondel, P.M., Chess, L. and Schlossman, S.F., Cellular Immunol. (1975)
127. Sorenson, S., Freisleben and Hawkes, S.P., Transplant. Proc., 5 (1973) 1361.
128. Sorensen, S.F., Transplant. Proc., 5 (1973) 1657.
129. Stastny, P., Tissue Antigens, 4 (1974) 571.
130. Staub-Nielsen, L. and Thorsby, E., Tissue Antigens, 1 (1971) 81.
131. Staub-Nielsen, L., Mayr, W., Sandberg, L., Piazza, A. and Svejgaard, A., In: Histocompat. Testing (1975) (in press)
132. Svejgaard, A., Bratlie, A., Hedin, P.J., Hogman, C., Jersild, C., Kissmeyer-Nielsen, F., Lindblom, B., Lindholm, A., Low, B., Messeter, L., Moller, E., Sandberg, L., Staub-Nielsen, L. and Thorsby, E., Tissue Antigens, 1 (1971) 81.
133. Svejgaard, A., Platz, C., Ryder, L.P., Staub-Nielsen, L. and Thomsen, M., Transplant. Reviews, 22 (1975) 3.
134. Teisberg, P., Olaisen, B., Geede-Dahl, T., Jr. and Thorsby, E., Tissue Antigens, 5 (1975) 257.
135. Thompson, J.S., Parmely, M.J., Flink, R.J. and Severson, C.O., J. Exp. Med., 135 (1972) 596.
136. Thorsby, E., Hirschberg, H. and Helgesen, A., Transplant. Proc., 5 (1973) 1523.
137. Thorsby, E., Hirschberg, H., and Helgesen, A., Transplant. Proc., 5 (1973) 1523.
138. Trinchieri, G., Bernoco, D., Curtoni, S.E., Miggiano, V.C. and Ceppellini, R., Histocompat. Testing 1972, (1972) 509.
139. Tweel, J.G. van den, Blusse, S. van oud Alblas, A., Keuning, J.J., Goulmy, E., Termijtelen, A., Bach, M.L. and Rood, J.J. van, Transplant. Proc., 5 (1973) 1535.
140. Walford, R.L., Finkelstein, S., Hanna, C. and Collins, Z., Nature, 224 (1969) 74.
141. Ward, F.E., Southward, J.G. and Amos, D.B., Transplant. Proc., 1 (1969) 352.
142. Widmer, M.B., Peck, A.B. and Bach, F.H., Transplant Proc., 5 (1973) 1501.
143. Winchester, R.J., Dupont, B., Wernet, P., Hansen, J., Fu, S.M., Ojea, F., Laursen, N., Kunkel, H.G., Proc. Sixth Intern. Histocompat. Workshop, (1975) (in press).
144. Winchester, R.J., Fu, S.M., Kunkel, H.G., Dupont, B., Jersild, C., J. Exp. Med., 141 (1975) 924.
145. Yunis, E.J., Plate, J.M., Ward, F.E., Seigler, H.F. and Amos, D.B., Transplant. Proc., 3 (1971) 118.
146. Yunis, E.J. and Amos, D.B., Proc. Nat. Acad. Sci., 68 (1971) 3031.
147. Yunis, E.J., Krivit, W., Reinsmoen, N. and Amos, D.B., Nature, 248 (1974) 517.

148.Yunis, E.J., Amos, D.B. and Blumenthal, M.N., Transplant. Proc.,
     7 (1975) 49.
149.Ziegler, J.B., Alber, Ch.A. and Balner, H., Nature, 254 (1975)
     609.

ALLOANTIBODIES IN RELATION TO THE REJECTION OF SKIN ALLOGRAFTS IN THE MOUSE

T. N. HARRIS and S. HARRIS

The Children's Hospital of Philadelphia and School of
Medicine, University of Pennsylvania, Philadelphia,
Pennsylvania (USA)

The rejection of skin allografts is accepted as a cell-mediated phenomenon, presumably by host lymphocytes invading the graft, but the specific effector molecule by which the lymphocytes bring this about has not been identified. Data will be presented here suggesting, in the mouse, an association of antibodies with either accelerated rejection or prolonged retention, respectively, depending on the 7S subclass of the antibodies involved.

The developments which will be described in this paper arose from cell transfer studies in the rabbit. Following the demonstration of immune responses transferred by lymph node cells of the guinea pig (4), a series of studies was done with transfer of lymph node cells from antigen-injected donor rabbits to recipient rabbits, in which the corresponding antibody subsequently appeared (8,9). It was shown that the antibody which was found in the serum of the recipients was due to the transfer, within the cells, of the mechanism for the synthesis of antibody (4,8).

In studies directed at the function of these transferred cells, use was made of the then recent observations by Medawar (21), in the genetically heterogeneous rabbit, that second-set skin grafts were rejected at an accelerated rate, in comparison with first-set grafts. The application undertaken to our studies was to see the effect of transferring lymph node cells into recipient animals which had been immunized with leukocytes from the prospective donors. If recipient rabbits had previously been injected with blood leukocytes of prospective donors, or with pooled rabbit leukocytes, the production of antibody by the subsequently transferred lymph node cells in the recipients' tissues could be comp-

letely abolished (10). It was then found that this suppression of
function of transferred allogeneic lymph node cells could also be
carried out passively, by injection of serum of third party rabbits
which had been immunized with pooled rabbit leukocytes, or merely
by incubation of the cells, before transfer, with rabbit anti-
rabbit leukocyte serum (5).

The observation of a serum antibody which suppressed the
function of allogeneic cells in a normal recipient raised a question of studying the effect of this antibody on other transplanted
allogeneic cells, including those contained in solid tissue grafts,
such as skin grafts. In such a study, accelerated rejection of the
grafts would be the analogue of the suppression of allogeneic spleen
cells which we had observed. This necessitated working in a species
in which animals of inbred strains are available, such as the
mouse, developing in this species the appropriate conditions of
cell transfer, and the demonstration of both actively induced and
passively transferred suppression of allogeneic spleen cells (11,6).
Methods were worked out for immunization of inbred mice with allogeneic spleen cells, for the production of ascitic fluid globulin
(12), and for measuring the suppressive antibody in vitro (suppression of the production of hemolytic antibody plaques by spleen
cells of SRBC-immunized mice) (7).

Various Effects of Alloantibody-Containing Globulins
on Allograft Retention

When enough globulin, of sufficiently high suppressive titer,
had been obtained from mice of three $H-2^k$ strains versus cells of an
$H-2^d$ strain (Balb/c), individual preparations were tested for their
effect on the retention of skin allografts, by injection into
normal mice grafted in the same strain combination (e.g., CBA anti-
Balb/c globulin injected into normal CBA mice grafted with Balb/c
skin). The titer of suppressive antibody was thus being used as a
tentative measure of the appropriate anti H-2 alloantibody, since
it was from the suppression of such allogeneic cells that the question arose. Early experiments were carried out with ascitic fluids
pooled from C3H or CBA mice between the 11th and 13th days after
a second injection of Balb/c spleen cells, when the titer of suppressive antibody in the ascitic fluids was at its highest.
Against a normal first-set rejection time of 14 days for skin
grafts in this strain combination, in the first three experiments,
21 of 23 antibody-injected mice did, in fact, show complete rejection of the graft by day 11, in comparison with zero among 19
control mice.

In the next experiments of this series, however, because of the limited supply of ascitic fluid globulin obtained at 11 to 13 days it was decided to use globulins obtained from the same mice later, relative to the immunizing injections of Balb/c spleen cells, when the suppressive titer had declined somewhat, but with concentrating such globulins to reach a titer of this antibody equal to that of the 11-13 day globulins used in earlier experiments. Globulins which were obtained at later times after the immunizing injection, and concentrated, showed a diminishing effect of rejection of Balb/c skin grafts, until, with fluids obtained later and concentrated up to 4-fold, no effect on the time of rejection of skin grafts was seen. It was thus clear that some factor other than the suppressive titer as measured in vitro by the hemolytic antibody plaque test was involved in the effects on skin allografts which we had seen.

TABLE I

Various Effects of CBA Anti-Balb/c Globulin Pools on the Time of Retention of First-Set Balb/c Skin Grafts in CBA Mice

Type of effect	No. of globulin pools	Grafts rejected by day 11/ total no.	Grafts retained after day 17/ total no.
Accelerated rejection	5	64/78 (82%)	
Accelerated rejection, partial	5	31/61 (51%)	
No effect	57	0/34* (0)	
Prolonged retention	5		43/46 (93%)

*Data obtained in the tests on 4 typical pools of this group.

All globulins obtained 11-13 days after second injection of Balb/c spleen cells.

A series of experiments was therefore undertaken involving a
larger number of anti-Balb/c ascitic fluid globulin pools obtained
under varying immunizing conditions, but in each case tested for
effects on skin grafts if the titer of suppressive antibody was
high enough. These experiments were all done with CBA anti-Balb/c
ascitic fluid globulins tested in CBA mice given Balb/c skin grafts.
All of these fluids were obtained between 11 and 13 days after the
second injection of Balb/c spleen cells and had relatively high
suppressive antibody titers for Balb/c cells. However, the injection of these fluids into allografted mice produced a range of
different effects on the skin grafts. Table I shows results of
four kinds: first, in a few of the globulin pools, accelerated
rejection of the great majority of the test grafts; second, rejection by day 11 of approximately 25-60% of test grafts; third, no
significant difference from normal rejection time (found in the
great majority of globulin pools); and fourth, prolonged retention of Balb/c skin grafts (13).

Skin Allografts in Immunologically Tolerant Mice

The clear effect of accelerated rejection of skin allografts
following the injection of anti graft strain alloantibody, in the
few instances in which this occurred, and the wide range of effects
produced by the injection of such antibodies, made it desirable to
examine the effects of such alloantibodies on skin grafts in a
situation where there could be no significant contribution by an
immune response of the host. Another study was therefore carried
out on the effect of such alloantibody-containing globulins on the
fate of the skin grafts in CBA mice which had been rendered
immunologically tolerant to Balb/c alloantigens by perinatal injection of spleen cells from (Balb/c x CBA) F$_1$ hybrid mice (14).

The first series of experiments in immunologically tolerant
CBA mice was carried out in the same general plan as the experiments
in the normal mice, with anti-Balb/c globulin of relatively high
suppressive titer injected, as before, in the first few days following the placing of the skin grafts. Observations were begun
on day 10 after grafting and continued until the rejection of the
graft or until the growth of white hair on the graft site was
noted, usually day 21-24. Four anti-Balb/c globulin preparations
obtained from CBA mice between days 11 and 13 after their second
injection of Balb/c spleen cells caused rejection of the grafts
in all of the mice tested, from five of five on days 10-12, in
the most effective of these globulin preparations, to five of six
within days 14-19 in the case of the least effective of the group.
Of the 16 control mice for these four globulin pools, 13 showed
retention of the graft to the stage of growth of white hair after
the third week, and three showed a late rejection, between days 19
and 22.

From some groups of mice ascitic fluid globulins were also collected later. These showed decreased effectiveness in causing rejection of the skin graft, or none, an observation consistent with the decrease or loss of effectiveness of late globulins in the normal allografted hosts.

In another series of experiments, Balb/c skin grafts were placed on immunologically tolerant CBA or C3H mice and, after two months, groups of mice bearing tolerated grafts were given a series of injections (over five days) of anti-Balb/c ascitic fluid globulin (11-13 day) raised in mice of the same strain. Each experiment included two control mice, littermates of the antibody-injected mice, also bearing tolerated grafts and injected with normal ascitic fluid globulin at the same dose. Of 17 mice injected with anti-Balb/c globulins in five experiments, 15 showed clear shrinking of the graft five to 14 days after the first injection and complete rejection of the graft between 13 and 22 days. The remaining two antibody-injected mice showed only a clear decrease in size of the graft, with some areas of scarring, but the graft remained in that state for the duration of the two-month period of observation. Control grafts all remained intact through this period, with no change in the size of the graft or the appearance of the hair.

The short interval between the injection of alloantibody and the rejection of the tolerated grafts indicated that this rejection could not have been due to breakdown of immunologic tolerance by attack of the alloantibody on the chimeric cells, followed by immunization of the host by graft strain antigens, since there was no time for breakdown of the chimeric cells, loss of tolerance, and then for a first-set graft rejection period of 14 days. Rather, it appeared that the rejection was due to a direct injury to the allogeneic tissue by the alloantibody.

Further, when these mice were re-grafted with Balb/c skin, the second grafts, while ultimately rejected, were retained for substantially longer period (21-45 days) than first set grafts between these strains, again indicating that the rejection of the first grafts was not due to an active process by the host, but to injury of the grafted tissue by the alloantibody injected.

Effect of Anti Mouse IgG_1 on the Titer of the Suppressive Antibody

It will be recalled that, in the normal hosts, among the globulin pools which failed to cause accelerated rejection of skin allografts were some which, on the contrary, caused prolonged retention. The observation of these opposite effects suggested the presence

in our globulin pools of two antibodies, of the same specificity
but of different biologic effect, one cell-injuring and the other
not, which could compete with each other for antigenic sites on the
cells of the graft. The net effect on the graft would then be
determined by the relative concentrations of the antibodies of the
two classes in the globulin injected. A likely pair of candidates
for such roles in our essentially 7S preparations would be anti-Balb/
c antibodies of the IgG_2 and IgG_1 classes, since it has been shown
in several species that IgG_2 antibodies, but not IgG_1, fix complement (1,22) and complement fixation is very probably an essential
step in the immunologic injury of cells by antibody. Also, in the
guinea pig, competition has been shown between IgG_1 and IgG_2 class
antibodies in the fixation of complement by fractions within a
given serum (19).

     In our approach to the resolution of this problem, because of
the great difficulty in separating mouse IgG_1 and IgG_2 in such
quantity that each could be tested separately, we undertook preparation of anti-mouse IgG_1 serum as a reagent. In the postulated
scheme, anti-Balb/c antibodies of IgG_2 class would be the cell-
injuring or rejecting class of antibody, and IgG_1 would be the competing class. Anti-mouse IgG_1 could then serve as a reagent to
remove IgG_1 and thus yield preparations of IgG_2 anti-Balb/c alloantibodies to test for a role in graft rejection. The preparation
of anti mouse IgG_1 was approached by preparing extracts of tumors
of MOPC 21 (an IgG_1 producing plasmacytoma), digesting with papain,
absorbing the digest with a glutaraldehyde immunoadsorbent of rabbit
anti mouse IgG_2 (from MOPC 173) and injecting rabbits for the anti
mouse IgG_1.

     A pool of rabbit anti-IgG_1 which gave no evidence of contami-
nation with anti-IgG_2, by immunodiffusion (IDF), at our threshhold
of detection, was tested for adequacy in removing the IgG_1 present
in our alloantibody preparations, by adding the anti-IgG_1 in increasing amounts to a constant amount of a CBA anti-Balb/c globulin and
removing the resulting precipitates. A portion of each supernatant,
concentrated to the original volume of anti-Balb/c globulin, was
tested to determine, by IDF, the level at which all IgG_1 had been
removed.

     A set of supernates which by this criterion included the end-
point of complete removal of IgG_1 from the anti-Balb/c globulin
preparation was now tested for suppressive titer vs. Balb/c cells.
The supernates of these mixtures were found to have suppressive
titers higher than the original globulin. These titers rose with
increasing amounts of the anti-IgG_1 added until they reached a
plateau, at which the titer remained constant through further in-

creases in the amount of anti-IgG_1. The level of addition of anti-IgG_1 at which the plateau was reached agreed with that at which complete removal of IgG_1 was indicated by IDF. Thus the plateau level was clearly the actual titer of the suppressive antibody present in each globulin preparation. The fact that the original titer is lower than the plateau would reflect the effect of competition by the anti-Balb/c antibody of IgG_1 class in the globulin, and the extent of that difference would give an estimate of the relative abundance of anti-Balb/c antibody of IgG_1 class in that globulin preparation (15).

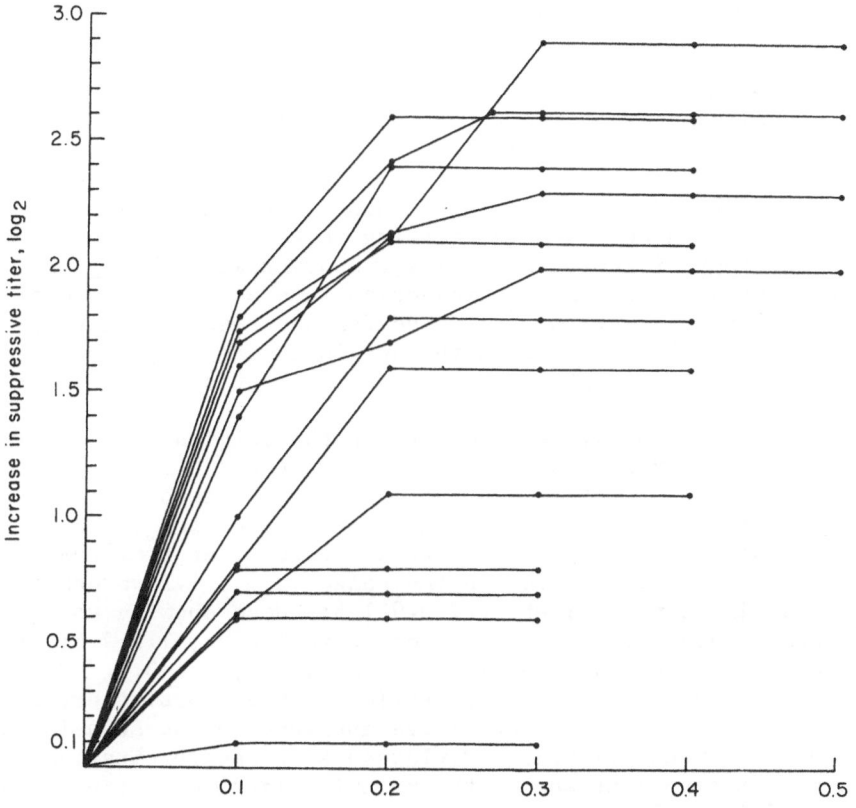

Amount of anti IgG_1 added per 0.1 ml of anti BALB/c globulin

Fig. 1. Increase in suppressive titer of anti-Balb/c globulin after addition of anti mouse IgG_1, in amounts of 0.1 to 0.5 ml/0.1 ml of anti-Balb/c globulin, and removal of precipitate. For clarity, titers of only two additional supernatants are shown after the plateau level was attained.

A systematic exploration of other preparations of CBA and C3H anti-Balb/c globulins was now undertaken. In all of these, it was again found that the suppressive titer for Balb/c spleen cells rose, on treatment with the anti-IgG$_1$, from the original level to a plateau level. As can be seen in Fig. 1, the anti-Balb/c globulin preparations varied in the amount of anti-IgG$_1$ required to raise the titer to plateau level, and in the total increase of suppressive titer from the original to plateau level (and therefore in the relative level of IgG$_1$ class of antibody).

Correlation of Graft Retention with the Relative Levels of IgG$_1$ and IgG$_2$ Subclass of the Injected Antibodies

We were now able to examine the effects of the various anti-Balb/c globulin pools on the retention of Balb/c skin allografts in relation to the relative levels of anti-Balb/c antibodies of the two subclasses, since all of the pools tested had titers of suppressive antibody (IgG$_2$ class) within a certain range, and since we now had a method of estimating relative amounts of anti-Balb/c antibody of IgG$_1$ class. The globulins which had given us effects of various kinds on injection into normal mice receiving primary skin grafts were now titrated for suppressive antibody against Balb/c spleen cells, in the original form and after treatment with anti-mouse IgG$_1$ to a plateau level of suppressive titer.

Fig. 2 shows the difference between the original and plateau levels for a number of pools, classified by their effects on the time of retention of the skin grafts. It can be seen that the globulin pools which had caused accelerated rejection (by day 11) showed an increase in suppressive anti-Balb/c titer from the original to the plateau level in the range of 0.1-1.2 powers of 2; those producing a partial effect, 0.9-1.6; those showing no significant degree of difference from control mice, 1.2-2.6; and those causing prolonged retention of the grafts, 2.1-3.1. Thus, on proceeding from accelerated rejection to the opposite extreme, of prolonged retention, a progressive increase can be seen in the differences between original and plateau suppressive titer, and thus a progressive increase in the relative level of anti-Balb/c antibody of IgG$_1$ class (16).

These data suggested that the retention of a skin allograft may be a consequence of the level of anti-graft antibody of IgG$_1$-class, within a set of preparations all having suppressive antibody in an adequate range of concentration, that is, on the relative levels of IgG$_1$ and IgG$_2$ class among the anti graft strain antibodies.

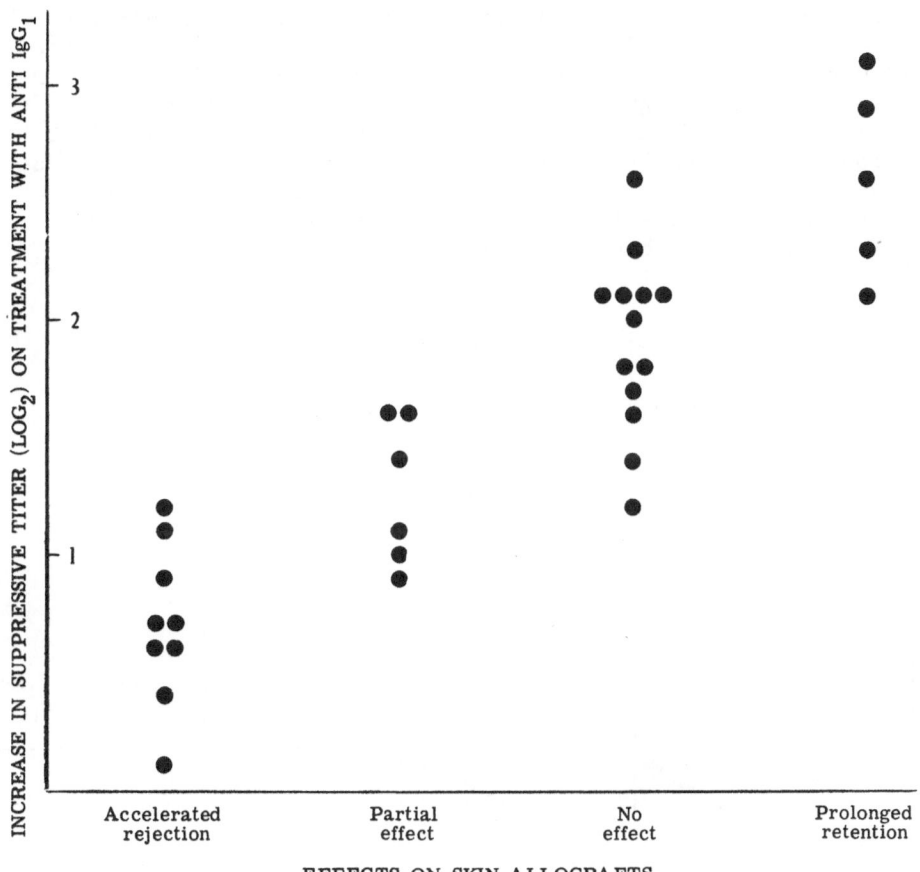

Fig. 2. Anti-Balb/c alloantibody-containing globulins grouped by their effects on the retention of Balb/c skin allografts. Each point shows the increase (log2) from the original anti-Balb/c suppressive titer to the plateau level attained on removal of IgG_1 class antibody by treatment of the globulin with anti-mouse IgG_1.

Confirmation of this relation was obtained by determining the original and plateau titers in the anti-Balb/c globulin pools which had given accelerated rejection, and in globulins obtained from the same mice a week later, since we had found that in the case of some earlier globulin pools which had caused accelerated rejection, globulins obtained 1 week later from the same mice had failed to

do so. This effect was now confirmed in a series of 19-23 day
globulins from mice which had given accelerated rejection in their
11-13 day globulins. When the increases in suppressive titer on
treatment with anti-IgG$_1$ were determined for these later-day
globulins, it was found that whereas the increases in the 11-13
day globulins ranged from 0.1 to 1.2 powers of two, the analogous
increases in the later globulins ranged from 1.4 to 2.5. This
indicated a higher level of alloantibody of IgG$_1$ class in the
later-day globulins than the earlier ones.

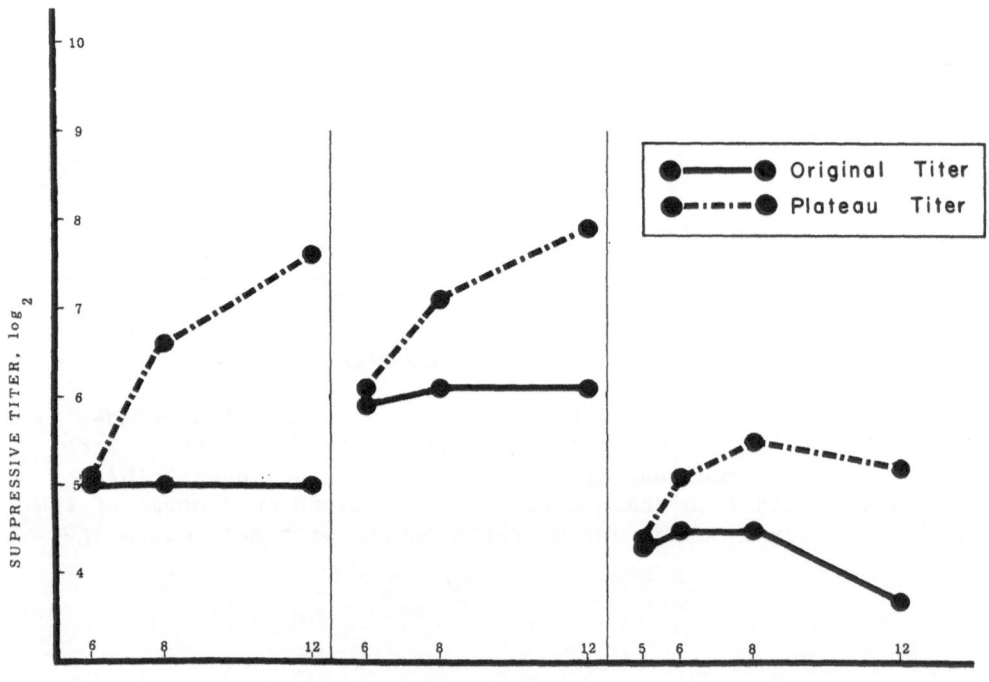

DAYS AFTER 1° INJECTION OF SPLEEN CELLS

Fig. 3. Suppressive antibody in sera of C3H and CBA mice given
an injection of 5 x 10^8 Balb/c spleen cells. Plateau titers
express the full titer of IgG$_2$ (plus IgM class) antibody, which
is attained by the anti-Balb/c serum on addition of increasing
amounts of rabbit anti mouse IgG$_1$ serum.

Shifts in Subclass of Antibodies in Mice
Injected with Allogeneic Spleen Cells

The observation summarized above were all made by injection of grafted mice with antibody prepared in other mice of the same strain. In studies directed at the possible significance of these observations to the sequence of events by which a normal mouse brings about the active rejection of a skin allograft, we examined CBA and C3H mice at intervals following a primary and secondary injection of Balb/c spleen cells. Since the methods described above make possible the determination of relative changes or differences in the levels of antibody of a given specificity of IgG_1 class and IgG_2 class in whole, unfractionated, serum or globulin preparations, we examined sera of mice following the primary injection of allogeneic spleen cells, and in the later days following a secondary injection. Fig. 3 shows such data, from the primary injection. On day five or six suppressive antibody was detectable, and of this about one-half was of IgG_2 class, since treatment with 2-ME reduced the titer by about one power of two. On that day, however, the fact that the plateau titer was no higher than the original indicated that antibody of IgG_1 class had not yet appeared. Thereafter, the rising plateau level indicated that the level of IgG_2-class antibody was increasing, and the appearance and progressive increase of a difference between plateau and original titer indicated that IgG_1 class antibody had appeared and was also increasing in concentration (17).

At the other end of the immunization we found that whereas the plateau titer of anti-Balb/c suppressive antibody had reached its maximum level by day 11-13 after the secondary injection of Balb/c spleen cells, and remained constant for the following week, the titer of the original globulin fell during this period, as shown in Fig. 4. This increase in the difference between original and plateau titers indicated that the IgG_1-class antibody was continuing to increase after that of IgG_2 class had stopped increasing. Thus, anti-Balb/c antibody of IgG_1 class begins to appear later than that of IgG_2 class, and continues to increase later.

Returning to the early stages of the primary response, if we assume that the timing of the response to alloantigens of the skin graft is similar to that after the primary injection of spleen cells, the IgM and IgG_2 class of antibodies by day six would provide an early population of complement fixing, cell injuring, antibodies, which could cause sufficient injury to the cells of the graft to ensure graft rejection six or seven days later, despite the later appearance of competing antibodies of IgG_1 class. However, at the time of full rejection of the graft, there would be a substantial serum level of alloantibody of IgG_1 class. This could be sufficient

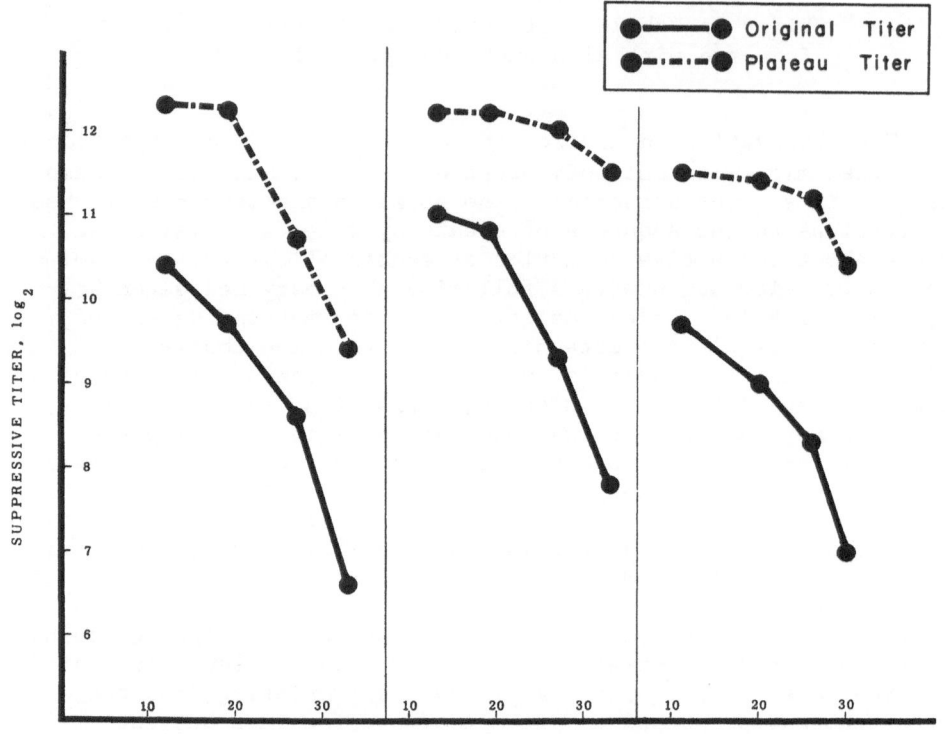

Fig. 4. Suppressive titers of secondary C3H or CBA anti-Balb/c
globulins. Original and plateau titers as in Fig. 3.

to account for the failure of passive transfer of accelerated rejection of allografts by serum (2,3), and indeed for the failure of
the great majority of our own anti-Balb/c globulins obtained at
11 to 13 days after secondary immunization to cause accelerated
rejection of allografts.

It may be noted that the association suggested by these
studies of IgG_2-class antibody with rejection, and IgG_1-class with
retention or enhancement, would appear to be at variance with the
position of most workers in the tumor graft area, in which enhancing
antibody is generally considered to be of IgG_2 class, if one
assumes similar mechanisms of rejection or retention of normal
tissue allograft and tumor. A discussion of this point has been
presented elsewhere (16).

Effects of IgG_2 Preparations of Alloantibody

A critical set of experiments on the question of the roles of antibodies of these subclasses in allograft rejection would be to test the effects of preparations of alloantibodies which would express the effect of one subclass only, or be restricted to that subclass. In our first approach to this, CBA anti-Balb/c globulin preparations which did not cause accelerated rejection of Balb/c skin grafts in their original form were brought to their plateau level of suppressive titer by precipitation with appropriate amounts of anti-IgG_1. When these were injected into CBA mice given Balb/c grafts, accelerated rejection of the skin grafts was observed. However, this required levels of suppressive titer substantially greater than those which had been found in the native preparations which had caused accelerated rejection in our earlier study. It was considered that the requirement for the high titer might be due to inefficient use of the IgG_2-class antibody, perhaps because of binding of complement in the mouse by the soluble complexes produced by the specific precipitation. When tested for this, these IgG_2-expressing preparations were indeed found to bind complement, even over a number of two-fold steps of dilution.

An IgG_2 fraction of alloantibody-containing globulin was therefore prepared by the use of an immunoadsorbent of Sepharose-bound anti-IgG_1. This IgG_2-restricted preparation was also found to cause accelerated rejection, and at considerably lower levels of the alloantibody, as judged by the suppressive titer, than the preparation produced by precipitation. IgG_1-restricted preparations of the alloantibodies have also been made, by treatment of our anti-Balb/c globulins with staphylococcal protein A which, in the mouse, has been found to bind IgG_2 globulin (20). These IgG_1 preparations have produced prolonged retention of skin allografts, to 17 or 18 days.

Very recently Jansen et al., (18) have reported on the effects on skin allografts of IgG_1 and IgG_2 restricted preparations of alloantibody in the mouse. The subclass fractions were prepared by Sepharose-bound anti-IgG_2 and anti-IgG_1. The IgG_1 preparation was found to cause prolonged retention of the skin allografts. The IgG_2-restricted preparation caused a hyperacute destruction of grafts if injected with rabbit complement. However, in the absence of the rabbit complement, injection of this preparation led to prolonged retention of the graft. This last observation is at variance with our observations, in which prolonged retention was observed only with the IgG_1-restricted preparations.

In summary, we have shown accelerated rejection of skin allografts in normal mice, and rejection in immunologically tolerant mice, by a few preparations of alloantibody-containing globulins. We have also found that alloantibody preparations which did not show this property gave evidence of having more alloantibody of IgG_1 class, and could be converted to preparations causing accelerated rejection if they were allowed to express fully their IgG_2-class antibody, or were restricted to this class. Finally we have shown a shift from IgG_2 class toward IgG_1 class in the allo-antibodies produced in CBA and C3H injected with Balb/c spleen cells.

In view of these observations, and of the as yet unidentified specific effector by which lymphocytes cause allograft rejection, we think that the question could be considered of whether a factor in the rejection process could be an early production of IgM and IgG_2 class antibodies by the lymphocytes involved, followed by a shift toward IgG_1-class antibody, but with the competing, IgG_1-class, antibody reaching effective levels at a time when irreversible injury to the grafted cells or tissues has already occurred.

ACKNOWLEDGMENTS

The work described in this report was supported by Grants CA 14483 and AI 11466 of the National Institutes of Health, and Grant IM-3A of the American Cancer Society.

Fig. 1. is reproduced by permission from the Journal of Immunology.

Fig. 2. is reproduced by permission from Immunology.

Table I and Figs. 3 and 4 are reproduced by permission from Transplantation.

REFERENCES

1. Benacerraf, B., Ovary, Z., Bloch, K. J., and Franklin, E.C., J. Exp. Med., 117 (1963) 937.
2. Billingham, R.E., Brent, L. and Medawar, P.B., Proc. Roy. Soc. B 143 (1954) 58.
3. Billingham, R.E. and Brent, L., Brit. J. Exp. Path., 37 (1956) 566.
4. Chase, M.W., Immunological reactions mediated through cells, in: The Nature and Significance of the Antibody Response (Ed. A. M. Pappenheimer, Jr.) Columbia University Press, New York, New York (1953) 156.
5. Harris, S., Harris, T.N. and Farber, M.B., J. Exp. Med., 108 (1958) 411.
6. Harris, S., Harris, T.N., Ogburn, C.A. and Farber, M.B., J.

Immunol. , 99 (1967) 447.
7. Harris, S. and Harris, T.N., J. Immunol., 96 (1966) 478.
8. Harris, T.N., Harris, S. and Farber, M.B., J. Immunol., 72 (1954) 148.
9. Harris, T.N., Harris, S. and Farber, M.B., J. Immunol., 72 (1954) 161.
10. Harris, T.N., Harris, S. and Farber, M.B., J. Exp. Med., 108 (1958) 21.
11. Harris, T.N., Harris, S. and Farber, M.B., J. Immunol., 99 (1967) 438.
12. Harris, T.N., Harris, S. and Ogburn, C.A., Transplantation, 12 (1971) 448.
13. Harris, T.N., Harris, S., Bocchieri, M.H., Farber, M.B. and Ogburn, C.A., Transplantation, 14 (1972) 495.
14. Harris, T.N., Harris, S. and Farber, M.B., Transplantation, 15 (1973) 383.
15. Harris, T.N. and Harris, S., J. Immunol., 109 (1972) 1096.
16. Harris, T.N. and Harris, S., Immunology, 25 (1973) 409.
17. Harris, T.N. and Harris, S., Transplantation, 19 (1975) 318.
18. Jansen, J.L.J., Koene, R.A.P., v Kamp, G.J., Tamboer, W.P.M. and Wijdeveld, P.G.A.B., J. Immunol., 115 (1975) 387.
19. Kourilsky, F.M., Bloch, K.J., Benacerraf, B. and Ovary, Z., J. Exp. Med., 118 (1963) 699.
20. Kronvall, G., Grey, H.M. and Williams, R.C., Jr., J. Immunol., 105 (1970) 1116.
21. Medawar, P.B., J. Anat., 48 (1944) 176.
22. Nussenzweig, R.S., Merryman, C. and Benacerraf, B., J. Exp. Med. 120 (1964) 315.

IMMUNOSUPPRESSION OF ROSETTE-FORMING CELLS

S. S. LEFKOWITZ and D. NEMETH

Texas Tech University School of Medicine
Lubbock, Texas (USA)

Park and Brody (13) have shown that phenobarbital markedly affects cellular immunity even at therapeutically prescribed dosages. A number of the narcotic analgesics and psychotomimetic drugs has been shown to markedly reduce production of interferon in random-bred mice (6,10). Although many investigators have concentrated their efforts on the effects of various drugs on cellular immunity, some attention has recently been directed toward humoral immunity. It has been reported that morphine as well as Δ-9-tetrahydrocannabinol (THC) markedly affect the development of plaque forming cells (PFC) in the spleens of mice immunized with sheep erythrocytes (SRBC) (8,9). Impairment of cellular immunity in human chronic marijuana smokers was also reported (5). The drug caffeine has also been shown to be immunosuppressive in mice (7, 9).

The primary purpose of this work was to screen a number of frequently abused drugs and determine their immunosuppressive effects on mice and to correlate these effects with lymphoid cell toxicity. The principal method used to study the effects of certain of these drugs upon immunity was the immunocytoadherence assay which was initially reported by Zaalberg (14) and Nota, et al. (12). They reported that SRBC would readily adhere to spleen cells from mice immunized with SRBC. These cells were termed rosette forming cells (RFC) and consisted of 5 or more SRBC physically attached to each immune spleen cell. This assay has been utilized as a measure of immunosuppression (1,5).

The experiments reported herein showed that morphine, methadone, pentobarbital, phenobarbital and caffeine caused a marked reduction of splenic RFC. Amphetamine sulfate did not affect the number of RFC under the same experimental conditions. This reduction in RFC was not paralleled by a reduction in splenic cellularity indicating that a general toxicity to lymphoid cells may not be the primary cause of the reduced number of RFC. These results emphasize the possibility that the administration of certain drugs could result in a reduction of immunocompetent cells thereby affecting host immunity.

MATERIALS AND METHODS

Random-bred, male Swiss-Webster mice weighing 18 to 20 gms were obtained from Sprague-Dawley, Madison, Wisconsin. The animals were immunized intraperitoneally with 0.2 ml of a 50% dilution of washed SRBC. This contained approximately 2×10^9 cells/ml. Five mice were used for each group, unless otherwise stated.

The drugs employed included morphine sulphate (Mallinkrodt), methadone hydrochloride (Mallinkrodt), sodium phenobarbital (Winthrop), sodium pentobarbital (Elkins-Sinn), amphetamine sulfate (Sigma) and caffeine (Sigma). These drugs were diluted with Hanks balanced salt solution (BSS) and injected subcutaneously in a volume of 0.2 ml daily. Control animals received an equal volume of BSS.

Various dosages and schedules were employed to ascertain the levels as well as the conditions under which these drugs would affect rosette formation. The mice were sacrificed by cervical dislocation 8 to 10 days after immunization. Their spleen cells were then removed and pressed through a 60 mesh screen. The spleen cells were then passed through a 22 gauge needle to obtain single cell suspensions.

The methods employed were essentially those of McConnell, et al. (11) and are described as follows: Spleen cells were washed 4 times by centrifugation at 1000 rpm for 10 min using cold phosphate buffered saline (PBS). The number of cells was determined by counting in a Coulter Counter, model ZBI. Approximately 5×10^6 spleen cells in 0.2 ml were mixed in a plastic tube with 30 to 40 $\times 10^6$ SRBC in 0.1 ml. After centrifugation of the cell suspension in the cold at 1200 rpm for 10 min, 0.7 ml of cold PBS was added to each tube and the rosettes were resuspended by gentle tapping of the tubes. The resulting suspensions were gently aspirated through a Pasteur pipet. The RFC were counted in a hemocytometer and expressed as the amount per 10^6 leukocytes.

RESULTS

Several experiments were done to ascertain the time-course relationship of rosette production. Fig. 1 illustrates that following a single injection of SRBC, peak rosette formation occurred between days 6 and 10 with a maximum at about day 8. Because of this, most of the experiments involved the assessment of splenic rosettes between days 6 and 10.

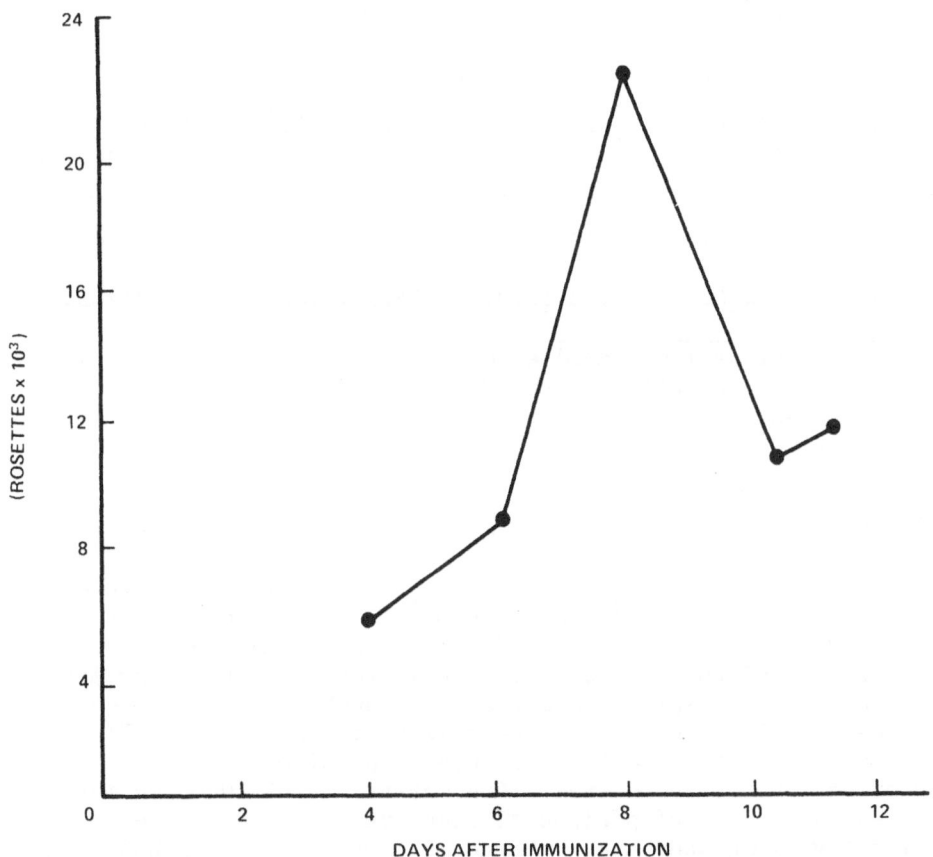

Fig. 1. Rosette formation by sheep erythrocytes (SRBC) following a single injection of SRBC.

The injection of morphine at 75 mg/kg had a major effect on the behavior of mice. This was characterized by hyperactivity and a Straub tail. After 7 to 9 injections of morphine, there was approximately a 60% inhibition of rosettes (Table I). This dosage also resulted in a diminution of spleen weight as well as the num-

ber of leukocytes per spleen. When 50 mg/kg of morphine were injected for up to 11 days, the body weight, spleen weight and number of RFC were not significantly altered. However, this dosage
did reduce the number of splenic leukocytes by approximately 30%.

TABLE I

The Effects of Morphine on Rosette Forming Cells*

Dosage** (mg/kg)	Body Wt. (gms)	Spleen Wt. (gms)	Cells x 10^7 Spleen	Rosettes x 10 10^6 leukocytes
75	23.52±1.33	0.096±0.038	10.69±0.49***	11.58±1.66****
Control	23.94±0.97	0.124±0.014	14.95±1.45	26.92±2.33

*All calculations ± standard error of the mean.
**7 injections starting 3 days prior to SRBC
***Significant at $P \leq 0.05$
****Significant at $P \leq 0.01$

     Methadone also had similar effects on both the behavior and
immune competence as measured by rosette formation in mice. Table
II (Exp. 1) illustrates the effects of 3 injections containing 30
mg/kg methadone on the number of RFC. It should be noted that the
number of rosettes in the methadone treated animals was approximately half that of the control animals, whereas the other parameters,
i.e. body weight, spleen weight and the number of leukocytes/spleen
were essentially the same as equivalent controls. Table II (Exp.
2) illustrates the effects of 7 injections of methadone on the number of peripheral leukocytes. Concentrations of 20 mg/kg markedly
reduced the number of peripheral leukocytes. This dosage also significantly diminished the number of RFC while 10 mg/kg had only
peripheral effects. Only marginal toxicity was apparent in a few
of the experiments.

     Loss of equilibrium and lethargy occurred following injection
of phenobarbital. This drug was also effective in reducing the
number of rosettes under the present experimental conditions. In

Table III, there was approximately a 50% inhibition of RFC using
75 mg/kg. However, this difference was not reflected in the number of leukocytes/spleen when compared to the control. With the
lower dose of 25 mg/kg, a greater inhibition was noted than with
75 mg/kg phenobarbital. This was shown in a number of experiments.
An experiment was designed to determine whether exposure to phenobarbital prior to immunization with SRBC would have the same effect
as exposure following immunization. It was noted that if the animals were injected with phenobarbital the same number of times, i.e.
7, prior to injection with erythrocytes, the number of RFC was comparable to that of the control animals suggesting that inhibition
of RFC occurred only when mice were exposed to phenobarbital following immunization.

Phenobarbital was also used to see what effect it might have
on the number of peripheral leukocytes. It can be seen in Table
IV that a marked diminution of splenic leukocytes was not reflected
in the peripheral leukocyte count. The numbers of peripheral leukocytes were essentially the same whether or not the mice were injected with phenobarbital.

Another barbital compound, pentobarbital, was evaluated for
its effect on rosette formation (Table V). After 40 mg/kg was administered to mice for 10 days, approximately 50% inhibition of
RFC was observed. Slight toxicity was apparent, but this was not
statistically significant.

Amphetamine sulphate was administered for 7 days at a dosage
of approximately 5 mg/kg. Following injection of this compound,
the animals became hyperactive and extremely nervous. In spite of
the very marked physical and behavioral effects, including an occasional death, no significant differences in the above parameters
were noted between the experimental and control groups.

Hyperactivity was characteristic of mice injected with caffeine. It also dramatically affected the number of RFC. One hun-
dred mg/kg of caffeine administered for 8 days markedly suppressed
the number of RFC by approximately 50%. The number of leukocytes/
spleen was suppressed even further by approximately 65%, whereas
the body weight was not affected. Table VI illustrates that similar results were obtained with a greater suppression of both body
and spleen weight, whereas the suppression of the number of rosettes
was only about 30%. It should be noted that in this latter experiment, only 5 injections were employed. The administration of caffeine for as long as 11 days using 50 mg/kg did not significantly
affect the above parameters when compared with the control groups.

TABLE II

The Effect of Methadone on Rosette Forming Cells and Peripheral Leukocyte Count*

Dosage (mg/kg)	Body Wt. (gms)	Spleen Wt. (gms)	$\dfrac{\text{Cells x } 10^7}{\text{Spleen}}$	$\dfrac{\text{Rosettes x } 10^3}{10^6 \text{ leukocytes}}$	Peripheral Leukocytes/cc x 10^7
Exp. 1**					
30	23.02±1.57	0.120±0.02	8.50±1.43****	24.00±6.76****	ND
Control	23.91±1.23	0.152±0.01	12.30±0.73	44.00±1.03	ND
Exp. 2***					
20	23.28±0.71	0.090±0.01	17.28±1.99	4.30±0.72****	2.49±0.39****
10	24.06±1.18	0.090±0.01	15.36±1.65	6.96±0.61	3.83±0.46****
Control	24.73±0.87	0.080±0.02	18.48±1.09	7.80±0.86	9.61±1.56

*All calculations ± standard error of the mean.
**3 injections started 1 day prior to SRBC
***7 injections started 1 day prior to SRBC
****Significant at $P \leq 0.05$

TABLE III

The Effects of Phenobarbital on Rosette Forming Cells*

Dosage** (mg/kg)	Body Wt. (gms)	Spleen Wt. (gms)	Cells x 10^7 Spleen	Rosettes x 10^3 10^6 leukocytes
75	29.70+1.83	0.180+0.02	12.15+0.63	13.60+3.65****
25	28.78+1.62	0.190+0.04	7.40+1.20***	4.40+1.20****
Control	32.00+2.83	0.175+0.03·	14.37+1.69	27.30+2.29

*All calculations $\pm$ standard error of the mean.
**5 injections starting 1 day prior to SRBC (3 animals/group)
***Significant at P $\leq$0.05
****Significant at P $\leq$0.01

TABLE IV

The Effect of Phenobarbital on Splenic
and Peripheral Leukocyte Counts*

Dosage (mg/kg)	Body Wt. (gms)	Spleen Wt. (gms)	Peripheral Leukocytes/cc x 10^7	Splenic Leukocytes x 10^7
75**	24.15+1.73	0.078+.006****	4.83+0.11	5.05+0.89***
Control	23.26+0.74	0.128+0.012	4.57+0.24	6.80+0.60

*All calculations $\pm$ standard error of the mean.
**8 injections starting 1 day prior to SRBC
***Significant at P $\leq$0.1
****Significant at P $\leq$0.01

TABLE V

The Effect of Pentobarbital on Rosette Forming Cells*

Dosage** (mg/kg)	Body Wt. (gms)	Spleen Wt. (gms)	Cells x 10^7 Spleen	Rosettes x 10^3 10^6 leukocytes
40	28.48±1.40	0.056±0.002	16.29±1.83	6.16±1.03***
Control	30.27±1.00	0.066±0.004	17.21±0.84	12.56±1.45

*All calculations ± standard error of the mean.
**10 injections starting 1 day prior to SRBC
***Significant at P ≤0.01

TABLE VI

The Effect of Caffeine on Rosette Forming Cells*

Dosage** (mg/kg)	Body Wt. (gms)	Spleen Wt. (gms)	Cells x 10^7 Spleen	Rosettes x 10^3 10^6 leukocytes
100	19.97±.97***	0.068±0.006	9.03±.83***	16.68±2.01***
Control	22.80±1.67	0.096±0.020	20.76±5.30	23.28±3.67

*All calculations ± standard error of the mean.
**5 injections starting 1 day prior to SRBC
***Significant at P ≤0.05

DISCUSSION

The cell types which are responsible for rosette formation are not clearly defined. In man, RFC in the peripheral circulation are considered to be thymus dependent leukocytes or T cells. It has also been reported that human blood lymphocytes may form "non-immune" rosettes with SRBC (2). In mice, the majority of RFC are lymphoid cells and probably synthesize "antibody like" molecules on their surface (3). In spite of this, both T and B (bone marrow) lymphocytes form rosettes (4). Since the number of RFC increased following immunization approximately ten-twenty-fold, it is apparent that the development and production of rosettes does represent a measure of the immune potential of a given host (1).

A number of differences was noted between the effects of many of these abused drugs on PFC (9) as compared to the results reported in these studies. In general, the number of PFC following exposure to many of these drugs paralleled the splenic cellularity. In cases of overt drug toxicity, both cell numbers and PFC were reduced approximately the same proportion with some exceptions. This was not the case when measuring RFC as reported in this study. Marked reduction in RFC was not always a reflection of total spleen cellularity, but rather appeared to be independent of it. This was especially evident when methadone was used. This drug had no apparent effect on spleen cellularity, but markedly reduced the number of RFC. This difference indicates the probability that different types of cells are involved in the production of PFC and RFC. This is further emphasized by the timing of the peak response, i.e. day 4 to 5 for the PFC and 6 to 10 days for the RFC. It should be emphasized that the peak rosette response occurs about day 8 which is about the same time as the production of indirect PFC which produce IgG.

It was noted that some of the drugs employed, such as morphine, methadone, etc. had a very critical dosage response. With 75 mg/kg of morphine, marked inhibition of RFC was obtained as well as some toxicity as exemplified by weight loss and reduction of splenic cellularity. However, 50 mg/kg did not affect these parameters. This toxicity, such as the loss of splenic cellularity, etc. has been reported previously (8). In general, the greater the dosage of the drug the greater the inhibition. This was not the case with phenobarbital. In 3 out of 4 experiments, the lower dosage used, i.e. 25 mg/kg, was much more effective in reducing the number of splenic rosettes. A possible reason for this is a more efficient uptake of the lower dose resulting in a greater suppression of the immunocompetent cells.

The administration of phenobarbital prior to immunization with SRBC did not affect the number of RFC. Only when the drug was administered after immunization was it effective in reducing the number of RFC. This would tend to indicate a recovery from the effects of phenobarbital and/or a lack of residual effect following discontinuation of the drug. The lack of a generalized effect of phenobarbital on circulating lymphoid elements does not detract from its properties as a potentially immunosuppressive drug. Methadone did markedly affect the number of leukocytes in the blood indicating that its effect may be on different populations of lymphocytes.

The results reported in this study, along with others, tend to indicate the broad effects of some of the abused drugs on various immune parameters. These types of studies tend to substantiate certain clinical observations on the incidence of morbidity and/or mortality associated with drug use. Whether the use of these drugs has a direct and significant effect on disease processes remains to be determined.

ACKNOWLEDGEMENTS

This investigation was supported in part by a grant from Eli Lilly and Company, Indianapolis, Indiana.

REFERENCES

1. Bach, J.F., Dardenne, M. and Fournier, C., Nature, 22 (1969) 988.
2. Coombs, R.R.A., Gurner, B.A., Wilson, A.B., Holm, G. and Lindgren, B., Int. Arch. Allergy, 39 (1970) 658.
3. Greaves, M.F., Europ. J. Immunol., 1 (1971) 186.
4. Greaves, M.F. and Möller, E., Cell Immunol., 1 (1970) 372.
5. Gupta, S., Greico, M. and Cushman, P., New Eng. J. Med., 291 (1974) 874.
6. Hung, C.Y., Lefkowitz, S.S. and Geber, W.F., Proc. Soc. Exp. Biol. Med., 142 (1973) 106.
7. Laux, D.C. and Klesius, P.H., Proc. Soc. Exp. Biol. Med., 144 (1973) 633.
8. Lefkowitz, S.S. and Chiang, C.Y., Res. Comm. Chem. Path. Pharm. (1975) in press.
9. Lefkowitz, S.S. and Chiang, C.Y., Res. Comm. Chem. Path. Pharm (1975) in press.
10. Lefkowitz, S.S., Hung, C.Y. and Geber, W.F., Res. Comm. Chem. Path. Pharm., 5 (1973) 885.
11. McConnell, I., Munroe, A., Gurner, B.N. and Coombs, R.R.A., Int. Arch. Allergy, 35 (1969) 1209.
12. Nota, N.R., C. R. Acad. Sci., 259 (1964) 1277.
13. Park, S.K. and Brody, J.I., Nature, New Biol., 233 (1971) 181.
14. Zaalberg, O.B., Nature, 202 (1964) 123.

BIOLOGICAL AND PATHOLOGICAL CHARACTERISTICS IN MOUSE LINES WITH

LARGE DIFFERENCES IN LEUKOCYTE COUNTS

C. K. CHAI

The Jackson Laboratory
Bar Harbor, Maine (USA)

Experimental inquiry of quantitative characters of polygenic basis has been seen to contain many practical implications for agricultural programs, and to provide explanations for microevolution. Depending on the traits and organisms under investigation, study of quantitative genetic variations can also have practical applications in medicine. Gowen (6), for instance, may have been the first to experimentally demonstrate the genetic influences of disease resistance and susceptibility by selectively breeding mice for differences in resistance to typhoid organisms. Recently, Biozzi, et al. (1) selectively bred for differences in response to sheep red blood cells in mice, and they produced lines differing in a number of immunological parameters. These differences were attributed to the action of a large number of genes. This work provided a strong basis for the presence of the polygenic system in the immune response.

This present study, in parallel to the breeding experiments mentioned above, deals with our selection for leukocyte level differences in mice. This work was initiated 18 yr ago, and the lines have gone through 32 generations, including 22 generations of selection and 10 generations of inbreeding. The lines appear to have been stabilized for a larger difference in leukocyte level and in a number of biological, immunological, and pathological characteristics. The genetic basis of leukocyte level variations (2,3) and differences in immune responses have been reported earlier (8,9). Besides a brief description of the biological characteristics of the high (HCL) and low leukocyte count (LLC) lines, we report here 2 pathologic conditions in the LLC mice: reticular cell hyperplasia and amyloidosis. The origin of the amyloid fibrils on the basis of electron microscopic studies and the etiology of amyloidosis on the basis of biological characteristics and immune responses will be discussed.

MATERIALS AND METHODS

Beginning from a hybrid population which was produced by intercrossing 6 inbred strains (C57BL/6, C57BR/cd, A, BALB/c, LG/Ckc and SM/Ckc), the LLC and HLC mice were derived by selective breeding for leukocyte count differences. The selection procedure and cell counting methods were given in previous papers (2,4).

Serum immunoglobulins of the HLC and LLC mice at about 2 months and 1 yr of age were tested by the method of radial immunodiffusion. The immunodiffusion plate embedded with different goat anti-mouse immunoglobulins and the standard for each immunoglobulin were purchased from Meloy Laboratories, Springfield, Virginia. Four ml of each serum from each mouse were used and the diffusion time was 18 hr. Five young and 5 old mice were used for each strain.

Histological preparations were made according to conventional procedures for inguinal lymph node, kidney, spleen, liver and adrenal gland tissues. Hematoxylin and eosin stain was used for each slide, supplemented by crystal violet and Congo red stains to confirm the presence of amyloid.

For electron microscopy, we used spleen tissue fixed in Karnofsky's fixative, embedded in Epon and stained with uranyl acetate and lead citrate.

RESULTS

The average total leukocyte counts for each generation of each line are shown in Fig. 1 for 22 generations of selection and 4 generations of inbreeding. Approximately a six- or seven-fold difference between the HLC and LLC line persisted through the tenth generation of inbreeding, indicating fixation of genes which affect leukocyte production.

There were marked differences in the percentages of lymphocytes and neutrophils between the LLC and HLC mice. We examined 40 LLC and 42 HLC mice at the second generation of inbreeding. The average lymphocyte and neutrophil percentages were 83.2 and 22.4 for the LLC and 90.0 and 7.0 for the HLC, respectively. These values were outside the range of all inbred strains. In some LLC mice, lymphocytes reached 55%, a value approximating lower human measurements. Thus, the relative difference in the number of lymphocytes between LLC and HLC was even greater than the relative difference in total leukocyte counts.

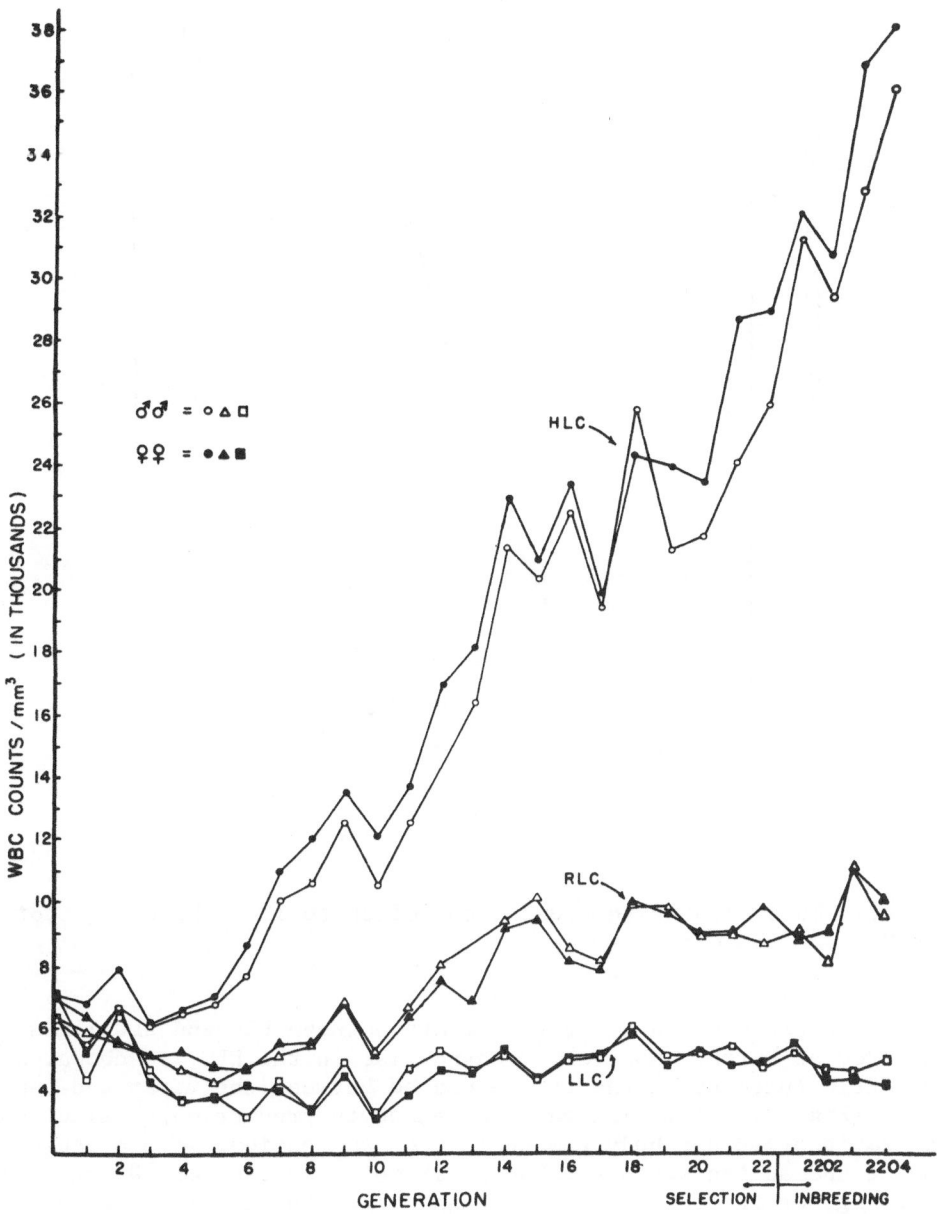

Fig. 1. Mean total leukocyte counts at each generation of selection and inbreeding for the high leukocyte count (HLC), low leukocyte count (LLC) and the random (RLC) lines.

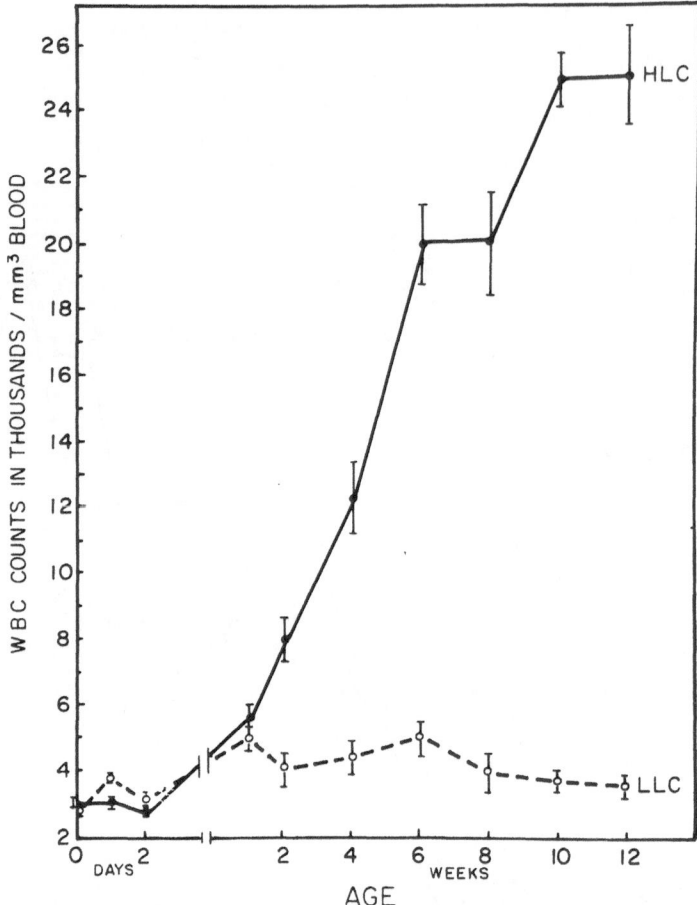

Fig. 2. Total leukocyte counts from birth to 12 weeks of age for the HLC and LLC mice.

        No difference in leukocyte counts between LLC and HLC lines occurred at birth. However, a sharp rise in the HLC, as compared to little increase in the LLC, appeared between the first and tenth weeks (Fig. 2). This increase in leukocyte production paralleled the rapid organ and body growth in the same period. The average thymus and spleen weights taken at 10 weeks of age were 89 mg, and 116.4 mg for the females and 62.1 and 112.8 for the males of HLC mice, respectively. In the LLC mice, the averages were 41.9 and 108.4 for the females and 25.6 and 88.3 for the males, respectively. Twelve to 18 mice for each sex of each line were used. The differences in leukocyte counts must be a reflection of increasing phys-

iological variations between the 2 lines in certain biological systems. The direct contributor to the difference was probably the
thymus.

The levels of immunoglobulins are shown in Table I. There was
a significant increase in each immunoglobulin with age in the LLC
but not in the HLC mice.

TABLE I

Amount of Immunoglobulins (mg/ml) in the Serum of
HLC and LLC Mice Tested by Radial Immunodiffusion

Strain	Age	IgM	IgA	IgG$_1$	IgG$_2$
HLC	2 mo	0.29+.07	0.25+.06	0.49+.24	3.04+.39
HLC	1 yr	0.36+.06	0.58+.06	0.41+.04	2.48+.31
LLC	2 mo	0.14+.02	0.33+.07	0.37+.17	1.90+.23
LLC	1 yr	0.69+.05	0.87+.32	1.06+.27	4.80+.27

Reticular cell hyperplasia. Histological examination was made
of the inguinal lymph nodes of 43 mice between 2 and 3 months of age.
Reticular cell hyperplasia appeared in either a diffuse or localized
form, or a combination of both was found in 33 mice. The proliferation of fusiform histiocytes, surrounded by fibrous materials of
a light pinkish hue on hematoxylin and eosin stains, was characteristic. Lesions occurred in the cortex as well as in the medulla,
exhibiting a mosiac appearance with pinkish stained patches or strips
when the entire node was involved. Dunn (5) expected that these lesions would progress to a malignant form. But, to our surprise, none
of the inguinal nodes of mice autopsied between 10 and 20 months of
age showed any malignant development. Instead, the nodes in general
exhibited atrophy with a marked depletion of lymphocytes. The fibrous and the connective tissues appeared fused and less well
defined than in younger mice.

Amyloidosis. A total of 44 mice at the age of 10 to 18 months
was examined for amyloidosis. Amyloid deposition was found in the
spleen and kidney of each mouse. Out of a total of 44 mice, 2 were
found with no amyloid deposition in the liver and 3 in the adrenal
glands. An arbitrary grading according to the relative amount of

amyloid was given for the spleen, liver and adrenal glands (Table II). Traces of amyloid were found in some of the other organs which were examined.

TABLE II

Number of LLC Mice with Different Degrees of
Amyloidosis in Spleen, Liver and Adrenal Tissues

	No. of Mice	Degree of Amyloidosis			
		+++	++	+	-
Spleen*	44	10	28	6	0
Liver**	44	6	22	14	2
Adrenal***	44	8	18	15	3

*Spleen: +++=amyloid deposition throughout the whole spleen; ++= circling around the spleen follicles; +=focal or localized at the periphery of the follicles.
**Liver: +++=amyloid deposition throughout the whole liver; ++= circling around the central veins; +=focal or localized in the periphery of the central veins.
***Adrenal: +++=amyloid deposition throughout all layers of the cortex with or without the medulla involved; ++=in the zona fasciculata and reticularis: +=only in the zona reticularis.

    Ultrastructure of amyloid. We examined various sections of spleen using an electron microscope (from a mouse with amyloid deposition graded as +++). Both intra and extracellular amyloid fibrils were revealed. Fig. 3 shows the presence of fibrils in the cytoplasm of a reticular cell; they were oriented more or less at right angles to the cytoplasmic membrane, and more densely packed at the end of the cytoplasmic membrane than at the end of nuclear membrane. Such patterns of distribution gave the impression that the fibrils grew out from the cytoplasmic membrane and extended toward the cell nucleus. Some areas of this cytoplasmic membrane appeared undefined and they were assumed to be in the process of undergoing disintegrative changes.

    Fig. 4 is an electron micrograph showing the relationship of a bundle of amyloid fibrils to a reticular cell of the spleen. Most fibrils are exterior to the cell surface, but some appear to originate from the cytoplasm, and stretch out into the intercellular space, presumably the venous sinus. Note that the fibrils are closely packed at their origin and gradually separate from each other after leaving the cell. They form beautiful smooth sigmoid curves not often observed in biological fibrils. The pattern of distribution of the fibrils and their relationship to the cell indicate that they originated from the cell. In the background is a mass of amyloid fibrils, presumably old ones, and some are detached from the cells from which they were secreted.

The longest individual fibrils measured at least 2 μ, or about 2 times the length of the long axis of the nucleus of the reticular cell. These were the longest amyloid fibrils reported in the tissue sections. The diameter of 50 filaments was measured from an electron micrograph print of higher magnification in a different field. We observed an average of 59.3 Å with a range between 43 and 83 Å.

DISCUSSION

In view of the data provided above, 2 basic concepts concerning amyloidosis in the LLC mice require discussion: etiology of amyloidosis and the origin of amyloid.

The low leukocyte level, of the lymphocytes in particular, together with the increase in immunoglobulin levels in the old LLC mouse group reflects an abnormal lymphoid system. One of the characteristics of the immune response in LLC mice concerned the production of IgM primarily as a result of repeated injections of sheep red blood cells; this differed from the general pattern of immune response involving a shift from IgM to IgG production following continuous immunization. Secondly, the LLC mice (in spite of low leukocyte counts) produced greater hemagglutinin titers than HLC mice having leukocyte counts 6 times higher. We have interpreted this finding as due to the lack of suppressor T cells in the LLC mice. But, whatever the underlying factor, such immune response may be regarded as abnormal. Much subsequent investigation of the immune response of the LLC and HLC mice is now in progress. Nevertheless, the overall immunological and biological evidence to date suggests that the production of reticular cell hyperplasia and amyloidosis represents a consequence of immune deficiency in LLC mice, and supports Teilum's (9) two-phase theory of amyloidogenesis.

The present series of electron microscopic observations seem to support the view that amyloid fibrils are formed in the reticular cells producing through the disruption of the cytoplasmic membrane, resulting meantime in the death of the cells, a degenerative change which begins with the immunological stimulation of the reticular cell system.

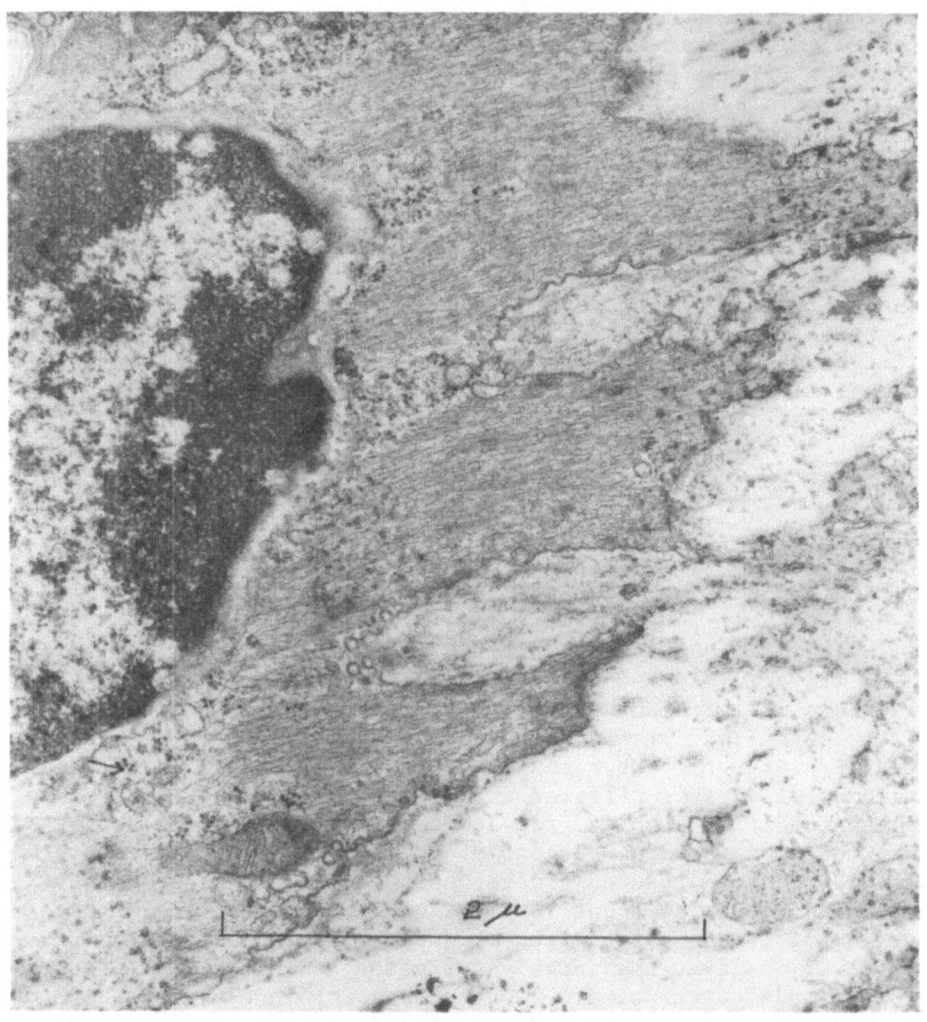

Fig. 3. Intracytoplasmic amyloid fibrils in a reticular cell of
the LLC mouse spleen. Notice the fibrils relatively densely packed
at the end toward the cytoplasmic membrane comparing with the end
toward the nucleus and the rather parallel distribution patterns,
conveying the impression that they are originated from the cytoplasmic membrane. The cytoplasmic membrane is well defined, causing no confusion with extracellular fibrils which are practically
nonexistent in this place. N = nucleus, M = mitochondria, GR =
granular reticulum. (45,800 x)   (reduced 30% for reproduction)

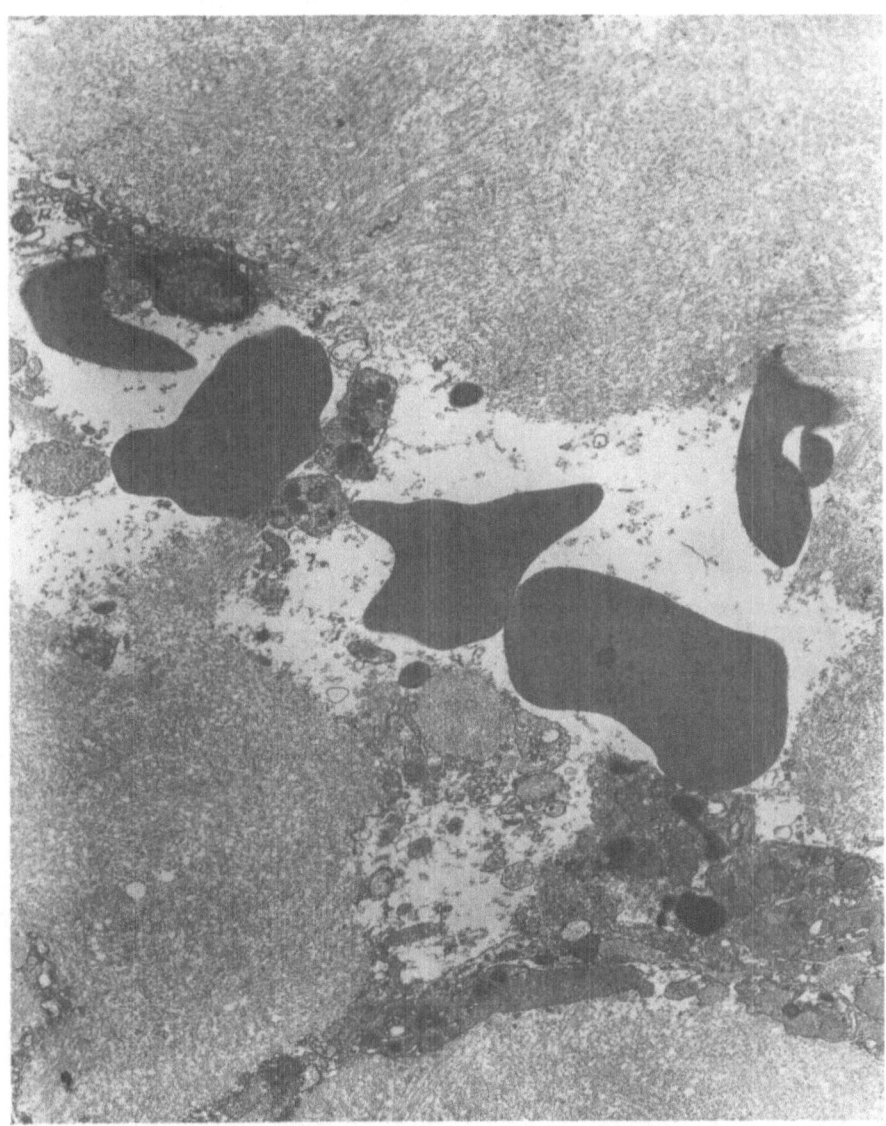

Fig. 4. An electron micrograph showing the association of amyloid fibrils and a reticular cell of the spleen. The fibrils appear to originate from within the cell. They are distributed in curves into the extracellular space which is filled with randomly lined or cross sectioned amyloid fibrils. (13,700 x)
(reduced 30% for reproduction)

In summary, selective breeding for leukocyte level differences resulted in 2 strains of mice with six- or seven-fold differences in leukocyte counts. They differed in percentages of different leukocyte types, lymphoid organ weights and response to sheep red blood cells. Two pathological conditions, reticular cell hyperplasia and amyloidosis, occurred in the low leukocyte count line.

At the age of 3 to 6 months, 70% of the mice developed reticular cell hyperplasia in the inguinal lymph nodes, and at 10 to 18 months, about 100% of them developed amyloidosis in the spleen, kidney, liver, adrenal glands and various other organs. Both intra- and extracellular amyloid fibrils were found in the spleen sections by electron microscopy. Our interpretations on the basis of these findings is that the pathologic conditions observed represent manifestations of immunological events characteristic of the LLC mice with immune deficiency or abnormality in the immune system. We would like to suggest that the origin of amyloid fibrils is at the reticular cells.

ACKNOWLEDGEMENTS

This investigation was supported by NIH Grant No. CA-14036 from the National Cancer Institute. Many thanks to Thomas Farley for help in managing the mouse strain, autopsy and blood counts and to Lester Bunker for assistance in histological preparations and electron microscopy. The Jackson Laboratory is fully accredited by the American Association for Accreditation of Laboratory Animal Care.

REFERENCES

1. Biozzi, G.C., Stiffel, C., Mouton, D., Bouthillier, Y. and Decreusefon, C., J. Exp. Med., 135 (1972) 1071.
2. Chai, C.K., Genet. Res., 8 (1966) 125.
3. Chai, C.K., J. Hered., (1975) in press.
4. Chai, C.K., Farley, T. and Morang, P., J. Lab. Clin. Med., 72 (1968) 165.
5. Dunn, T.B., J. Nat. Cancer Inst., 14 (1954) 1281.
6. Gowen, J., J. Infect. Dis., 73 (1943) 40.
7. Janigan, D.T., Amer. J. Pathol., 47 (1965) 159.
8. Kaliss, N. and Chai, C.K., J. Reticuloendothel. Soc., 16 (1974) 330.
9. Silver, D. and Chai, C.K., J. Immunol., 115 (1975) 462.
10. Teilum, G., Acta. J. Pathol. Microbiol., 61 (1964) 21.

CELL-MEDIATED IMMUNITY IN LONG TERM TRANSPLANT PATIENTS

J. M. THOMAS, A. M. KAPLAN, F. T. THOMAS and H. M. LEE

Medical College of Virginia, Virginia Commonwealth
University, Richmond, Virginia (USA)

We recently reported 8-12 yr follow-up studies of the earliest
renal transplant recipients at the Medical College of Virginia (44).
At a mean follow-up of 10 yr, about 50% of these early transplant
recipients are alive, all with functioning kidneys. Tissue typing
was not available when these patients were transplanted and retrospective typing has shown that most patients were non HL-A identical
with their recipients. Therefore, the immunological mechanisms
which permitted successful transplantation for a decade or longer
are of great interest.

The present study was designed to determine whether differences
in specific cellular immunity against donor and indifferent allogeneic lymphocytes could be demonstrated in long term (LT) renal
transplant recipients.

MATERIALS AND METHODS

Patients

The 12 individuals selected for this study were long-surviving
renal transplant recipients with related donors. These patients
had a functioning primary kidney allograft 8-12 yr (mean = 10.0 yr)
after transplantation and a satisfactory course throughout this
period. Of these, 10 had received their transplant from a parent

and 2 from a sibling. Donor and recipient lymphocytes were tissue-
typed at the time of the present study and 92% (11/12) of the recipients were shown to be haploidentical with the donor and mismatched for 1-2 defined HL-A antigens.

Lymphocyte Preparation

Prior to obtaining blood samples, the recipients were withdrawn from Prednisone and azathioprine (Imuran) for 24 hr. Heparinized peripheral blood was obtained from the recipient (A),
donor (B) and unrelated third (C) and fourth party (D) normal volunteers. Lymphocytes were separated on a Ficoll-Hypaque gradient,
washed 3 times in medium RPMI-1640 and counted in a hemocytometer.
Tissue culture medium RPMI-1640 (Gibco) containing 100,000 U/ml
penicillin and 50,000 µg/ml streptomycin was used in all of the
assay systems.

Mixed Lymphocyte Culture

One-way mixed lymphocyte cultures (MLC) were carried out
according to the micro-method of Hartzman, et al.(19). Mitomycin
C (Sigma) was used to treat stimulator cells for 20 min at 37 C at
a concentration of 50 µg per 2×10^6 cells. After 3 washes in RPMI-
1640, 3×10^5 stimulator cells were seeded in quadruplicate cultures
with 1.5×10^5 responder cells in micro-culture plates (Linbro).
The culture medium was supplemented with 25% heat-inactivated pooled normal human serum (NHS). Incubation time was 5 days at 37 C
in the presence of 5% CO_2 in 95% air. Tritiated thymidine (2 µCi
per culture, s.a. = 1.9 Ci/m Mole, Schwartz, Mann) was added for
the final 18-20 hr.

Recipient lymphocytes were tested both as responder and stimulator cells with cultures of B, C and D individuals, respectively.
Response to phytohemagglutinin (PHA-P, Difco) was also included.
Results are expressed in mean counts per min (cpm) and as a stimulation index (SI), which is the ratio of mean cpm in allogenic cul-
tures/mean cpm in autologous cultures.

For serum blocking studies, recipient serum was inactivated
at 56 C for 30 min and diluted in five- to ten-fold dilutions from
1/5-1/5000 using NHS as the diluent. Cultures were set up in recipient serum for the entire incubation period, and results were
compared to control cultures in NHS. Significant blocking activity
was considered present if the mean cpm in recipient serum were at
least 2 standard deviations lower than the mean cpm in normal plasma (greater than 25% inhibition).

Cellular Cytotoxicity

Direct lymphocyte mediated cytotoxicity (LMC) was performed according to the method of Garavoy, et al.(16). Freshly drawn, washed recipient peripheral blood lymphocytes were used as effector cells in a 100:1 ratio with ^{51}Cr-labelled target cells. Incubation was carried out in triplicate cultures in RPMI-1640 containing 25% NHS for 4-6 hr at 37 C in a CO_2 incubator. The cultures were centrifuged and supernatant aliquots were counted for released ^{51}Cr in a gamma scintillation counter.

Cell-mediated lympholysis (CML) assays were performed according to the method of Bonnard, et al.(8). For effector cell generation, 1×10^7 washed peripheral blood lymphocytes were incubated with 3×10^6 mitomycin-treated stimulator cells (B or C subjects) for 6 days in culture medium containing 20% NHS in the presence of 5% CO_2 in 95% air. On day 6, the effector cells were centrifuged, placed in fresh culture medium and incubated with ^{51}Cr-labelled target cells in an E/T ratio of 100:1. Target cells were PHA-induced lymphoblasts which had been cultured for 6 days. The assay incubation time was 6 hr under the same conditions as described for LMC.

Antibody-dependent cell-mediated cytotoxicity (ADCMC) was performed essentially according to the method of Trinchieri, et. al (49). For ADCMC, effector cells were taken from healthy normal volunteers, and target cells were PHA-induced lymphoblasts tagged with ^{51}Cr.

In all 3 cell-mediated cytotoxicity assays with ^{51}Cr labelled target cells, cytotoxic activity was tested on target cells of A, B, C and D subjects. The percent lysis was calculated according to the equation:

$$\frac{\text{mean cpm experimental} - \text{mean cpm spontaneous}}{\text{mean cpm maximum} - \text{mean cpm spontaneous}} \times 100$$

Spontaneous release of ^{51}Cr was obtained from cultures of target cells in the absence of effector cells, and maximum release was obtained from target cells frozen and thawed repeatedly.

RESULTS

The response of LT transplant recipients to PHA and to indifferent allogeneic lymphocytes in one-way MLC was not significantly different from the response of normal control volunteers included in each assay. The mean cpm of PHA-stimulated cultures of recipient lymphocytes was $113,638.9 \pm 12,355$ compared to a mean of $116,080 \pm 8,412$ in normal controls. The mean MLC SI of LT transplant recipients to random unrelated individuals (AX_m) was 20.6 ± 4.1 (SE) com-

pared to a SI of 21.5 $\pm$ 2.5 (SE) for MLC between normal unrelated
individuals (Table I). Moreover, the one-way MLC response of recipient to the haploidentical donor (AB$_m$) in NHS was positive in
all cases, except in the single HL-A identical sibling-sibling combination. The mean SI for haploidentical AB$_m$ cultures was 7.3 $\pm$
1.0 (SE). This SI was not significantly different from the mean
SI of 10.4 $\pm$ 1.7 (SE) obtained for normal haploidentical offspring
to parent (OP$_m$) responses in one-way MLC.

TABLE I

One-Way Mixed Lymphocyte Culture (MLC) Studies
in 8-12 Year Transplant Recipients

	MLC Combinations**			
	AB$_m$	AX$_m$	XX'$_m$	OP$_m$
mean stimulation index (SI)* ($\pm$SE)	7.3$\pm$1.0	20.6$\pm$4.1	17.7$\pm$2.1	10.4$\pm$1.7

*SI = <u>cpm in allogeneic mixed culture</u>
       cpm in autologous mixed culture
**A = recipient; B = donor; X = normal unrelated individuals C+D;
O and P = normal offspring and parent; m = mitomycin treated

    Serum blocking activity in one-way MLC was detected in 88.8%
of the recipient sera tested. The degree of inhibition by recipient serum in AB$_m$ cultures was variable, (25-80%) and never achieved
100% inhibition at any dilution in any of the sera. Of the sera
which exhibited MLC blocking activity, 63% were specific for donor
cells and 37% appeared to be nonspecific in that they inhibited one-
way MLC stimulation between random unrelated individuals, not sharing
donor HL-A antigens. Nonspecific blocking sera also inhibited PHA
stimulation, but this inhibition (10-15%) was weaker than the inhibition of one-way MLC responses.

    When the serum of LT transplant recipients was assayed for AD-
CMC with normal indifferent effector cells, 50% (6/12) of the recipients exhibited specific ADCMC activity against ^{51}Cr-tagged donor target lymphocytes (Table II). In positive sera, the percent specific
cytolysis was always greater than 10% (mean 18.5% $\pm$ 62 [SE]) in contrast to negative sera in which mean percent specific cytolysis was
1.3% $\pm$ 2.8 (SE). The number of mismatched donor HL-A antigens

was the same (mean 1.2 mismatched antigens per recipient) in both ADCMC positive and negative groups. A striking association was observed between the incidence of detectable ADCMC against donor cells and the clinical signs of chronic renal allograft rejection, especially persistent proteinuria. All of the ADCMC negative recipients exhibited urinary protein values within the normal range (mean 0.29 ± 0.07 grams/24 hr) over the last 3 yr. In contrast, the mean urinary protein values for the ADCMC positive group were significantly elevated with a mean of 1.81 ± 0.35 grams/24 hr ($t=2.764$, $p< 0.01$).

No active T cell cytotoxicity against ^{51}Cr-tagged donor target lymphocytes was detected in 4 hr LMC assays in any of the LT transplant recipients tested. The next question was whether their lymphocytes, which were MLC responsive to the HL-A non-identical donors, could be stimulated in vitro in the MLC-CML system to generate cytotoxic effector cells against the donor and unrelated individuals. The results of CML tests indicated that the peripheral blood lymphocyte population of this group of patients was selectively deficient in ability to generate cytotoxic effector cells against the donor (Table III). The mean percent lysis against B targets in AB_m effector cultures was $5.1 \pm 2.1\%$ (SE). This figure is sharply contrasted by a mean percent lysis of $42.1 \pm 8.4\%$ (SE) obtained against C targets in AC_m effector cultures. The percent lysis in AC_m effector cultures was lower than that obtained against the specific targets in CB_m ($53.2 \pm 10.5\%$ [SE]) and DC_m ($63.5 \pm 9.7\%$ [SE]) effector cell cultures, respectively. This difference, however, was not statistically significant ($p > 0.10$). Therefore, the depressed AB_m effector activity could not be simply explained as an effect of residual immunosuppressive drugs.

TABLE II

Antibody Cell-Mediated Cytotoxicity (ADCMC) and Progressive Proteinuria in 8-12 Year Renal Transplant Recipients*

	ADCMC Positive	ADCMC Negative
Mean Urinary Protein (g/24 hr)	1.81 ± 0.35 (SE)	0.20 ± 0.07 (SE)

*t = 2.764; p = <0.02

TABLE III

Cell-Mediated Lympholysis (CML) Studies in
8-12 Year Renal Transplant Recipients

Effector Cells	Mean Percent Lysis to ^{51}Cr Target Cells	
	Specific Allogeneic Target	Autologous Target
AB_m	(B) 5.1$\pm$2.1% (SE)	(A) 2.4$\pm$1.9%
AC_m	(C) 42.1$\pm$8.4%	(A) 4.6$\pm$1.5%
CB_m	(B) 53.2$\pm$10.5%	(C) 7.4$\pm$3.8%
DC_m	(C) 63.5$\pm$9.7%	(D) 9.1$\pm$5.2%
BA_m	(A) 41.6$\pm$13.9%	(B) 7.2$\pm$5.0%

A = recipient; B = donor; C and D = unrelated normal individuals;
m = mitomycin treated

Since the mean MLC-SI of the LT transplant recipient group to
haploidentical donor stimulating cells was only 41.8% of the MLC SI
obtained toward stimulating cells from C and D parties, the possibility had to be considered that the proliferating helper cell
activity generated during the 6 day MLC incubation phase of the
effector cultures was inadequate to generate cytotoxic effector
cells against the 1-2 HL-A antigen mismatched B cells. To test
this possibility, we compared the CML response of the LT transplant
recipients to that of normal adult offspring to parental (haploidentical) stimulating cells, since the OPm SI in MLC was comparable to
that of the LT transplant recipient to donor response. The mean
percent lysis to specific parent target cells in OPm effector cultures was 34.7% $\pm$ 6.7 (SE). Thus, it is unlikely that inadequate
MLC stimulation could alone explain the lack of cellular cytotoxicity in AB_m effector cultures.

DISCUSSION

This report presents data on the cellular immune responsiveness of successful LT transplant recipients, 92% (11/12) of whom
have tolerated haploidentical kidney allografts, free of late acute
rejection episodes, for a mean of 10.0 yr. They were transplanted

without benefit of established tissue typing or crossmatching techniques and have received standard immunosuppressive therapy. The
precise immunological adaptation mechanisms which have contributed
to such long term graft acceptance remain undefined. We have applied some of the currently available techniques for monitoring
cellular immune responses in an effort to identify possible clues
which might explain their optimal post transplant course.

The lymphocytes of the LT transplant recipients demonstrated
a normal response to PHA and to unrelated allogeneic lymphocytes
in one-way MLC. Moreover, they were MLC-responsive to specific
donor lymphocytes. Thus, a decade of chronic immunosuppressive
therapy has not eliminated lymphocyte clones capable of recognition
and blastogenic response to donor specific or unrelated LD alloantigens in this patient population. Other investigators have reported specific one-way MLC responsiveness to donor lymphocytes in
transplant patients with functioning renal allografts within the
first 5 yr after transplantation (2,11,15,21,39). Bach, et al.
(2) noted an absence of MLC stimulation in only 25% of recipients
with HL-A non-identical transplants at 1-4 yr post-transplant.
Studies reported from other laboratories indicated that only a
small, but significant, proportion of 1-5 yr recipients are specific non-responders in MLC (11,15,39).

The original contention (6) that transplantation tolerance
involves a central inhibition of specifically reactive clones has
gleaned support from observations showing that lymphoid cells from
tolerant animals are poorly reactive to donor alloantigens in MLC
(4,22,41,52,53) and in graft versus host (GVH) assays (1,14). On
the other hand, in some experimental models there is clear evidence
that lymphoid cells of tolerant animals can respond to variable
degrees in MLC (3,4,31) and GVH assays (7,34,43,46,53). This dis-
crepancy has been attributed by some investigators to the degree of
operational allograft tolerance, i.e. complete or partial (4,22).
With the human LT transplant recipients it is not possible to determine the degree of operational tolerance involved while they are
maintained on chronic low dose immunosuppressive therapy. However,
reports of cases where immunosuppressive drug lapses have occurred
indicated that a significant number of patients continue to maintain good allograft function suggesting that immune modulation factors, in addition to immunosuppressive therapy, are operational in
maintenance of LT allograft function.

Wright, et al.(53) have shown that although host GVH reactivity and MLC reactivity to donor lymphocyte antigens may be detected in operationally tolerant rats, the degree of MLC reactivity is quantitatively decreased compared to controls. The relative
degree of specific lymphocyte reactivity in one-way MLC assays needs
to be considered. We found that with these assays, the mean SI

(7.3) of LT transplant patients to their haploidentical related
donor was within the range of the mean SI (10.4) obtained in normal offspring to their haploidentical parent. These SI values are
similar to the peak distribution of those in a large series of single LD-1 locus-allele-disparate related combinations reported by
Thorsby, et al.(45). We also calculated MLC relative stimulation
ratio (RSR) for the LT transplant recipient group. As defined by
Segall, et al.(42), the RSR is a ratio of stimulation of recipient by donor cells to stimulation of recipient by cells of unrelated individuals. A mean RSR of 0.42 + 0.07 (SE) (range 0.20-
0.89) was found for the LT transplant recipients. Thus, they fell
equally into categories of low to moderate responders, respectively,
according to the classification outlined by Segall, et al.(42).

Most (88%) recipients with long functioning kidneys had clearly
demonstrable serum blocking factors in MLC. Of the patients with
serum blocking activity, 63% showed specific serum blocking against
their donor while the rest had non-specific blocking factors. Since
all patients were withdrawn from Prednisone and Imuran for 18-24 hr
prior to obtaining blood samples, the blocking was probably not due
to a drug effect. Clearly, however, the specific nature of these
blocking factors needs to be better characterized, and experiments
are currently in progress in our laboratory to study this question.
There is evidence from numerous laboratories that both in human
and animal renal transplant recipients the recipient serum may
block MLC responsiveness (15,21,25,39). As in our studies the
blocking effects of some sera are specific for donor stimulating
cells while others exhibit a lack of specificity (15,21).

The significance of MLC blocking antibody in allotransplantation is not clear. In some patients the serum blocking effect seen
in both MLC studies was present at serum dilutions out to 1/3,000.
The presence of high titer blocking factors would tend to support
the concept that these factors might be operational in vivo. In
a rat cardiac allograft model of passive enhancement, Gordon and co-
workers (18) have demonstrated that all alloantisera with in vivo
enhancing activity could inhibit MLC responses. On the other hand,
recent studies of Larner and Fitch (29) failed to demonstrate a
correlation between in vitro inhibition of MLC by rat alloantisera
and in vivo kidney allograft enhancement. In their experience all
antisera tested inhibited MLC, but only few had enhancing activity.
Hasek, et al.(20) demonstrated that serum which blocked cytotoxic
activity of lymphocytes in vitro had no effect in vivo on induction
of tolerance in neonatal rats. The role of blocking factors was
not studied, however, in relation to induction and maintenance of
allograft acceptance in immunosuppressed adults.

Quadricci, et al.(38) have shown that about 35% of transplant
recipients who have well tolerated allografts possess blocking fac-

tors demonstrable by the Hellstrom microcytotoxicity assay. These authors found a correlation of clinical nonreactivity to allografts with the development of blocking factors in the 1-12 month post-transplantation period. During the first post-transplant month and after the first year no such clear-cut correlation existed suggest-ing that effective blocking factors develop after the first month, but that after the first year other factors are necessary to explain the benign clinical course of surviving patients.

The studies reported here on 8-12 yr LT transplant recipients demonstrate a specific defect in in vitro generation of cytotoxic T cells against donor cells in all cases. This phenomena could not be explained by a drug effect or nonspecific blocking since recipi-ent cells showed normal in vitro cytotoxic effector cell generation against indifferent donors. Furthermore, donor cells were shown to be adequate immunogens in terms of eliciting a normal in vitro cyto-toxic T cell response by indifferent cells. Therefore, in the face of normal stimulation by donor cells and normal responsiveness of recipient cells, a state of donor-specific hyporesponsiveness in CML is apparent. This phenomena has not, to our knowledge, been previously reported in LT transplant patients. Our preliminary studies of transplant recipients at the 1-4 yr interval did not re-veal often this specific deficiency in cytotoxic effector cell gen-eration suggesting that the long term durability of allograft func-tion in man may be related to development of a specific defect in generation of in vitro cytotoxic effector cells against donor anti-gens. There are several possibilities to explain these phenomena: a) suppressor T cell development, b) clonal deletion of precursor effector cells, c) development of anti-idiotypic antibody, and d) in vitro production of blocking antibody or antigen-antibody com-plexes.

Waldmann, et al.(49) have demonstrated the role of suppressor T cells in the pathogenesis of variable hypogammaglobulinemia but little is known of the role of suppressor T cells in other human immune responses. Suppressor T cell development might specifically inhibit the generation of cytotoxic T cells by T-T cell interaction as proposed by Gershon (17). There are recent data available from animal studies suggesting that active suppressor cells may play a role in allograft tolerance. Lymphoid cells from tolerant recipi-ents can suppress the anti-donor GVH reactivity of normal lympho-cytes of the recipient strain (40). Moreover, specific cardiac allograft tolerance in mice has been transferred with lymphoid cells from tolerant recipients (28). The mechanism of active suppression of effector cell immunity in allograft tolerance remains unresolved, but according to studies of Wegmann and Drell (50) the block prob-ably occurs between the effector precursor and actual cytotoxic ef-fector cell.

Clonal deletion of reactive cell groups (or "clones) was first proposed by MacFarlane Burnett as an explanation for acquired tolerance. Again, despite a plethora of work in lower animals on this phenomena, little is yet known of its importance in clinical allotransplantation. Our findings are compatible with a hypothesis of deletion of specific clones of precursor cytotoxic effector cells perhaps due to the influence of chronic Imuran therapy. Imuran is known to be most active against blastogenic and differentiating cells. Therefore, the clonal cells programmed for anti-donor cytotoxicity might be preferentially eliminated by Imuran in vivo as they differentiated into cytotoxic effectors. The precise nature of this effector cell defect should be of interest, since the successful LT transplant recipients are specifically deficient in response to donor cells but otherwise exhibit an intact T cell system to guard against infection. Clearly, the factors promoting development of this specific immune defect and the time course of development are worthy of more detailed studies.

Brevity does not permit doing justice to the theories of anti-idiotypic antibody production or in vitro generation of blocking antibodies. In recipients with specific serum blocking antibodies, anti-idiotype antibodies might be present. If so, the anti-idiotypic antibodies might interfere with recipient T cell activity in CML. In vitro generation of anti-donor blocking antibodies similarly could interfere with generation or expression of T cell cytotoxicity. It is known that alloantibodies can block CML assays at both the initiation stage and at the effector stage (9,36).

Chronic rejection is thought to be a leading cause of graft loss in LT renal transplant patients (51). The principal characteristics of chronic rejection include progressive deterioration of renal function, and gradual onset of proteinuria, hypertension, and decreased glomerular filtration (23). Pathologically, kidney damage is manifest by glomerular and arterial changes (37). The glomerular lesions of chronic rejection have been postulated to be a result of chronic immunological damage mediated by humoral antibodies (23, 26, 35).

In the present study all long term recipients having clinical symptoms of chronic rejection as defined by persistent proteinuria at 8-12 yr post-transplant demonstrated significant anti-donor ADCMC activity in their serum. In contrast, 75% of the LT transplant recipients with negative ADCMC had urinary protein values in the normal range.

At present, little is understood about the role of specific anti-donor ADCMC activity in clinical organ transplantation. Some investigators have associated pre-transplant and early post-transplant ADCMC activity with poor post-transplant courses marked by severe early acute rejections (27,33,47). Others, however, have not observed such an association (10,12,30). Our studies of short term (1-12 yr) and LT renal transplant recipients have demonstrated an association between proteinuria and other signs of chronic renal allograft rejection and anti-donor ADCMC. More detailed studies and longitudinal follow-up studies of ADCMC may be of importance in further understanding of the possible role of ADCMC in chronic allograft rejection.

Clinically, these results suggest that LT transplant patients should be serially monitored for ADCMC activity. We are now testing a patient with a previously positive ADCMC test who has only recently begun to show signs of chronic rejection. Another patient had clear ADCMC activity 4 months prior to the onset of irreversible chronic rejection. Our results also suggest that the ADCMC test may be useful in matching donor-recipient pairs pre-transplant to rule out the presence of pre-existing ADCMC activity which may lead to later chronic rejection. We currently are following a patient who in retrospect had a positive ADCMC test pre-transplant but who up to now has no signs of chronic rejection.

The question now arises as to what therapy, if any, is indicated when patients develop chronic rejection and a positive ADCMC test. Because of the generally poor prognosis of chronic rejection, it would seem reasonable to consider pharmacological techniques to block or decrease the level of ADCMC in chronic rejection patients. Little is known concerning the effects of various immunosuppressive drugs on ADCMC although at least anti-lymphocyte globulin (ALG), Prednisone and Imuran are known to inhibit effector cell activity in ADCMC (5,13,32). Since most of our patients are receiving maximally tolerated doses of Imuran, increase in the dose of this drug is not possible. Most LT transplant patients, however, are on very low (5-15 mgm) doses of Prednisone and, therefore, it might be possible to decrease ADCMC activity by increasing the Prednisone dosage. Descamps, et al.(13) have suggested that Prednisone is a far more effective inhibitor of ADCMC activity than Imuran. Biberfeld, et al.(5) have shown that ALG can inhibit ADCMC in vitro, but the need for chronic therapy in humans with chronic rejection would make this agent a less than optimal one, since long term ALG therapy is not well tolerated. It would be important also to study the in vivo and in vitro effect of other immunosuppressive agents (e.g. cyclophosphamide) on post-transplant serum and effector cell activity in ADCMC.

Immunologically, the association of ADCMC and chronic renal allograft rejection should be of great interest. Some previous studies have suggested an association of humoral antibody with chronic renal allograft rejection although the relationship has continued to be an elusive one. The major features of chronic re-

jection, glomerular and blood vessel injury, have long been thought
to be due to an immunological reaction; although, in the case of
the glomerular injury, other pathology such as recurrent glomerulonephritis or antigen-antibody complex disease must be ruled out.
In previous studies from our institution, a very low rate of recurrent glomerulonephritis (about 6%) has been documented (24). Furthermore, most of the ADCMC positive patients in this study did
not have glomerulonephritis as their original disease. The presence of both glomerular and arterial lesions seen in 2 patients
who were ADCMC positive strongly suggests that the kidney disease
represents chronic rejection and not recurrent glomerulonephritis.
Since renal allograft rejection is currently responsible for the
large share of allograft loss from rejection in our unit, the role
of ADCMC in development of the syndrome is worthy of more detailed
studies.

REFERENCES

1. Atkins, R. and Ford, W., Transplantation, 13 (1972) 442.
2. Bach, M.L., Engstrom, M., Bach, F., Etheridge, E. and
    Najarian, J.S., Cell. Immunol., 3 (1972) 161.
3. Bernstein, I., Hamilton, B., Wright, P., Burnstein, R.
    and Hellstrum, R., J. Immnnol., 114 (1975) 320.
4. Beverly, P., Brent, L., Brooks, C. and Medawar, P., Transplant.
    Proc., 5 (1973) 679.
5. Biberfeld, P., Biberfeld, G., Perlmann, P. and Holm, G., Cell.
    Immunol., 7 (1973) 60.
6. Billingham, R., Brent, L. and Medawar, P., Nature, 172 (1953)
    603.
7. Bilsdoe, P., Ford, W.L., Pettirossi, O. and Simonsen, M.,
    Transplantation, 12 (1971) 189.
8. Bonnard, G., Lemon, L. and Chappius, M., Scand. J. Immunol.,
    3 (1974) 7.
9. Cordier, G., Betuel, H. and Revillard, J., Transplant. Proc.,
    5 (1973) 1855.
10. D-Apice, A.J.F. and Morris, P.J., Transplantation, 18 (1974)
    20.
11. Debray-Sachs, M. , Dimitriu, A., Bach, A.M. and Hamburger, J.,
    J. Cell. Immunol., 7 (1973) 181.
12. Descamps, B., Gagnon, R., Debray-Sachs, M., Barbanel, C. and
    Crosnier, J., Transplant. Proc., 7 (1975) 635.
13. Descamps, B., Gagnon, R. and Crosnier, J., Abstr. 10th Leuco.
    Cult. Conf., Amsterdam, September (1975).
14. Elkins, W., Transplant. Proc., 5 (1973) 685.
15. Etheredge, E., Shons, A.R. and Najarian, J.S., 7th Leucocyte
    Culture Conference (Ed. F. Daguillard) Academic Press, New York,
    New York (1973) 399.
16. Garavoy, M., Granco, V., Zschaeck, D., Strom, T., Carpenter,

C. and Merrill, J., Lancet, 1 (1973) 573.

17. Gershon, R., Contemporary Topics in Immunobiology (Ed. M. D. Cooper and N. L. Warner) Plenum Publishing Co., New York, New York, 4 (1974) 1.

18. Gordon, R., Stinson, E., Souther, S. and Oppenheimer, J., Transplantation, 12 (1971) 484.

19. Hartzman, R., Segall, M., Bach, M. and Bach, F., Transplantation, 11 (1971) 268.

20. Hasek, M., Chunta, J., Sladecek, M., Machockova, M., Bubenik, J. and Matousek, V., Transplantation, 20 (1975) 95.

21. Hattler, B.G., Jr., Karesh, C. and Miller, J., Tissue Antigens, 1 (1971) 270.

22. Heron, I., Transplantation, 15 (1973) 534.

23. Hume, D., Sterling, W., Weymouth, R., Siebel, H., Madge, G. and Lee, H.M., Transplant. Proc., 2 (1970) 361.

24. Hume, D. and Bryant, C.P., Transplant. Proc., 4 (1972) 108.

25. Ippolito, R., Mahoney, R. and Murray, I., Transplantation, 17 (1974) 89.

26. Jeannet, M., Pinn, V.W., Flaz, M.H., Winn, H.J. and Russell, P.S., New Eng. J. Med., 282 (1970) 111.

27. Jeannet, M., Vassalli, P. and Botella, F., Transplant. Proc., 7 (1975) 631.

28. Jirsch, D., Kraft, N. and Diener, E., Transplantation, 18 (1974) 155.

29. Larner, B. and Fitch, F., Transplantation, 16 (1973) 54.

30. Lordon, R.E., Garovoy, M.R., Ball, E. and Thompson, A., Western Dialysis and Transplant Society Meeting, Las Vegas, Nevada, Sept. 19-21 (1975).

31. Lucas, Z., Markley, J. and Travis, M., Fed. Proc., 29 (1970) 2041.

32. MacLennan, I.C.M., Transplant. Rev., 13 (1972) 67.

33. McConnachie, P.R. and Dosseter, J.B., Tissue Antigens, 3 (1973) 303.

34. Mullen, Y., Takasugi, M. and Hildemann, W., Transplantation, 15 (1973) 238.

35. Pierce, J., Kay, S. and Lee, H.M., Surgery, 78 (1975) 14.

36. Phillips, S., Strom, T., Corson, J., Transplant. Proc., 5 (1973) 577.

37. Porter, K.A., Transplant. Proc., 6 (1974) 79.

38. Quadricci, L., Tremann, J., Marchioro, I. and Striker, G., Transplantation, 17 (1974) 361.

39. Revillard, J.P., Robert, M., Beteul, H. and Rifle, G., 7th Leucocyte Culture Conference (Ed. F. Daguillard) Academic Press, New York, New York (1973) 415.

40. Rouse, B. and Warner, N., J. Immunol., 113 (1974) 904.

41. Schwarz, M.R., J. Exp. Med., 127 (1969) 879.

42. Segall, M., Bach, F.H., Bach, M.L., Hussey, H.L. and Ushling, D.T., Transplant. Proc., 7, Suppl. 1 (1975) 41.

43. Sumerska, T., Betel, I., Balner, H. and Warren, H., Trans-

plantation, 17 (1974) 1.
44. Thomas, F., Lee, H.M., Wolf, J.S., Pierce, J.C. and Hume, D.
    C., Transplant. Proc., 8 (1975) 707.
45. Thorsby, E., Bondevik, H., Helgesen, A. and Hirschberg, H.,
    Transplant. Proc., 7 (1975) 87.
46. Tilney, N.L. and Bell, P., Transplantation, 18 (1974) 31.
47. Ting, A. and Terasaki, P.I., Transplantation, 18 (1974) 371.
48. Trinchieri, G., DeMarchi, M., Mayr, W., Saur, M. and Ceppellini,
    R., Transplant. Proc., 5 (1973) 1631.
49. Waldman, T., Broder, S., Durm, M., Blaese, D.M., Blackman,
    M. and Strober, W., Lancet, 2 (1974) 609.
50. Wegmann, T. and Drell, D., Suppressor Cells in Immunity (Ed.
    S. Singahl and N. Sinclair) University of Western Ontario
    Press, London, Ontario (1975) 76.
51. Williams, G.M., Transplant. Proc., 6 (1974) 71.
52. Wilson, D.B., Silver, W. and Novell, P., J. Exp. Med., 126
    (1967) 655.
53. Wright, P., Bernstein, I., Hamilton, B. and Hellstrom, K.,
    Transplantation, 19 (1975) 437.

The RES in Tumor Immunology

THE INTERACTIONS OF A LEUKEMIA VIRUS WITH CELLS OF THE RES

W. S. CEGLOWSKI, A. A. MASCIO and R. P. CLEVELAND

The Pennsylvania State University, University Park,
Pennsylvania (USA)

The ability of a variety of microorganisms to interact with cells of the reticuloendothelial system (RES) is well documented. Recent reviews (14,34,42) have evaluated the experimental evidence concerning the ability of microorganisms, including oncogenic and non-oncogenic viruses, to suppress host immune mechanisms. The role of immune mechanisms in the induction and maintenance of the neoplastic state has also been the subject of much discussion and experimentation (6,36,43). The isolation and characterization of several viruses (16,18,27,38) capable of inducing a leukemia-like disease in mice has provided systems in which one can study the effects of well defined oncogenic agents on the cells of the RES. These oncornaviruses also provide an opportunity to study the influence of RES functions on the induction and maintenance of the neoplastic state. We have selected Friend leukemia virus as the oncogenic agent of primary interest in our studies which have been primarily concerned with determining the ability of the leukemia virus-infected host to mount humoral and cell mediated immune responses. We have been concerned with characterizing the suppressive events with the intent of eventually elucidating the precise mechanism(s) by which leukemia virus infections mediate their immune suppression.

MATERIALS AND METHODS

Our stock Friend leukemia virus (FLV) (16) was originally obtained from the American Type Culture Collection. It has been passaged in Balb/c mice by intraperitoneal inoculation of a 1:10 dilution of a 10% homogenate of splenic tissue from infected mice.

For use in experiments the virus is now passaged once in newborn
Sprague-Dawley rats in order to remove contaminating lactic dehydrogenase virus, LDV (40). This is necessary since LDV has been shown
to affect thymic-dependent areas of lymphoid tissues (44).

Friend leukemia virus has been shown (45) to be a complex of
at least two viruses. One component of the complex is a lymphatic
leukemia virus (LLV) that produces a disease similar to the Gross
virus-induced disease (46). The other component is called the spleen
focus forming virus (SFFV) because of its ability to form macros-
copic spleen foci of transformed erythroid precursor cells (1).
The LLV component appears to serve as a helper virus in the production of spleen foci by SFFV by supplying the genome for coat
formation (13,15). For this reason it is necessary to quantitate
each of the components of the complex in the virus inoculum. The
spleen focus forming virus (SFFV) component is assayed by the
method of Axelrad and Steeves (1). The lymphatic leukemia virus
(LLV) component is assayed by means of the XC cell assay (41)
using either primary Balb/c embryo cells or the SC-1 cell line (19).
As a consequence of the forced passage in Balb/c mice this virus
is now N-b tropic. The stock virus preparations usually contain
$3.0-6.0 \times 10^2$ FFU/ml and $4.0-8.0 \times 10^4$ PFU/ml.

Even in the early studies with FLV it was apparent that there
was a genetic component to susceptibility and resistance to the
pathogenicity of the virus. Recent investigations have shown that
two independently segregating genes control susceptibility to FLV.
The two loci are designated FV-1 which governs relative resistance
to the LLV component (35), and FV-2 which controls absolute resistance to spleen focus formation by SFFV (25). The two alleles at
FV-1 are designated as FV-1^n and FV-1^b; mice being either N-type
(DBA/2) or B-type (Balb/c) (26,35). The relative resistance at
FV-1 to LLV is reflected in the in vitro titration patterns of the
virus. N-tropic viruses show one-hit, dose response patterns in
cells from N-type mice, whereas they show multiple-hit patterns
in B-type mice; on the other hand B-tropic viruses show single-hit
patterns in B-type mice and multiple-hit patterns in N-type mice
(35). The original FLV isolate (N-tropic) can be adapted to a
stable NB-tropic form by repeated forced passage in B-type (Balb/c)
mice (26). Mice homozygous or heterozygous for susceptibility at
the FV-2 allele are susceptible to SFFV (Balb/c and DBA/2),
whereas mice homozygous for the FV-2^r allele (C57Bl/6) are resistant (25). The FV-2 phenotype masks the FV-1 genotype in the
intact animal and thus controls the outcome of FLV infection.

In our studies inbred mice of the Balb/c strain (FV-1^b, FV-
2^s) have been used as the principal susceptible experimental animal.
C57Bl/6 (FV-1^b, FV-2^r) mice are used as the leukemia virus-resistant
host. Female weanlings are obtained from commercial sources.

Following an appropriate quarantine period, the animals are tested for LDV and then maintained in groups of five and provided water and food ad libitum.

Humoral immune status was assessed by immunizing control and FLV-infected mice by intraperitoneal (i.p.) inoculation of 0.5 ml of a 10% suspension of washed sheep red blood cells (SRBC). At appropriate time intervals after immunization, mice were killed by cervical dislocation. Their spleens were removed, weighed, and then teased into a single cell suspension with dissecting needles and forceps. Aggregates were disrupted by aspirating with a Pasteur pipette. The cell suspension was filtered through a double layer of gauze and washed three times with Hank's balanced salt solution.

Single antibody forming cells (AFC) were detected by plaque formation in agar gel (21). A 0.1 ml volume of the spleen cell suspension was rapidly mixed with 2.0 ml of melted 1% Noble agar, 0.1 ml DEAE dextran and 0.1 ml of a washed 20% SRBC suspension maintained at 48 C. This mixture was carefully poured onto the surface of a previously prepared Petri plate containing 15 ml of solidified 1.4% Noble agar in HBSS. Each spleen cell suspension was plated in triplicate. After the plates were incubated at 37 C for one hour, 2 ml of guinea pig complement diluted 1:10 in veranol buffer, was added to each plate. The plates were then reincubated for 30 minutes at 37 C.

The number of antibody plaque forming cells (PFC) per plate was counted and the total number of plaques per spleen and the number of plaques per million leukocytes were calculated. Total splenic leukocyte counts were determined with a hemocytometer using 4% (v/v) acetic acid as the diluent.

In the studies concerning the effect of FLV infection on macrophage migration inhibition, mice were sensitized with Mycobacteria 21 days prior to infection with FLV by subcutaneous injection at five sites on the dorsum with complete Freund's adjuvant (CFA) supplemented with dried Mycobacterium tuberculosis.

At appropriate time intervals dispersed cell suspensions were prepared by teasing excised spleens in HBSS with rat tooth forceps. The dissociated cells were washed twice at 4 C in 10 ml of HBSS. Erythrocytes in the suspension were lysed by a single wash with an ammonium chloride (0.83%) - Tris (0.15 M) buffer, pH 7.4 (5). The cells were then resuspended and washed once in Medium 199. Approximately 15-25% of the cells in these preparations were capable of rapidly phagocytizing carbon particles and were presumably macrophages. Phagocytosis was measured by incubating a cell suspension (10^6 cells/ml) in Medium 199 containing india ink (1:100) for

60 minutes at 37 C and microscopically counting the fraction of cells darkened by the uptake of the carbon (32). Adherent cells were separated from non-adherent cells by incubating the cell suspension in Medium 199 for 30 minutes at 37 C on glass petri plates (31). The adherent cells were eluted from the glass petri plates with 0.15 M phosphate buffered saline (PBS) (pH 7.4) containing 0.1 M ethylenediaminotetracetic acid (EDTA) and were then washed once in medium. These cells were predominantly macrophages (70-90%) as judged by phagocytosis. The non-adherent cell fraction was centrifuged once and resuspended in Medium 199 plus 10% heat inactivated (56 C, 12 minutes) fetal calf serum (FCS).

For the direct assay of migration inhibitory factor (MIF), spleen cells were placed in capillary tubes plugged at one end with soft paraffin and then centrifuged at 100 x g for four minutes. Two capillaries, containing approximately 5×10^6 cells each, were cut at the cell-fluid interface and placed into a Sykes-Moore Chamber. At least four chambers per mouse were prepared, two containing Medium 199 plus 10% FCS, the remaining two receiving the same medium to which the antigen had been added (3). The antigen employed was PPD at a final concentration of 50 ug/ml. At this concentration the antigen did not inhibit the migration of spleen cells from non-sensitized mice.

The production of MIF-rich supernatants for use in the indirect assay was achieved by incubating 2×10^7 non-adherent spleen cells/ml in Medium 199 plus 10% FCS to which PPD at a concentration of 100 ug/ml had been added. The total volume of each culture was 2 ml. The spleen cells from each mouse were cultured separately. Following 24 hours of incubation at 37 C, the cell-free supernatants from these cultures were assayed for MIF activity against normal mouse spleen cells prepared in the same manner as described for the direct assay. Control fluids consisted of supernatants from similar cell cultures incubated without antigen.

The chambers for both direct and indirect assays were incubated for 24 hours at 37 C and the area of migration from each capillary tube was determined by projecting and tracing an enlarged (13X) image of the migrating cells onto paper. The areas were corrected for the magnification and expressed in terms of the actual size of the migration area (mm^2). Migration inhibition was expressed as the following percentage:

(1.00 - average area with antigen/average area without antigen) x 100%: inhibition of the area of migration by 20% or greater is considered by most investigators to be an indication of delayed hypersensitivity (2).

Lymphocyte mediated cytotoxicity was assessed by the procedure

described by Henney (20). In brief, a lymphoma, EL-4, of the
$C_{57}Bl/6$ mouse was maintained by weekly serial passage in the
syngeneic strain. Target cells were prepared from the ascitic
fluid by suspending 10^7 washed cells in 1 ml of Eagle's minimal
essential medium and incubating with 0.5 ml of sodium ^{51}chromate
for 30 minutes at 37 C. The target cells were then washed three
times in medium and adjusted to 10^6 cells per ml.

The effector spleen cells from Balb/c mice sensitized with
3×10^7 EL-4 cells (i.p.) 11 days prior to the assay were pressed
through a stainless steel mesh (size 100) in HBSS. The resulting
cell suspension was centrifuged at 300 x g for five minutes.
Erythrocytes were lysed by a single wash in 0.83% ammonium
chloride-Tris buffer at 37 C. Cell debris was removed by centri-
fuging for 20 seconds at 200 x g, discarding the pellet, and
recentrifuging the supernatant at 300 x g for five minutes. The
resulting pellet was washed twice in Eagle's Medium containing 10%
fetal calf serum, resuspended in the same medium and the number of
viable lymphocytes determined by dye (0.1% trypan blue) exclusion.
Viability was usually 90-95%. The assay is performed in the
following manner. Labelled target cells (10^5 in 0.1 ml) were
added to 1×10^5 to 5×10^6 spleen lymphocytes and the total
volume was adjusted to 1.0 ml in 10 x 75 mm plastic tubes to
obtain lymphocyte/target cell ratios of 1:1 to 50:1. The mixtures
were allowed to settle in upright tubes. Following incubation for
four hours in five percent CO_2, 95% air at 37 C the tubes were
centrifuged at 1000 x g for five minutes and a 0.5 cc aliquot of
the supernatant counted for ^{51}Cr content on a γ-ray spectrometer.
Control mixtures consisted of spleen cell suspensions from unsensitized mice. The amount of ^{51}Cr released from these control
mixtures is most frequently the same as that released in similar
tubes without any lymphocytes. Thus, the spontaneous release is
the amount of ^{51}Cr released in the absence of any lymphocytes and
did not exceed five to ten percent of the total ^{51}Cr content.
The percentage specific cytolysis was calculated by subtracting
from the CPM of ^{51}Cr released in the absence of lymphocytes
(spontaneous release) and then dividing by the total CPM of ^{51}Cr
present in 0.5×10^5 target cells minus the CPM of ^{51}Cr spontaneously released. The lysis in this sytem has been shown to be a result
of a single-hit interaction between an effector lymphocyte and a
target cell.

In some studies effector splenocytes were treated with either
anti-theta or anti-FLV serum. Anti-theta serum was prepared by
the method of Rief and Allen (39). In brief, C_3H mouse thymocytes were inoculated into AKR mice at weekly intervals. Four days
after the last injection, the mice were bled and the resulting
serum was absorbed with C_3H erythrocytes. The cytotoxic capacity
of the serum was assessed by incubating serial diltuions of heat

inactivated antiserum with target cells (spleen or thymus) which had been labeled with ^{51}Cr. After 30 minutes incubation the cells were washed to remove the excess antiserum and incubated in the presence of guinea pig complement (which had been absorbed with mouse spleen cells) for an additional 30 minutes. The release of ^{51}Cr into the supernatant was then determined on a gamma counter. The theta antigen has been extensively utilized as a marker for thymus-derived lymphocytes in mice (37). Antiserum to Friend leukemia virus and virus infected cells was prepared in the follow-ing manner. Balb/c mice were vaccinated with a 10% homogenate of FLV infected Balb/c splenic tissue inactivated with formalin. This was followed by weekly inoculations of splenocytes from FLV infected Balb/c mice. After a total of six inoculations the mice were bled and the serum separated. The serum was assayed for its ability to neutralize FLV infectivity and by the indirect immuno-fluorescent technique using both control and FLV-infected spleen cells.

In other studies spleen cells from both normal and infected Balb/c mice were treated with Vibrio cholera neuraminidase (VCN): 50 units of VCN in 0.5 ml of sterile saline was added to 10^8 (2 ml) spleen cells in medium and the mixture was incubated for 15 minutes at 37 C. The cells were then centrifuged at 300 x g, for five minutes and washed three times with medium before determining either susceptibility to anti-theta serum of cytolytic capacity.

RESULTS

Infection of susceptible Balb/c mice with FLV leads to a rapid and prolonged decline in humoral immune competence as measured by the ability to generate PFC to an antigen such as SRBC (7,8). Suppression of both the primary and secondary responses is observed (Tables I and II). Peak numbers of PFC are reduced in mice infected with FLV at either the time of immunization or prior to immunization. Mice inoculated with FLV two days after immunization exhibit peak numbers of PFC equal to or greater than those of appropriate cont-rols. As the time interval between infection and assay increases the magnitude of the depression increases. Late in the primary immune response conversion of PFC formation from IgM type to IgG type plaques is observed in control mice as well as mice infected two days after immunization. An apparent impairment in the "switch" from IgM to IgG production is observed in animals infected on the day of immunization or earlier. Studies concerning the anamnestic response to sheep erythrocytes confirmed this observation by demon-strating that the reduction in the number of IgG producing cells was much greater than that observed for IgM producing cells (Table II). The degree of reduction of both the primary and secondary immune response was also dependent on the dose of virus adminis-

TABLE I

Effect of Friend Leukemia Virus (FLV) Infection on the
Primary Immune Response to Sheep Erythrocytes

Treatment	4 Days Post Immunization		17 Days Post Immunization	
	IgM PFC/Spleen	IgM PFC/10^6 Splenocytes	IgM PFC/Spleen	PFC/Spleen
Control	6.8×10^4	4.2×10^2	2.1×10^2	4.0×10^3
Infected with FLV 2 Days After Immunization	6.2×10^4	1.9×10^2	1.8×10^2	1.8×10^3
Infected with FLV Same Day as Immunization	1.9×10^4	2.7×10^1	2.0×10^2	3.2×10^2
Infected with FLV 3 Days Before Immunization	5.8×10^3	9.3×10^0	1.7×10^2	2.4×10^2
Infected with FLV 8 days Before Immunization	1.2×10^3	1.5×10^0	2.6×10^2	1.0×10^3

tered.   In an attempt to determine if FLV had a direct effect on antibody forming cells or could indeed infect antibody forming cells, three separate but related studies were performed (9,11,23). One study demonstrated that antibody forming cells adoptively transferred to control and FLV-infected mice persisted in the FLV-infected animals in a manner similar to that observed in control mice.   The second study demonstrated that direct addition of FLV to short term in vitro cultures of antibody forming cells did not inhibit plaque formation.   The third study demonstrated that a substantial percentage (55%) of the antibody plaque forming cells studied from SRBC-immunized, FLV-infected mice appeared to be producing both FLV particles and anti-SRBC antibody.   Other studies were performed to study the ability of spleen, bone marrow and thymus cells from FLV-infected mice to restore immune competence in lethally irradiated syngeneic recipients (10,12).   This was accomplished by intravenous inoculation of   appropriate cell numbers into recipients followed immediately by immunization with sheep erythrocytes and assaying for antibody PFC nine days later (Table III).   These studies suggested that the decline in PFC observed was the result of a decline in the number of precursor cells.   These data also suggest that the leukemic environment as represented by the FLV-infected mouse exerts a further depressive effect on the proliferative potential of the precursor cells. We have also demonstrated that bone marrow cells but not thymus cells are impaired very early after infection with FLV in their ability to interact with their normal counterpart and reconstitute immune competence as was previously demonstrated by Bennett and Steeves (2).

Our more recent studies have been concerned with the study of effects of FLV-infection on cell mediated immune competence. Recent studies (28,30) were concerned with the effect of FLV on the production of MIF by mice sensitized with complete Freund's adjuvant prior to infection. These studies are summarized in Table IV and contrast with previous studies concerning humoral immunity in that an ongoing cell mediated immune response is diminished following infection of the susceptible DBA/2 and Balb/c mice. The leukemia virus resistant C57Bl/6 mice exhibit no appreciable diminution in MIF production following administration of FLV. Utilizing appropriate cell separation techniques one can demonstrate that the FLV-induced defect resides in the MIF secreting cell and not in the migrating indicator cell. Unfortunately, the macrophage migration inhibition technique does not permit one to distinguish between two possible causes of suppression. The first is that all cells are being suppressed; the second is that a given fraction or proportion of the cells are being completely inhibited while the remainder are producing normal levels of MIF. The experimental technique which appeared capable of resolving this question was the lymphocyte mediated cytotoxicity method (20).

TABLE II

Effect of Infection with Friend Leukemia Virus
(FLV) on the Secondary Immune
Response to Sheep Erythrocytes

| Group | PFC/Spleen | | Ratio |
	IgM	IgG	IgG/IgM
Control	8.9×10^3	9.2×10^4	10.3
Infected with FLV same day as immunization	9.5×10^3	9.0×10^4	9.4
Infected with FLV 3 days before immunization	5.0×10^3	0.8×10^4	1.6
Infected with FLV 8 days before immunization	2.2×10^3	0.2×10^4	0.9

TABLE III

Splenic Plaque Forming Cells (PFC) Resulting from
Transfer of Splenocytes to Irradiated Recipients
Compared with Splenic Plaque Forming Cells
Resulting from Direct Immunization

Group	Calculated Number* of Potential PFC in Donor Spleen Based on Cell Transfer Studies	Observed Number** of PFC/Spleen at the Peak PFC Response (4 Days Post Immunization)
Control	4.7×10^4	5.8×10^4
Infected with FLV on the same day	2.1×10^4	3.4×10^4
Infected with FLV 3 days before	1.0×10^4	8.8×10^3
Infected with FLV 7 days before	9.4×10^3	1.6×10^3
Infected with FLV 14 days before	7.2×10^3	2.3×10^2

* Groups of control or FLV-infected mice were killed and single
cell suspension of a portion of the spleen was transferred to
immunologically incompetent syngeneic recipients. All animals
were assayed nine days after immunization. The calculated number
represents the total of PFC expected if the entire donor spleen
was transferred.
** Groups of control or FLV-infected mice were immunized and
assayed for the number of PFC at four days post immunization.

TABLE IV

Effect of Inoculation of Friend Leukemia Virus (FLV)
on Migration Inhibition Factor (MIF) Production
in Leukemia Susceptible and Resistant Mice

Experimental Group*	% Migration Inhibition		
	DBA/2	Balb/c	$C_{57}B1/6$
Sensitized, Control	48	48	43
Sensitized, Infected with FLV one day prior to assay	45	43	36
Sensitized, Infected with FLV 3 days prior to assay	17	21	43
Sensitized, Infected with FLV 7 days prior to assay	15	3	47
Sensitized, Infected with FLV 10 days prior to assay	13	5	53

*Groups of mice were sensitized with supplemented complete Freund's
adjuvant 21 days prior to infection with FLV. Mice were infected
with FLV at the time indicated prior to performing the indirect
assay to assess the production of MIF.

TABLE V

Cytolytic Activity of Splenocytes to Alloantigen Bearing
Target Cells Following Administration
of Friend Leukemia Virus (FLV)

Experimental Group*	Day FLV Adminis- tered in Relation to Day of Assay	Cytolytic Activity** Splenocyte Source	
		Balb/c	$C_{57}B1/6$
Control	−	1.00	1.00
FLV-infected	−1	0.62	0.91
	−5	0.22	0.83
	−11	0.01	1.00
	−19	0.01	0.59

* Groups of mice were sensitized by i.p. inoculation of alloantigen bearing tumor cells 11 days prior to assay. Balb/c ($H-2^d$) mice
were immunized with the EL-4 lymphoma propagated in $C_{57}B1/6$ ($H-2^b$)
mice while $C_{57}B1/6$ mice were immunized with the P-815 mastocytoma
tumor passaged in DBA/2 ($H-2^d$) mice. Groups of mice were infected
with FLV on the days indicated.
** Cytotoxic activity is expressed as the ratio of the number of
splenocytes (FLV infected group) required to lyse 10% of the alloantigen bearing target cells/number of control splenocytes required
to lyse 10% of the alloantigen bearing target cells.

For these studies (29) Balb/c mice were sensitized with .an alloantigen bearing lymphoma (EL-4) cell which was maintained by passage in the ascites form in $C_{57}B1/6$ mice. At varying time intervals after sensitization mice were infected with FLV. At the peak day (day 11) of the immune response, the ability of spleno-cyte suspensions from control and FLV-infected mice to lyse the target cells was assessed by the release of ^{51}Cr. The cytotoxic potential of splenocytes from control and infected mice are present-ed in Table V. The efficiency of cytolysis decreases precipitously following infection of susceptible Balb/c mice with FLV. In contrast no marked depression in cytolytic activity is observed un-der similar experimental conditions in the resistant $C_{57}B1/6$ mouse.

In more recent studies we have attempted to elucidate the mechanism of this suppression of cell mediated immune function. Several experimental approaches have been utilized. We have studied the ability of FLV-infected spleen cells to influence the cyto-toxic potential of sensitized lymphocytes from normal mice. In these studies splenocytes from normal animals immunized with allo-antigen have been diluted with splenocytes derived from nonimmunized FLV-infected mice. Inhibition in excess of that caused by dilution with normal cells is induced by dilution with spleen cells from FLV-infected animals. Suppression increased as the time interval of infection increased; the degree of suppression also increased as the infecting dose of virus was increased. The suppression in this case might have been caused by a direct effect of the virus on the effector cell population. In order to test this possibility, dilutions of stock virus preparations were incubated with suspen-sions of splenocytes from normal sensitized animals. These studies demonstrated that the direct addition of infectious FLV into the assay system appeared to have no suppressive effect on lymphocyte mediated cytotoxicity under the experimental conditions employed. In order to further study the suppressive event, we have prepared extracts of FLV-infected spleens which are free of infectious virus as measured by the spleen focus forming virus assay (1) as well as the X C assay (41). These virus free extracts when incu-bated with effector cells from normal animals were capable of markedly inhibiting cytotoxic activity. The active component of the extract of virus infected cells has not yet been isolated or identified. In more recent studies we have determined that the impairment in the cytolytic potential of splenocytes from FLV-infected mice can be reversed in part by treatment of the effector cells with neuraminidase.

DISCUSSION

Studies in a number of laboratories have documented an impairment of host immune competence in leukemia virus infected mice (14). The concept of immunologic surveillance (43) might be invoked to explain the observed differences in immune competence in leukemia virus susceptible and resistant mice. Others (24,36) have raised questions concerning this concept and experimental evidence has been presented by Stutman and Dupuy (47) demonstrating that absolute resistance to Friend leukemia virus is not abrogated by pretreatment of such mice with anti-lymphocyte serum. However, evidence has been presented (49) suggesting that Statalon preparations can suppress established Friend virus induced leukemia in a susceptible strain of mouse by stimulating an immune response to both the virus and transformed cells.

In view of the relative importance assigned to cell mediated immune mechanisms in viral infections and neoplasia our observations concerning decreased cell mediated immune functions early in the disease process are of interest. Our studies utilizing the macrophage migration inhibition techniques which was believed at that time to be an *in vitro* correlate of cell mediated immunity demonstrated a rapid suppression of the production and release of MIF by FLV-infection of susceptible strains of mice. The effect of virus infection was clearly on the lymphocyte component of the reaction and not on the migrating indicator cells. Appropriate controls tended to rule out any contribution of either virus particles or virus-antibody complexes to the observed depression.

However, more recent studies by Bloom et al., (4) have demonstrated that the production of MIF in the guinea pig is not exclusively a T cell product since PPD sensitized spleen cell suspensions depleted of T cells can indeed produce MIF as well as lymphotoxin. As a consequence the previous interpretation of our studies with MMI may require revision if indeed a similar situation prevailed in the mouse.

Our most recent studies have been performed utilizing the lymphocyte mediated cytotoxicity technique which is currently believed to be exclusively a T cell mediated phenomenon. These studies have demonstrated a rapid suppression of cytolytic activity of sensitized cells shortly after infection of susceptible mice with FLV. Such suppression does not at present appear to be mediated by a direct virus-effector cell interaction. Our current studies suggest the possibility that spleen cells from FLV-infected mice secrete a "factor" which is capable of inhibiting lymphocyte mediated cytolysis. The relation of this factor to the suppressor cells observed in normal and virus-infected mice (17,22) or suppressor substances (32,48) observed in malignant cells is at this time not known.

At present it is not known if this factor is produced by

either B cells, T cells, or macrophages. Also it is not at present
clear whether the virus alters the recognition step or the cytolytic
phase of the cytotoxic reaction. Studies are now in progress to
resolve some of these questions.

ACKNOWLEDGMENTS

    We thank Mrs. G. U. LaBadie for excellent technical assistance
in the performance of these studies. This work was supported in
part by research grants from the American Cancer Socity, Inc. and
by Public Health Research Grant No. CA-15643 from the National
Cancer Institute.

REFERENCES

1. Axelrad, A.A. and Steeves, R.A., Virology, 24 (1964) 513.
2. Bennett, M. and Steeves, R.A., J. Nat. Cancer Inst., 44 (1970)
    1107.
3. Bloom, B.R. and Bennett, B., In: In Vitro Methods in Cell-
    Mediated Immunity, (Eds. B.R. Bloom and P. Glade) Academic
    Press, New York, New York (1971) 235.
4. Bloom, B.R., Stoner, G., Gaffney, J., Shevach, E. and Green,
    I., Eur. J. Immunol., 5 (1975) 218.
5. Boyle, W., Transplant., 6 (1968) 761.
6. Burnet, M.F., Transplant. Rev., 7 (1971) 3.
7. Ceglowski, W.S. and Friedman, H., J. Immunol., 101 (1968) 594.
8. Ceglowski, W.S. and Friedman, H., J. Immunol., 102 (1969) 338.
9. Ceglowski, W.S. and Friedman, H., J. Immunol., 103 (1969) 460.
10. Ceglowski, W.S. and Friedman, H., J. Immunol., 105 (1970) 1406.
11. Ceglowski, W.S. and Friedman, H., In: Proc. Fourth Leukocyte
    Culture Conference, (Ed. O.R. McIntyre), Appleton-Century-
    Crofts, New York, New York (1971) 309.
12. Ceglowski, W.S., LaBadie, G.U. and Friedman, H., In: Micro-
    environmental Aspects of Immunity, (Eds. B.D. Jankovic and K.
    Isakovic), Plenum Press, New York, New York (1973) 499.
13. Dawson, P.J., Tacke, R.B. and Fieldsteel, A.H., Br. J. Cancer,
    22 (1968) 569.
14. Dent, P.B., Prog. Med. Virol., 14 (1972) 1.
15. Eckner, R.J. and Steeves, R.A., J. Exp. Med., 136 (1972) 832.
16. Friend, C., J. Exp. Med., 105 (1957) 307.
17. Gershon, R.K., Cohen, P., Hencin, R. and Liebhaber, S.A., J.
    Immunol., 108 (1972) 586.
18. Gross, L., Proc. Soc. Exptl. Biol. Med., 76 (1951) 27.
19. Hartley, J.W. and Rowe, W.P., Virology, 65 (1975) 128.
20. Henney, C.S., J. Immunol., 107 (1971) 1558.
21. Jerne, N.K. and Nordin, A.A., Science, 140 (1963) 405.
22. Kirchner, H., Chused, T.M., Herberman, R.B., Holden, H.T. and

Lavrin, D.H., J. Exp. Med., 139 (1974) 1473.

23. Koo, G.C., Ceglowski, W.S., Higgins, M. and Friedman, H., J. Immunol., 106 (1971) 815.

24. Kripke, M.L. and Borsos, T., J. Nat. Cancer Inst., 52 (1974) 1393.

25. Lilly, F., J. Nat. Cancer Inst., 45 (1970) 163.

26. Lilly, F. and Pincus, T., Adv. Cancer Res., 17 (1973) 231.

27. Moloney, J.B., J. Nat. Cancer Inst., 24 (1960) 933.

28. Mortensen, R.F., Ceglowski, W.S. and Friedman, H., J. Immunol., 111 (1973) 1810.

29. Mortensen, R.F., Ceglowski, W.S. and Friedman, H., J. Immunol., 122 (1974) 2077.

30. Mortensen, R.F., Ceglowski, W.S. and Friedman, H., J. Nat. Cancer Inst., 52 (1974) 499.

31. Mosier, D.A., Science, 158 (1967) 1574.

32. Murgita, R.A. and Tomasi, T.B., Jr., J. Exp. Med., 141 (1975) 440.

33. Nelson, D.S., Macrophages and Immunity, North Holland Publishing Company, Amsterdam, Holland (1969).

34. Notkins, A.L., Mergenhagen, S.E. and Howard, R.J., Ann. Rev. Microbiol., 24 (1970) 525.

35. Pincus, T., Rowe, W.P. and Lilly, F., J. Exp. Med., 133 (1971) 1234.

36. Prehn, R.T., J. Reticuloendothel. Soc., 10 (1971) 1.

37. Raff, M.C., Nature, 224 (1969) 378.

38. Rauscher, F.J., J. Nat. Canc. Inst., 29 (1962) 515.

39. Reif, A.E. and Allen, J.M.V., J. Exp. Med., 120 (1964) 413.

40. Riley, V., In: Methods in Cancer Research (Ed. H. Busch), Academic Press, New York, New York 4 (1968) 493.

41. Rowe, W.P., Pugh, W.E. and Hartley, J.W., Virol., 42 (1970) 1136.

42. Schwab, J.H., Bact. Rev., 39 (1975) 121.

43. Smith, R.T. and Landy, M., (Editors) Immune Surveillance, Academic Press, New York, New York (1970)

44. Snodgrass, M.J., Lowry, D.S. and Hanna, M.G., Jr., J. Immunol., 108 (1972) 877.

45. Steeves, R.A. and Eckner, R.J., J. Nat. Cancer Inst., 44 (1970) 587.

46. Steeves, R.A., Eckner, R.J., Bennett, M., Mirnad, E.A. and Trudel, P.J., J. Nat. Canc. Inst., 46 (1971) 1209.

47. Stutman, O. and Dupuy, M., J. Nat. Canc. Inst., 49 (1972) 1283.

48. Tanapatchaiyapong, P. and Zolla, S., Science, 186 (1974) 748.

49. Wheelock, E.F., Toy, S.T., Caroline, N.L., Sibal, L.R., Fink, M.A., Beverley, P.C.L. and Allison, A.C., J. Nat. Canc. Inst., 48 (1972) 665.

NEOPLASTIC INVASION AND METASTASIS WITHIN THE LYMPHORETICULAR SYSTEM

I. CARR and F. McGINTY

University of Sheffield and Weston Park Hospital
Sheffield (United Kingdom)

The common human cancers spread initially by lymphatic metastasis, but there are few accurate models of this phenomenon and no understanding of the way in which tumor cells invade lymphatic vessels. This report describes a defined and quantified model of lymphatic metastasis, an investigation of the lymph node reactions in this model and an analysis of the mechanism of invasion of lymphatic vessels. This material has been published in full elsewhere (1,3,4,5,9) or will be published (2). Reference should be made to these publications for full technical details and account of literature.

MATERIALS AND METHODS

The basic experiment carried out was as follows: 5×10^6 Rd/3 tumor cells (in 100 μl Hanks solution) were injected into the left footpad of inbred white rats. The footpad, the left popliteal lymph node and other organs were examined histologically after routine H and E staining of usually semiserial but often complete paraffin serial sections. The Rd/3 tumor is a transplantable tumor originally induced by injection of dibenzanthracene and carried by flank transplantation ever since. Transmission electron microscopy was carried out on material fixed in glutaraldehyde and osmium and embedded in Araldite by standard technique. Other animals received 50 μl BCG into the footpad or 10 μl BCG into the lymph node 7 days before giving tumor in an attempt to inhibit metastasis. Control animals were injected into the footpad with substances other than live tumor cells to study the lymph node response. These included 100 μg pneumococcal polysaccharide, 5×10^6 formolized tumor cells,

50×10^6 homologous RBC and a freeze dried extract of Rd/3 tumor. Some animals were injected with tumor extract after whole body irradiation with lead shielding of the left leg.

Quantitation of metastasis was carried out by counting tumor cells in paraffin sections of lymph nodes (9). It was then possible to assess the number of tumor cells necessary in a lymph node for successful metastasis, by injecting the standard dose (5×10^6) tumor cells, amputating the foot 24 hr to 7 days later and examining the popliteal lymph node for successful metastasis. Once an approximation had been reached of the tumor cell burden necessary for metastasis, tumor cells were injected directly into the popliteal node in appropriate dosage, after exposure of the lymph node under light ether anaesthesia; doses of 5×10^3 and 5×10^2 cells were injected in this way.

The idea that tumor cells might be destroyed in the lymph node was further explored by injecting low doses of tumor cells (5×10^5, 5×10^4 and 5×10^3) into the footpad and studying metastasis in the popliteal node 7, 14 and 21 days thereafter.

Since this and earlier work (5) suggested that a lymphoreticular response might be important in the rejection of tumor cells, tumor cell rejection across a species barrier was studied by injection of 5×10^6 Rd/3 tumor cells into normal mice, irradiated mice and nude athymic mice.

Next, the relationship between cell aggregation, penetration of lymphatics and metastasis was studied by examining metastasis in 2 mouse mammary tumors, a metastasizing tumor in CBA mice and a non-metastasizing tumor in C3H mice. Tumor cells were injected with and without prior incubation in 0.2% trypsin in Hanks solution for 30 min. This produced disaggregation of the tumor into a single cell suspension.

Lastly the ultrastructure of invasion of lymphatic vessels was studied, by examining with the electron microscope the footpads of rats which had received 5×10^6 and 20×10^6 Rd/3 cells up to 7 days previously.

RESULTS

In the initial group of experiments the model was established. Serial sections of popliteal lymph nodes immediately after and 6 hr after injection of tumor cells showed that no tumor cells were present and therefore that direct intralymphatic injections did not occur. This is most important; if this is not shown to be the case in a model of lymphatic metastasis then the first component of metastasis-invasion is missing from the system.

After injection of 5×10^6 Rd/3 tumor cells into the footpad of the rat, consistent lymphatic metastasis was present in the popliteal node (in over 95% of animals) (Fig.1). Twenty-four hr after injection a few tumor cells were present in the subcapsular sinus. Tumor cells traveled in the afferent vessels singly, sometimes in association with aggregated lymphocytes, settled in the subcapsular sinus and spread singly in the radial and medullary sinusoids (Fig. 2). There was a high rate of tumor cell mitosis in the lymph node and also continuous recruitment from the footpad. Two components were observed in the lymph node reaction, an early proliferation of sinus macrophages seen only with viable tumor and tumor extract (Fig. 3), followed by marked germinal center activity which was less specific. These reactions were later obliterated by tumor cell proliferation. When animals were allowed to survive beyond 14 days, they died with massive deposits of tumor in the para-aortic nodes and lungs. The injection of 10 µl BCG into either footpad or lymph node was effective in inhibiting metastasis - either partly or completely.

The macrophage proliferation response in the lymph node was not found after several nonspecific stimuli but was found after injection of a freeze-dried tumor extract and in animals which had received whole body irradiation with lead-shielding of the left lower leg, including the node. It was therefore probably a proliferation of resident lymph node macrophages.

When metastasis was quantitated by counting cells in semiserial sections it was found that the rate was somewhat inconstant. After injection of 5×10^6 cells into the footpad, the lymph node contained about 1.5×10^2 tumor cells in 1 day, 2.3×10^4 in 2 days, 4.5×10^4 in 3 days and 5.0×10^5 in 4 days (9).

The establishment of even approximate quantitation of this made it possible to determine whether tumor cell destruction could occur in the lymph node, or whether tumor cells once within the lymphatic vascular system were totally protected. If the footpad was removed any later than 24 hr after implantation of 5×10^6 tumor cells, progressive metastasis occurred; while of 20 animals whose feet were removed at 24 hr only 4 showed lymph node metastasis. When 5×10^3 tumor cells were injected directly into the popliteal node, growth in the lymph node always occurred. But when 5×10^2 tumor cells were similarly injected, growth resulted in the popliteal node in 3 animals out of 5. These findings allow the very approximate statement that the critical burden of tumor cells in the popliteal node in this experiment was 2.5×10^2 cells - remembering that the accuracy of cell counts at the low level involved is not better than $\pm$ 10%. When a dose of tumor cells lower than 5 x 10^6 was injected into the footpad (5 x 10^5, 5 x 10^4 and 5 x 10^3) and animals killed at 1, 2 and 3 weeks, broadly speaking, a metastasis appeared in the popliteal node by 14 days and disappeared by 21 days. It is clear from previous studies that tumor cells can be destroyed in the footpad. The present findings make it highly likely that they can also be destroyed in the lymph node.

Successful metastasis did not occur when Rd/3 tumor was injected in normal mice but occurred in 4/12 of the irradiated mice and all of 12 athymic mice. This indicates that T cells are important in tumor rejection at least across a species barrier. Previous work (5) has made clear the importance of a footpad lymphoreticular response in rejecting small numbers of Rd/3 tumor cells in the rat footpad.

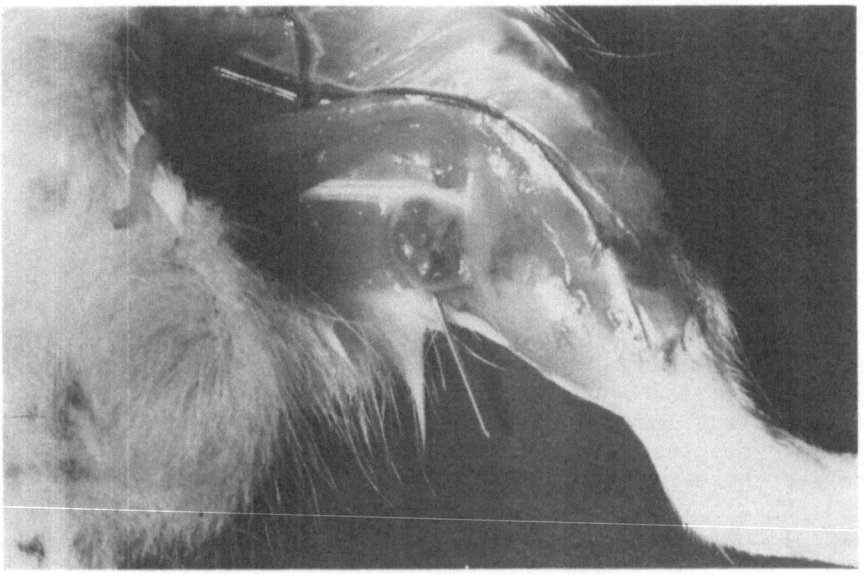

Fig. 1. Popliteal node enlargement 5 days after Rd/3 tumor into the footpad (x 0.7).

(Figures 1-5 have been reduced 10% for reproduction).

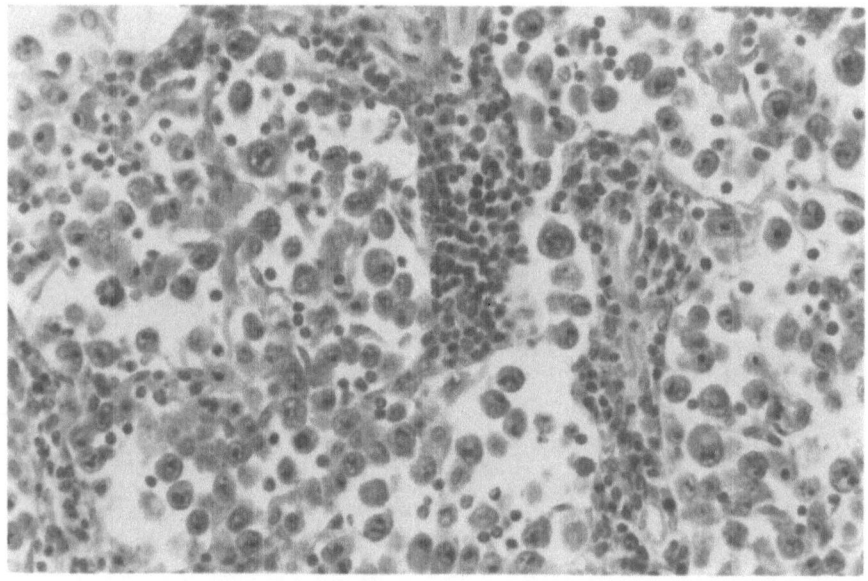

Fig. 2. Popliteal node 3 days after Rd/3 tumor into the footpad, showing tumor cells in medullary sinusoids (H and E x 275).

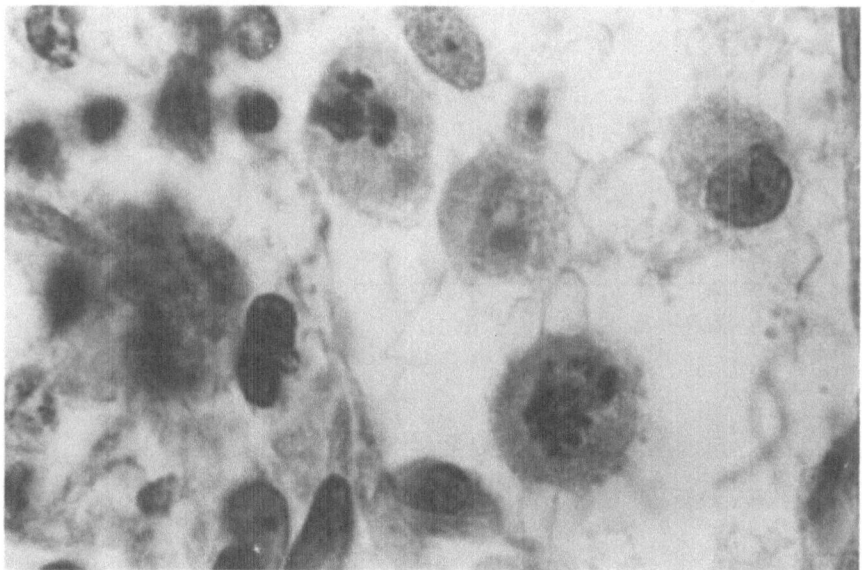

Fig. 3. Popliteal node 24 hr after Rd/3 tumor into the footpad showing macrophages in the medullary sinus in mitosis (H and E x 2000).

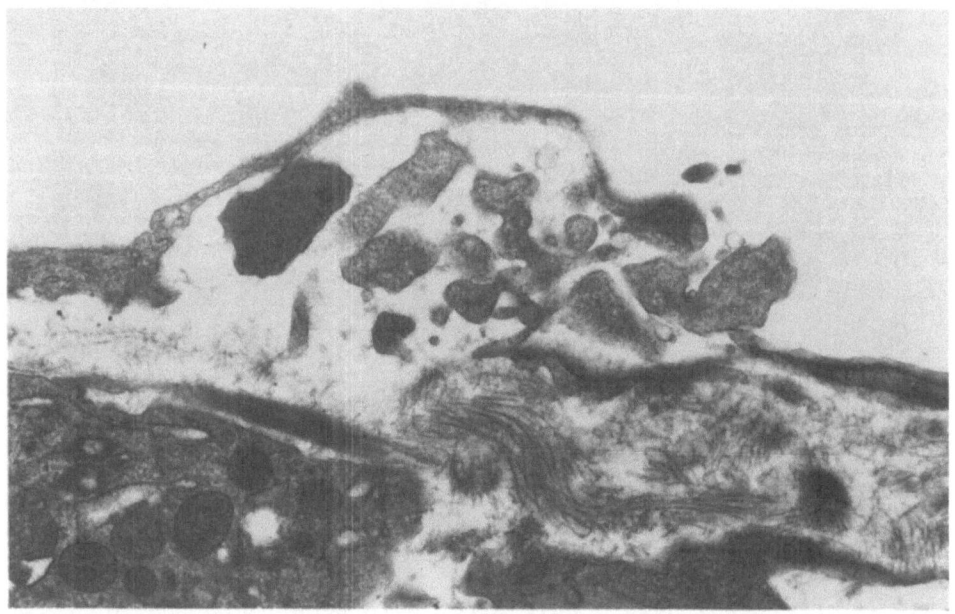

Fig. 4. Lymphatic wall showing fine cytoplasmic processes of a tumor cell protruding between entothelial cells (x 12,600).

TABLE I

The Effect of Trypsinization on Metastasis
from Footpad to Lymph Node

Tumor	Dose	Tumor in lymph node at 2-3 weeks
CBA mouse mammary	5×10^5 non-trypsinized	3 of 11
	3×10^5 trypsinized	9 of 11
C3H mouse mammary	4×10^5 non-trypsinized	0 of 9
	4×10^5 trypsinized	0 of 9

Mice were killed at 2-3 weeks when footpads were enlarged.

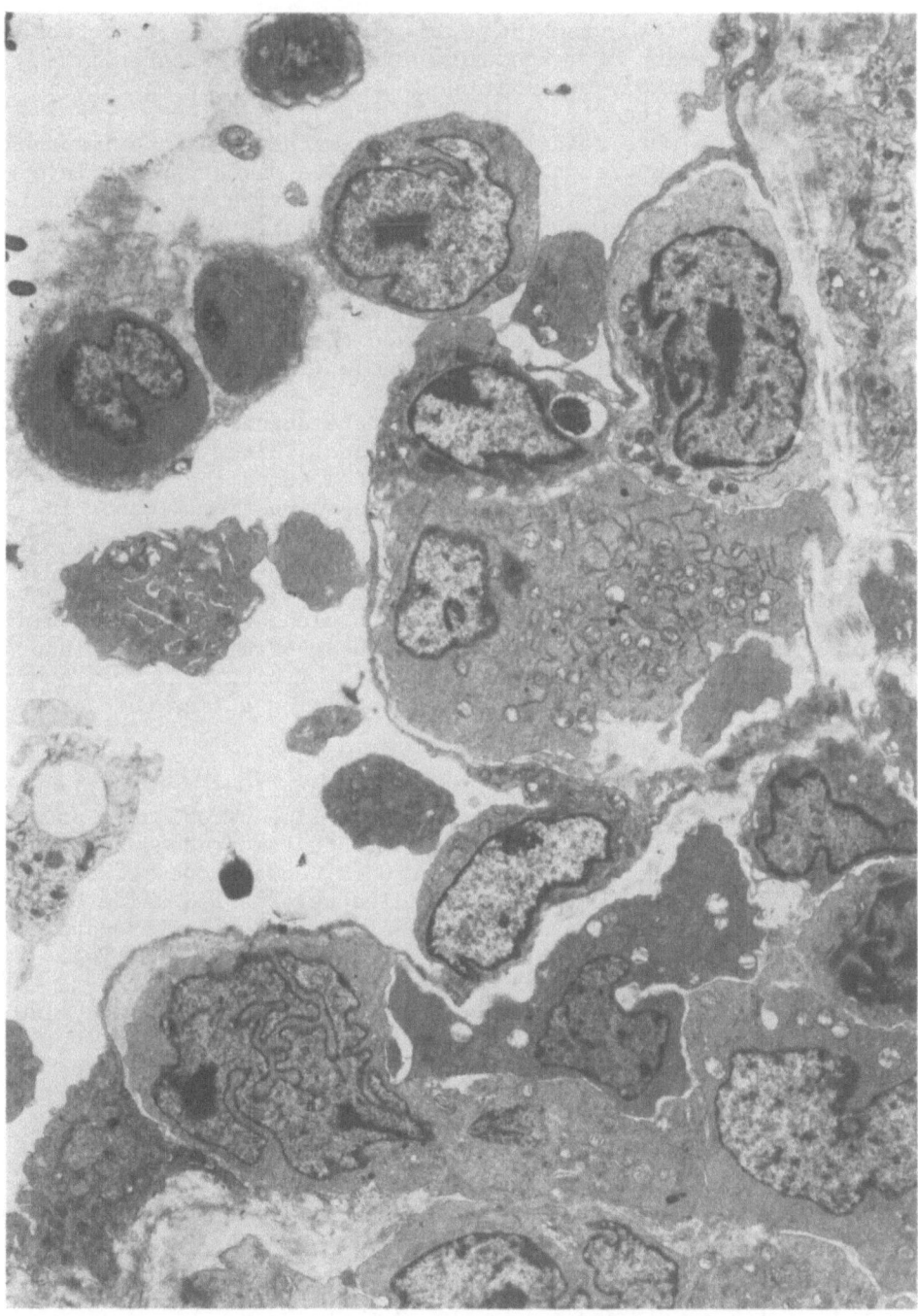

Fig. 5. Tumor cells passing between separated lymphatic endothelial cells into lymphatic vessel (x 4,300).

The results of the remaining experiments are related to the mechanism of invasion of lymphatics. If tumor cells enter lymphatics,as seems likely by individual active penetration between endothelial cells, then it is possible that disaggregated tumor cells will invade more readily. When the metastasis experiment was repeated with a spontaneous metastasizing CBA mammary tumor,injection of 5×10^5 non-trypsinized cells led to lymph node metastasis in 3 of 11 animals; when the cells were incubated in 0.2% trypsin in Hanks solution for 30 min, metastasis occurred in 9 of 11 animals. When the experiment was repeated with a non-metastasizing C3H mammary tumor, trypsinization did not however permit the non-metastasizing tumor to metastasize, even though it grew successfully on direct injection into the lymph node (Table I).

A search was subsequently made for the actual process of invasion of lymphatics as seen in the footpad. Since the process of metastasis was consistent, it could be assumed that all cells in transit across a lymphatic wall were passing from without in. However,of 500 Araldite blocks examined tumor cells invading lymphatic vessels were found in 6 blocks; these were serially sectioned. In the footpad neoplasm,elongated cells were common, notably at the periphery of the lesion. Such elongated cells frequently put out thick blunt processes in the neighborhood of both blood capillary and lymphatic capillary vessels. From these arose numerous long finger-like processes, some containing a fine core of 6 nm microfibrils.

Neoplastic cells were found alongside the lymphatic walls usually elongated along 1 axis and protruding fine cytoplasmic processes toward the vessel. Cytoplasmic processes pressed against and indented the endothelium and protruded between open endothelial junctions (Fig. 4). The collagen bundles were pushed aside but did not disappear over a wide area. Neoplastic cells lay apparently freely in the open gaps between endothelial cells (Fig. 5). Once within the lymphatic vessel the neoplastic cells reverted to a spherical shape. Sometimes the endothelial junctions were wide open and several cells could be seen migrating through; however, segmental disappearance of the lymphatic vessel wall could not be identified with the E.M. Over a stretch of 25 µm 8 tumor cells were identified migrating singly between endothelial cells. While the phenomenon was difficult to identify with the E.M. this is probably due to the low chance of a random section through a lymphatic hitting a tumor cell. Scrutiny of serial paraffin sections of a footpad tumor confirmed the presence of histological appearances consistent with those seen at E.M. at a rate of about 2/cu mm tissue. When, however,the incidence of such appearances in thick Araldite sections is compared with the incidence of true invasion as seen with the E.M. in thin sections, it is reasonable to suppose that only 1/10 of the appearances, which at light microscope level look like invasion are in fact invasion. An estimate of the incidence of a segment of a lymphatic which is actually being invaded might therefore be once in 5, cu mm of tumor tissue.

Macrophages and lymphocytes were seen migrating through the lymphatic wall in a similar way. It was on occasion impossible to identify an individual group of cytoplasmic processes as part of a tumor cell or macrophage. Serial sections usually settled the identity conclusively. The protrusion of cytoplasmic processes by tumor cells was occasionally accompanied by cell division in adjacent tumor cells. Simultaneous cell division and protrusion of processes was not identified.

Neither tumor cells nor macrophages were identified as passing into blood capillaries; occasional monocytes were identified passing through blood vessels. It seems likely that these were migrating out of the circulation. Numerous macrophages with well-developed processes were seen in the lymph, occasionally containing fragments of ingested tumor cell cytoplasm. The tumor cell seen in the subcapsular sinus of lymph nodes showed a circular profile once more.

The process of invasion of lymphatic vessels seems therefore to involve active migration of tumor cells between endothelial cells - reverse diapedesis. The direction of protrusion of cytoplasmic processes may determine the direction of movement of cells and may be related to the opening of previously closed endothelial cell junctions. No morphological evidence of secretion of toxins was seen.

DISCUSSION

Relatively few adequate models of lymph node metastasis had been presented at the point of first publication of this work (1). Additions to those quoted in that paper include the work of Hewitt, Kodama and Pollard (6,7,8). The present model has been shown to mimic well the natural process of detachment of cells, invasion of the lymphatic, embolization, survival and growth within the node, persistent recruitment from the primary site, passage to a tertiary site and ultimate production of lethal pulmonary metastasis.

The changes which occur in the draining lymph nodes include a follicular response, migration of cells from postcapillary venules, mitosis in paracortical areas, and mitosis in sinus macrophages. The latter response is interesting in that it has been clearly demonstrated in the present experiments to be tumor related and may be due to release of a soluble factor by the tumor which triggers the otherwise non-dividing sinus macrophages into mitosis. When the dose of cells is high (5×10^6) the host response is ineffectual, but when the dose is lower it seems very likely from the present experiments that metastasis occurs and then regresses. Destruction of tumor cells can probably occur in the lymph node but only when the burden is low - very approximately 2.5×10^2 cells in the node. The mechanism of this destruction and the mechanism of destruction of tumor after BCG injection are not certain but it seems likely that macrophages are involved. While clearly T cells are involved in rejection across a species barrier it is not certain that this is so within the same species.

These experiments showed that trypsinization improved the metastasizing properties of metastasizing tumors but did not permit non-metastasizing tumors to metastasize. This fits nicely with the electron microscopic demonstration that the cells passed between open junctions in lymphatic endothelium; protruding processes seemed to play an important part in this. Adjacent lymphatic junctions were tightly closed. There was no ultrastructural evidence of a secretory process, nor of tissue lysis at any distance from the tumor cells. It was however impossible to exclude the possibility that short acting lytic factors were being released. Since tumor cells penetrate the lymphatic vessel in the same manner as lymphocytes and macrophages it appears reasonable to suggest that the main property related to invasion of lymphatics in the present model is the motility of tumor cells.

ACKNOWLEDGEMENTS

We are grateful to the Cancer Research Campaign (Yorkshire Branch) for financial support, to Mr. P. Norris, Mr. S. Westby and Mr. P. Wood for much help and to the Journal of Pathology for permission to re-use illustrations.

REFERENCES

1. Carr, I. and McGinty, F., J. Path., 113 (1974) 85.
2. Carr, I., McGinty, F. and Norris, P., J. Path., in press, (1976).
3. Carr, I., McGinty, F., Potter, C. and Westby, S., Experientia, 30 (1974) 185.
4. Carr, I., Norris, P. and McGinty, F., Experientia, 31 (1975) 590.
5. Carr, I., Underwood, J.C.E., McGinty, F. and Wood, P., J. Path., 113 (1974) 175.
6. Hewitt, H.B., Blake, E., Brit. J. Cancer, 31 (1975) 25.
7. Kodama, T., Gotohda, E., Takeichi, N., Kuzumaki, N., Kobayashi, H., J. Nat. Cancer Inst., 52 (1974) 931.

8. Pollard, M., Luckert, P.H., J. Nat. Cancer Inst.,54 (1975)
    643.
9. Wood, P. and Carr, I., J. Path., 114 (1974) 85.

ADOPTIVE IMMUNOTHERAPY OF SPONTANEOUS LEUKEMIA-LYMPHOMA IN AKR MICE

M. M. BORTIN, R. L. TRUITT, W. C. ROSE, A. A. RIMM
and E. C. SALTZSTEIN
Mount Sinai Medical Center and the Medical College
of Wisconsin, Milwaukee, Wisconsin (USA)

Barnes, et al.(3) first proposed that the transplantation of
allogeneic bone marrow might serve as an aggressive antileukemic
treatment. They reasoned that following lethal irradiation, residual leukemia cells might be eliminated through a "reaction of
immunity" carried out by the transplanted cells. Mathé proposed
the term adoptive immunotherapy to describe this effect (10). New
insights in immunobiology have caused a resurgence of interest in
immunotherapy as an adjunct to currently available therapy of cancer. Many reasons exist which make it difficult to manipulate a
patient's own immune system so that it can become decisively effective against his own tumor. Transplantation of immunocompetent
cells from a normal individual to the tumor-bearing patient offers
an alternative immunotherapeutic approach. Unfortunately, transplanted immunocompetent cells also recognize and react against
histocompatibility antigens present on the normal tissues of the
host, causing potentially lethal graft-versus-host (GVH) disease,
the major complication of adoptive immunotherapy.

One approach to the use of adoptive immunotherapy while avoiding GVH disease was suggested by the work of Boranic (4). Boranic
transplanted immunocompetent cells from deliberately mismatched
donors to immunosuppressed leukemic mice. Following completion of
the desired graft-versus-leukemia (GVL) reaction, the effector cells
were killed to prevent the development of fatal GVH disease. Several hr later, in order to restore hematopoietic and lymphoid function, bone marrow from syngeneic donors was transplanted to the
"cured" hosts. He reasoned that "if there was a margin of time between eradication of the leukemia and irreparable damage to the host,
this interregnum should allow termination of the GVH reaction with-

out a recurrence of the leukemia" (5).

We used a similar treatment strategy in studies of a long-passage acute lymphoblastic leukemia (6) and spontaneous leukemia-lymphoma (SLL) in AKR mice (7). We now report the results of recent experiments using a temporary graft of H-2 mismatched cells as adoptive immunotherapy in combination with conventional chemoradiotherapy for the treatment of SLL, the experimental tumor which most closely simulated the refractory, "T cell" acute lymphoblastic leukemia-lymphoma of man (13). The results showed that adoptive immunotherapy, as an adjunct to chemoradiotherapy, significantly increased the survival rate and life span of leukemic AKR mice.

MATERIALS AND METHODS

Experimental design. A three-step treatment plan was employed in this study: a) AKR (H-2^k) mice bearing SLL were treated with chemoradiotherapy for immunosuppression and tumor cell reduction; b) to introduce the adoptive immunotherapeutic reaction against residual malignant cells, the mice were given bone marrow (BM) plus lymph node (LN) cells from H-2 mismatched DBA/2 (H-2^d) donors on day 8; c) to "rescue" the AKR hosts from GVH disease, the mismatched DBA/2 cells were killed after 6 days (day 14) with chemotherapy. Approximately 6 hr later, BM+LN cells from allogeneic but histocompatible RF (H-2^k) donors were administered to the AKR hosts to restore hematopoiesis. The schedules and doses for the 2 basic regimens tested are shown in Table I.

Three sets of tumor-bearing AKR mice were used as controls with each of the 2 treatment regimens. All treated control mice were given the same chemoradiotherapeutic treatments on days 0, 2, 8 and 14 as their corresponding experimental groups. "Chemoradiotherapy control" mice were given no cells on days 8 and 14; "syngeneic control" mice were given AKR BM+LN cells on days 8 and 14; and "rescue control" mice were given AKR BM+LN cells on day 8 and RF BM+LN cells on day 14. Four replicate experiments were done for each experimental group and 2 or 3 replicate experiments for each control group. Replicates were performed during separate weeks. Untreated AKR mice bearing SLL were included with each experimental and control group and the data have been combined with historical untreated control mice. Additional AKR mice bearing SLL received treatment similar to the experimental groups but the mice were not rescued from GVH disease (i.e. no treatment on day 14), or received the rescue treatments on day 17 instead of day 14 (i.e. a 9 day rather than a 6 day GVL treatment).

Mice. AKR/J, DBA.2J and RF/J mice were obtained from The Jackson Laboratory, Bar Harbor, Maine. AKR mice between 6 and 11 months of age were palpated once a week. Mice with splenomegaly

plus inguinal and axillary or brachial lymphadenopathy more than
3 times normal were randomized and entered into control or experimental groups.

TABLE I

Details of Treatment Plan for AKR Mice Bearing SLL

| | | | | Treatment* Given on Day | | | | | | |
| | 0** | | 2 | | 8 | | | 14 | | |
Treatment	Am-B	MeC	CY	TBR	BM	LN	CY	PCZ	BM	LN
Regimen A	10	16	175	300	10^7	2×10^6	150	100	10^7	5×10^5
Regimen B	10	16	175	200	10^7	4×10^6	150	100	10^7	5×10^5

*Abbreviations used: Am-B = amphotericin B; MeC = methyl cyclohexyl-
nitrosourea; CY = cyclophosphamide; TBR = total body radiation;
PCZ = procarbazine. Drug doses are expressed in mg/kg of body
weight, and TBR exposure is expressed in Roentgens.
**Day zero represents day of diagnosis.

Drugs and irradiation. AM-B, MeC and CY were provided by Drug
Research and Development, Chemotherapy Branch, National Cancer Insti-
tute. PCZ was the gift of Hoffman-LaRoche, Inc., Nutley, New Jersey.
All drugs were administered by intraperitoneal injection. TBR was
administered at a dose rate of approximately 43 R/min from a Picker
Vanguard X-ray therapy machine.

Cell suspensions. Femoral BM and mesenteric LN cells were
collected, processed into single cell suspensions and tested for
viability employing standard methods (8).

Mixed Leukocyte Culture (MLC) Assays. Reciprocal one-way MLC
assays (2) were performed to test 4 mice for chimerism (11) 134 to
155 days after treatment was completed.

Statistical analysis. Differences in survival were analyzed
by chi square analysis and the Wilcoxan rank test. Median survival
times (MST) and the survival rates at 60 and 90 days were calculated
for all mice entered into an experiment from the last day of treatment. The MST for untreated mice was calculated from the day of
diagnosis.

RESULTS

Untreated AKR Mice with Spontaneous Leukemia-Lymphoma. The median survival time (MST) for 363 untreated AKR concomitant and historical control mice diagnosed with SLL during the past year was 19.3 days (●, Fig. 1).   Only 2/363 mice (0.6%) survived 60 days after diagnosis.   Autopsy of untreated mice revealed the enlarged thymus, spleen and lymph nodes characteristic of SLL in AKR mice.

Chemoradiotherapy Controls. The survival data for AKR mice bearing SLL that were treated with the 2 chemoradiotherapy regimens (but without transplants) are shown in Table II.   Although the MST of 18 days for the chemoradiotherapy control mice treated with Regimen A (Group 2, Table I, and ■, Fig. 1) does not appear to be significantly different from the 19.3 days MST of the untreated controls, a remission of the disease was induced because the MST was 32 days when calculated from the day of diagnosis; a 66% increase over the life span of untreated mice.   Chemoradiotherapy using Regimen B (Group 5, Table II) was more effective than Regimen A as measured by the average duration of the remission.   Calculated from the day of diagnosis, Regimen B increased the life span by 164% over that of untreated mice.   Most of the early deaths in the chemoradiotherapy controls (Groups 1 and 5, Table II) were attributed to drug and/or radiation toxicity since most of the mice were leukemia-free on gross examination.

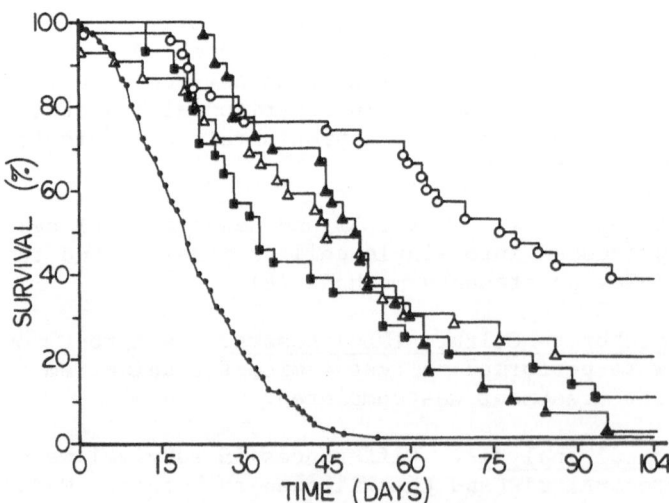

Fig. 1.   Survival curves for control and experimental AKR mice bearing SLL.   ●, untreated mice; ■, chemoradiotherapy controls (Group 1); ▲, syngeneic controls (Group 2); △, rescue controls (group 3); and O, experimental mice (Group 4).

TABLE II

Survival Data for Control and
Experimental AKR Mice Bearing SLL

Group Designation	Treatment Regimen	N	MST* (days)	Percent Surviving* (days)	
				60	90
1. Chemoradiotherapy Control	A	28	18	21	11
2. Syngeneic Control	A	30	39	13	3
3. Rescue Control	A	29	31	28	21
4. Experimental	A	38	64	53	40
5. Chemoradiotherapy Control	B	28	37	21	21
6. Syngeneic Control	B	30	40	33	30
7. Rescue Control	B	29	51	45	31
8. Experimental	B	38	71	55	29

*Calculated from last day of treatment (i.e. day 14).

Syngeneic Controls. The survival data for AKR mice bearing SLL treated with chemoradiotherapy plus syngeneic cells are shown in Table I (Groups 2 and 6). Some of the Regimen A chemoradiotherapy toxicity could be overcome by transplanting syngeneic BM+LN cells (cf. MST of Group 1 vs 2, Table II and the curves of ■ vs ▲ in the first 45 days, Fig. 1). The need for hematopoietic support was less evident in mice treated with the lower, less toxic dose of TBR in Regimen B (cf. MST of Group 5 vs 6, Table II).

Rescue Controls. The survival data for the rescue control mice are shown in Table II (Groups 3 and 7). The purpose of the rescue control groups was to evaluate a possible adoptive immunotherapeutic contribution of the RF cells. Cells from AKR donors were transplanted on day 8 (providing hematopoietic support, but presumably no GVL effect), and cells from RF donors were administered on day 14. Any improvement in survival of the rescue control groups over their counterpart syngeneic control groups (AKR cells on days 8 and 14) could be ascribed to an antileukemic (or other beneficial) effect of the RF cells. Cells from RF donors had no detectable antileukemic effect as measured by improved survival over syngeneic controls. There was no significant difference between rescue controls and syngeneic controls for Regimens A and B as measured by MST and proportion surviving 60 and 90 days. The survival pattern for Regimen A rescue control mice is shown in Fig. 1 (Δ).

Experimental Groups. A temporary graft of immunocompetent cells from mismatched DBA/2 donors was used for adoptive immunotherapy of SLL in AKR mice. Survival data for Regimens A and B are shown in Table II (Groups 4 and 8), and the survival pattern for mice treated with Regimen A is shown in Fig. 1 (0).

The 64 day MST for mice treated with Regimen A plus adoptive immunotherapy (Group 4, Table II) was significantly longer than the MST of untreated controls ($p < 0.001$), chemoradiotherapy controls ($p < 0.01$), syngeneic controls ($p < 0.01$), but not the rescue controls. The 53% and 40% survival rates at 60 and 90 days, respectively, were also superior to the 60 and 90 day survival rates of the control groups ($p < 0.06$ to 0.01) except for the rescue control at 90 days. In contrast, with the exception of the untreated controls ($p < 0.001$) and the chemoradiotherapy control mice surviving 60 days post-treatment ($p < 0.02$), the MST and survival rates for mice treated with Regimen B plus adoptive immunotherapy (Group 8, Table II) was not significantly different from their control groups.

Variations in Timing of Rescue Treatments. A total of 36 AKR mice bearing SLL (data not shown) received Regimen A treatments on days 0, 2 and 8, including DBA/2 BM+LN cells, but no treatments on day 14 (i.e. no rescue). The MST was only 28 days due to severe GVH disease. This control experiment demonstrated the need to treat the GVH disease or kill the graft of DBA/2 cells. The need to kill the DBA/2 cells in 6 days rather than 9 days was shown in a set of 8 experiments in which chemoradiotherapy regimens similar to Regimens A and B were employed, but in which the rescue treatments were given on day 17 rather than on day 14 (9 days rather than 6 days after transplantation of DBA/2 cells). All of these experiments (data not shown) were associated with unacceptably high mortality rates due to GVH disease.

MLC Results. In MLC tests, mitomycin-C treated spleen cells from long-term surviving experimental mice caused more than ten-fold stimulation of DNA synthesis in normal DBA/2 responder spleen cells but no stimulation of AKR or RF responder cells. Thus, the treatments applied on day 14 to "rescue" the AKR hosts from GVH disease appeared to have eliminated the DBA/2 cells from the spleens of tested mice. It was not possible to determine whether the survivors' spleen cells were of AKR or RF origin because AKR and RF cells are reciprocally non-reactive in MLC tests (11).

Incidence of Recurrent Leukemia. Almost all mice that died during the 14 day treatment, or within 14 days following completion of treatment were in remission and had no gross evidence of leukemia. For the purpose of this analysis we describe the incidence of recurrent leukemia in mice treated with transplants of BM+LN cells that lived more than 4 weeks following diagnosis and initiation of ther-

apy. Although significant prolongation of the survival time of
AKR mice was achieved by combining adoptive immunotherapy with
chemoradiotherapy, many of the mice in the experimental groups that
died more than 2 weeks after completion of therapy had gross and
histologic evidence of leukemia. The overall incidence of recurrent
leukemia in the transplanted control and experimental groups is presented in Table III. The lowest incidence of leukemic deaths (35%)
was found in mice from Group 4 that received Regimen A chemoradiotherapy plus adoptive immunotherapy. This was significantly lower
(p < 0.001) than the 91% incidence of recurrent leukemia which occurred in their syngeneic controls (Group 2, Table III). In contrast, there was no significant difference in the incidence of
leukemic deaths in the syngeneic control mice (Group 6) and experimental mice (Group 8) for Regimen B. Negative deaths were attributed
to infection, toxicity of the treatments, or GVH disease, and they
were evenly distributed throughout the 28 to 104 day period. It is
not now known whether the recurrent leukemia was in donor (RF) or
host (AKR) cells. The cytokinetic characteristics of AKR SLL suggest that deaths from leukemia occurring more than 60 days following
the end of treatment may represent viral re-induction of a new disease episode rather than recurrence of the original disease (12);
however, perturbation of the cytokinetic characteristics of SLL by
the treatments may have caused a prolonged remission in some mice.

TABLE III

Incidence of Recurrent Leukemia in
Control and Experimental Mice

Group*	Treatment Regimen	No. of Mice at Risk on Day 28**	Leukemic Deaths (%)***
2. Syngeneic Control	A	23	91
3. Rescue Control	A	21	48
4. Experimental	A	31	35
6. Syngeneic Control	B	26	62
7. Rescue Control	B	27	63
8. Experimental	B	35	54

*From Table II.
**Two weeks after completion of therapy
*** $\dfrac{\text{No. Leukemic deaths at 90 days post-treatment}}{\text{Mice at risk on day 28}} \times 100$

DISCUSSION

Although numerous investigators, including ourselves, have successfully applied adoptive immunotherapy to treat transplanted leukemias, there have been few successful reports using a spontaneous leukemia. Alexander (1) and Mathé, et al.(10) used adoptive immunotherapy to treat AKR SLL without success. Mathé, et al. treated mice at 6 months of age with C3H or C57BL/6 adult BM or fetal liver cells and reported mean survival times of 37 to 49 days after irradiation. Most of the mice died of GVH disease but free of leukemia. Truitt, et al.(14) treated germfree or decontaminated AKR mice bearing advanced SLL with 1000 R TBR and bone marrow from H-2 mismatched DBA/2 donors. The mice were maintained as germfree or decontaminated radiation chimeras throughout the study. In such an environment fatal secondary disease did not develop and 60% of the mice entered lived for 90 days (14). However, the success of their treatment was dependent upon rendering the AKR hosts germfree prior to clinical appearance of the malignancy. The same treatment procedures were not successful in conventional AKR mice.

We recently reported the first successful application of adoptive immunotherapy for SLL in conventional AKR mice (7). In those experiments, 37% of mice treated with a transient GVL reaction survived 60 days post-treatment. In the present studies, 53-55% 60-day survival and up to 40% 90-day survival was achieved. We believe that the current survival data surpass any previously reported for SLL in conventional AKR mice, employing any form of therapy, when treatment was not initiated until after clinical appearance of the disease.

Although numerous problems remain to be solved, the present experimental results in a most difficult model tumor system suggest that adoptive immunotherapy may prove useful as an adjunct to chemoradiotherapy for the treatment of advanced human leukemia that is refractory to "conventional" therapy.

In summary, adoptive immunotherapy in the form of a transient graft of mismatched DBA/2 bone marrow and lymph node cells was used in combination with chemoradiotherapy to treat AKR mice bearing advanced spontaneous leukemia-lymphoma. Leukemic mice treated in this manner had significant prolongation of their median survival times and significantly higher survival rates 60 and 90 days following the completion of treatment than untreated controls, chemoradiotherapy controls, or control mice that received chemoradiotherapy plus transplants of cells from syngeneic AKR donors. To our knowledge, the survival data are superior to those of any published report using any treatment modality initiated after clinical appearance of AKR spontaneous leukemia-lymphoma.

ACKNOWLEDGEMENTS

Supported by USPHS research contract No. N01-CB-33853 with the Division of Cancer Biology and Diagnosis, National Cancer Institute, the Patrick and Anna M. Cudahy Fund and the Board of Trustees, Mount Sinai Medical Center, Milwaukee, Wisconsin. R. L. Truitt is a Fellow of the Damon Runyon Memorial Fund for Cancer Research, Inc. We thank Dr. G. E. Rodey for performing the MLC tests. E. Allen, L. Crandell, C. Flanagan, E. Reynolds and B. Stephan provided expert technical assistance.

REFERENCES

1. Alexander, P., Cancer Res., 27 (1967) 2521.
2. Bach, F.H. and Voynow, N.K., Science, 155 (1966) 545.
3. Barnes, D.W.H., Corp, M.J., Loutit, J.F. and Neal, F.E., Brit. Med. J., 2 (1956) 626.
4. Boranić, M., J. Nat. Cancer Inst., 4 (1968) 421.
5. Boranić, M., Transplant. Proc., 3 (1971) 394.
6. Bortin, M.M., Rimm, A.A., Rodey, G.E.,Giller, R.H. and Saltzstein, E.C., Cancer Res., 34 (1974) 1851.
7. Bortin, M.M., Rose, W.C., Truitt, R.L., Rimm, A.A., Saltzstein, E.C. and Rodey, G.E., J. Nat. Cancer Inst., 55 (1975) 1227.
8. Bortin, M.M. and Saltzstein, E.C., Exp. Hematol., 10 (1966) 27.
9. Glucksberg, H. and Fefer, A., Cancer Res., 33 (1973) 859.
10. Mathé, G., Amiel, J.L., Schwarzenberg, L., Cattan, A. and Schneider, M., Cancer Res., 25 (1965) 1525.
11. Rodey, G.E., Bortin, M.M., Bach, F.H. and Rimm, A.A., Transplantation, 17 (1974) 84.
12. Schabel, F.M., Jr., Skipper, H.E., Trader, M.W., Laster, W.R., Jr. and Cheeks, J.B., Cancer Chemo. Rept., Part 2, 4 (1974) 53.
13. Senn, L. and Borella, L., New Eng. J. Med., 296 (1975) 828.
14. Truitt, R.L., Pollard, M. and Srivastava, K.K., Proc. Soc. Exp. Biol. Med., 146 (1974) 153.

SURFACE MORPHOLOGY OF LEUKEMIA VIRUS-INFECTED LYMPHOID CELLS

P. A. FARBER, S. SPECTER and H. FRIEDMAN

Albert Einstein Medical Center and Temple University
School of Dentistry and Medicine
Philadelphia, Pennsylvania (USA)

The malignant transformation of normal cells by tumor viruses, both in vivo and in vitro, is the subject of much current interest. Infection of mice with Friend Leukemia virus (FLV) produces marked changes to lymphoid cells which results in a generalized immunosuppression (3,10,11). Results of various studies in this and other laboratories indicate that the virus preferentially affects antibody precursor cells rather than antibody forming cells (6,8). Examination of spleen cells from FLV-infected mice with the transmission electron microscope (TEM), has shown the presence of C type particles as well as viruses budding from immature blastlike lymphoid cells (1,9). The present investigation deals with changes in the surface morphology of lymphoid cells which can be detected with the scanning electron microscope (SEM) in mice infected with FLV and correlates these with immunosuppression.

MATERIALS AND METHODS

Male Balb/c mice were infected with 0.2 ml of stock FLV homogenate containing 500 ID_{50} units of virus. The preparation contained both the spleen focus forming virus (SFFV) and lymphatic leukemia virus (LLV). No detectable lactic dehydrogenase or choriomeningitis viruses were present. This dose of virus resulted in a marked splenomegaly and most of the mice succumbed by 35 days post-infection. At various time intervals mice were killed, their spleen, bone marrow cells, thymus, and inguinal lymph nodes removed, and placed in cold Mc Coy's tissue culture medium. Thymus and lymph node cells were obtained by mincing with scissors. Spleen cells were purified by Ficoll-Hypaque centrifugation. The cells were

341

fixed with 1% glutaraldehyde in phosphate buffered saline (PBS) for
60 min and prepared for the SEM as described previously (4). The
numbers of antibody plaque forming cells (PFC) of mice immunized
with sheep red cells, both controls and FLV-infected, were determined by the standard localized hemolytic plaque assay in agar gel
(5).

RESULTS

Effect of FLV Infection on Splenocyte Cell Morphology

These results are similar to those reported previously (4).
Examination of spleen cells from normal Balb/c mice with the SEM
showed the expected spectrum of lymphoid cell types. The majority
of splenocytes had numerous villous projections. Lymphocytes with
smoother surfaces as well as cells with fewer villi were also observed (Fig. 1a). Large, irregular cells with ruffled surfaces
were evident in small numbers; these were considered to be macrophages. Within 5 days after FLV infection, marked changes in the
percentage of smooth and villous cell types became evident (Table
I). An increasing number of large, smooth-surfaced cells appeared
which were larger than the smooth-surfaced cells of uninfected
mice. This type of cell increased during the next few days so that
by 10 days post-infection they comprised nearly 30% of the spleen
cell population (Fig. 1b). By 17 days nearly 80% were of this
type (Fig. 1c) and few normal looking, villous lymphocytes were
present. By day 25 the splenocyte population was homogeneous,
comprised almost entirely of large smooth cells. Many of these
cells had blebs measuring 100 nM on their surface. These were
first noted on day 10 and cells bearing these surface modifications
increased in number as infection progressed. By day 30 degenerating
lymphocytes were often seen to coalesce and result in cell lysis
(Fig. 1d).

Lymph Node Cells, Bone Marrow and Thymus

In contrast to the spleen where changes in lymphocyte surface
morphology became evident within 5 days, lymph node cells manifested
few alterations until late in the infection. Both normal and infected lymph nodes had numerous villous-covered cells. The number
of smooth surfaced abnormal lymphocytes did not form an appreciable
percentage until 25 to 30 days post-infection. Even at this late
date, normal lymphocytes were still evident in significant numbers
(Fig. 2a).

TABLE I

Changes in the Spleen Cell Populations in Balb/c Mice
After Infection with Friend Leukemia Virus (FLV)

Time in days after infection*	PFC** per spleen (x10^3)	Percent cells*** staining for FLV	SEM topography****		
			S	I	V
0	97.5 $\pm$ 7.0	1.0	18	12	70
+5	12.5 $\pm$ 3.0	12.9	31	40	29
+10	4.5 $\pm$ 1.1	76.2	39	50	11
+15	1.1 $\pm$ 0.6	81.5	55	41	4
+20	0.6 $\pm$ 0.1	87.5	78	19	3
+25	0.5 $\pm$ 0.2	92.3	99	0	1
+30	0.1	95.0	98	1	1

*Groups of mice infected by intraperitoneal injection of 500 ID$_{50}$ of FLV on day indicated before assay; 5-10 mice tested at each time interval.
**Average antibody response for 5-10 mice 4 days after intraperitoneal immunization with 4 x 10^8 SRBC.
***Average percent of spleen cells positive for FLV antigen when stained with fluorescein-labelled antiserum.
****S = smooth surfaced lymphocyte; I = cells with an intermediate topography; V = cells with villi. Numbers indicate percent of splenic lymphocytes.

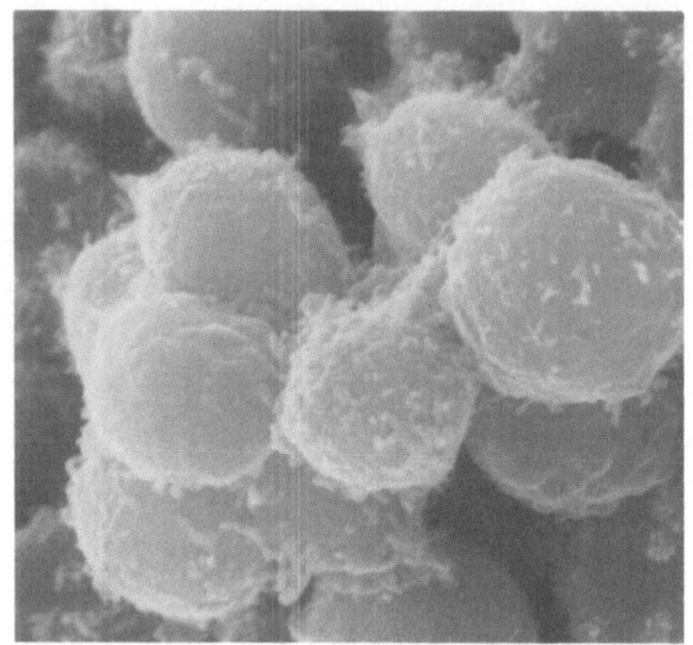

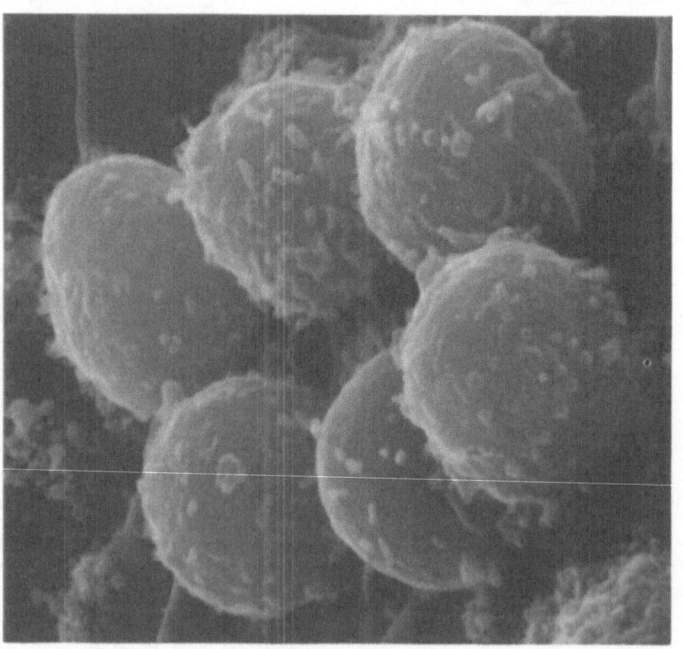

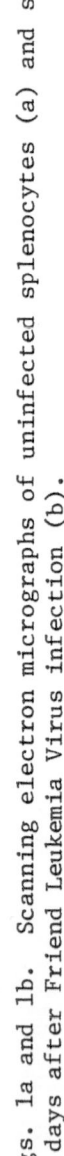

Figs. 1a and 1b. Scanning electron micrographs of uninfected splenocytes (a) and splenocytes 10 days after Friend Leukemia Virus infection (b).

a

b

c

d

Figs. 1c and 1d. Scanning electron micrographs of splenocytes 17 days (c) and 30 days (d) after Friend Leukemia Virus infection.

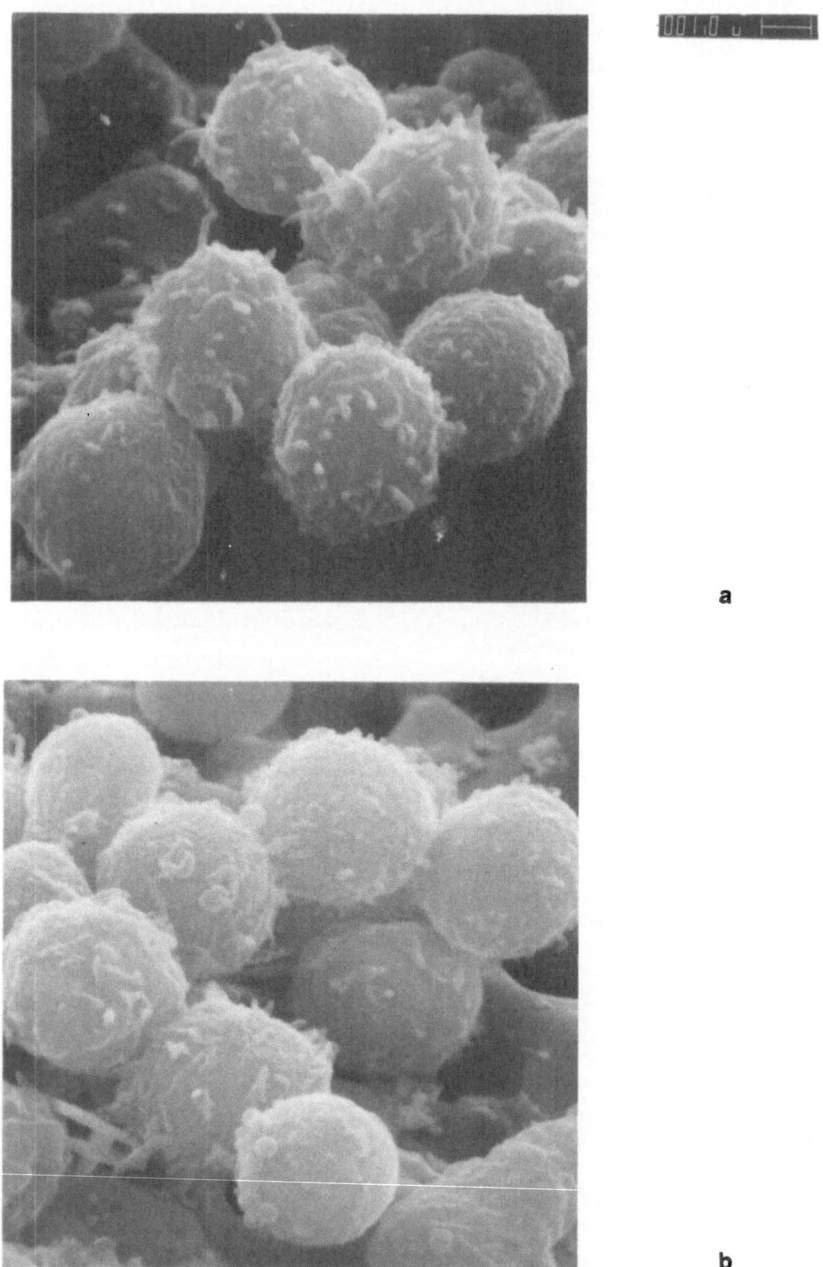

Figs. 2a and 2b. Scanning electron micrographs of uninfected lymph
node cells (a) and lymph node cells 30 days after Friend Leukemia
Virus infection (b).

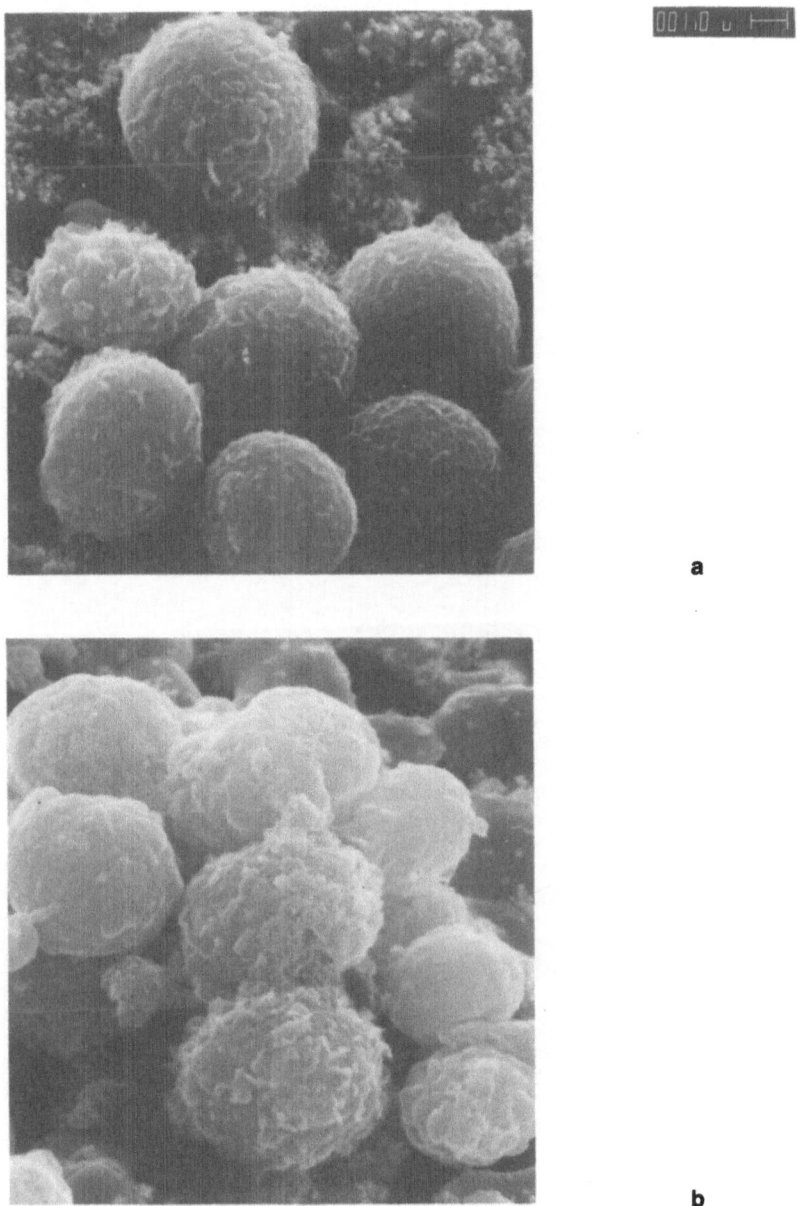

Figs. 3a and 3b.   Scanning electron micrographs of uninfected bone marrow cells (a) and bone marrow cells 30 days after Friend Leukemia Virus infection (b).

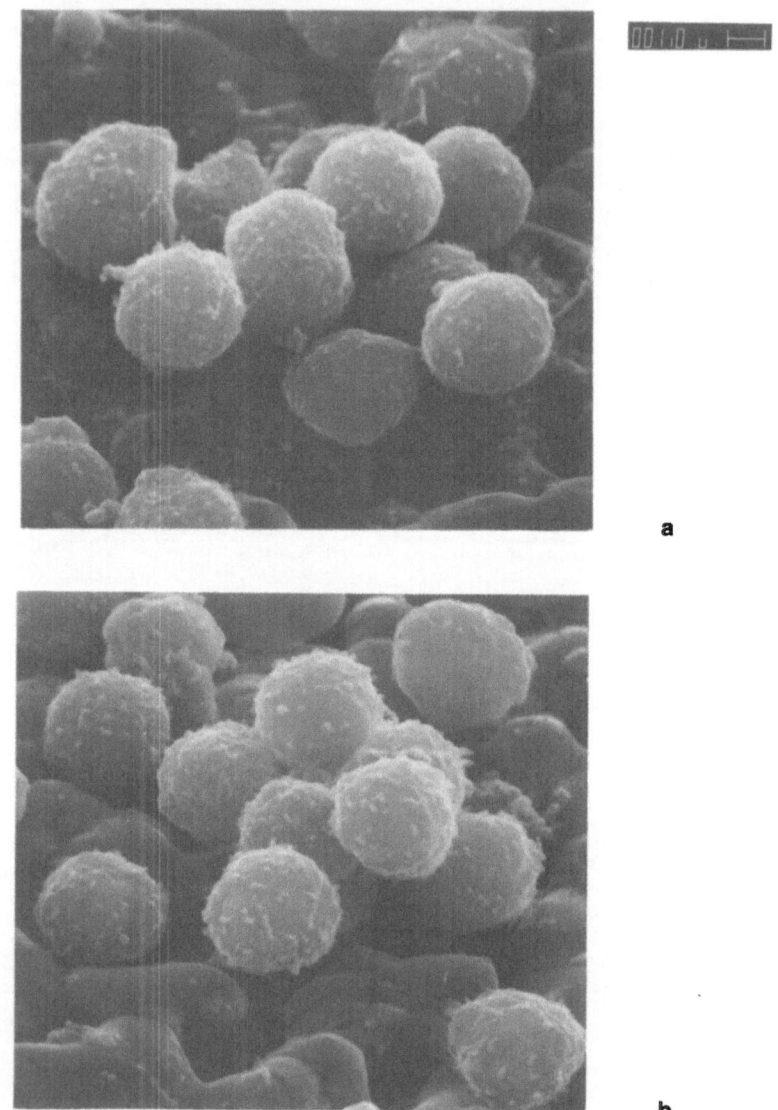

Figs. 4a and 4b. Scanning electron micrographs of uninfected thymus cells (a) and thymus cells 30 days after Friend Leukemia Virus in-fection (b).

   Both bone marrow (Figs. 3a and 3b) and thymus cell (Figs. 4a
and 4b) populations showed the least changes as a result of FLV infection. Few abnormal large cells with surface blebs or pores which
characterized the spleen (Figs. 1a-d) and lymph nodes (Figs. 2a and
2b) of FLV-infected mice were seen. Normal bone marrow cells were
smoother and larger than splenocytes with fewer and shorter villi.
In many respects these normal cells resembled the smoother cells
which characterized the FLV-infected spleen. Both cell-types seemed
to be blast cells, one apparently occurring naturally and the other
as a result of virus infection. There was little indication that
the cells in the bone marrow were changed by FLV infection. Similarly, the thymus cell surface morphology was not much different
than that of normal thymocytes. Thymocytes of normal mice were
smoother than splenic or lymph node lymphocytes with an occasional
villous-covered cell.

                          Immunosuppression

   As can be seen in Table I there was an 80% reduction in the
number of splenic antibody producing cells by 5 days. An even
greater suppression became evident as the infection progressed and
splenomegaly increased. On the other hand the number of PFC's produced by lymph node cells (not shown) is not significantly reduced
until much later in the infection.

                             DISCUSSION

   The results of the present study with the SEM are consistent
with, and extend results of earlier studies with the TEM regarding
ultrastructural changes in the spleen of FLV-infected mice (1,9).
The results of the present investigation with the SEM provided further insight regarding: a) morphologic surface alterations of lymphoid cells during FLV infection, and b) the relationship of these
changes to depressed immunologic responsiveness.

   The SEM has been used for the study of FLV-infected cells by
de Harven and Friend (2). They demonstrated that a continuous line
of FLV-transformed mouse cells was homogeneous in appearance with
surface blebs or knobs which were thought to represent budding
virus. Similar surface modifications were seen in this study.

   The most marked changes were evident in the spleen, where a
rapid reduction in the percent of villous-covered lymphocytes occurred after infection. A larger, smoother cell type appeared and
increased in number as infection progressed. These cells are
thought to represent virus-transformed lymphocytes and do not possess normal lymphocyte markers such as surface Ig or theta antigen,

but do have FLV-associated antigens. As infection progressed holes
developed in some of these cells which eventually underwent lysis.
The pores in the cell membrane may represent the point at which the
virus buds have pinched off. Alternatively, they may represent cells
with membrane damage as a result of immune lysis caused by antiviral, complement-fixing antibodies or the generation of cytotoxic
lymphocytes in other lymphoid organs. The alteration in the surface morphology of cells in the lymph nodes, which are considered
a secondary site for FLV replication, progressed at a slower rate.
No significant enlargement of lymph nodes occurred until late in
the disease. Some smooth-surfaced cells with blebs were noted later
in the infection but did not comprise more than 20% of the lymphocytes examined. The bone marrow and thymus cell populations showed
even less evidence of surface modifications. Even though fluorescent
staining has shown that up to 30% of bone marrow cells were infected
(7), SEM examination revealed only slight changes. Similar findings
were obtained with the thymus, however less than 10% of these cells
have virus-associated antigens.

REFERENCES

1. Chan, G., Rancourt, M.W., Ceglowski, W.S. and Friedman, H.
    Science, 159 (1968) 437.
2. de Harven, E., Lampen, N. and Sato, T., Virology, 51 (1973) 240.
3. Dent, P., Prog. Med. Virol., 12 (1972) 1.
4. Farber, P., Specter, S. and Friedman, H., Science, 190 (1975)
    469.
5. Friedman, H. and Ceglowski, W.S., Prog. Virol., 1 (1971) 815.
6. Friedman, H. and Ceglowski, W.S., Virus Tumorigenesis and
    Immunogenesis, Academic Press, New York, New York (1973) 299.
7. Friedman, H. and Kately, J.R., Amer. J. Clin. Path., 63 (1975)
    735.
8. Kately, J.R., Kamo, I., Kaplan, G. and Friedman, H., J. Nat.
    Cancer Inst., 53 (1974) 1371.
9. Koo, G.G., Ceglowski, W.S. and Friedman, H., J. Immunol., 106
    (1971) 799.
10. Notkins, A.L., Mergenhagne, S.E. and Howard, R.J., Ann. Rev.
    Microbiol., 24 (1970) 525.

HOST IMMUNOREACTIVITY TO CSA, A TISSUE-SPECIFIC ANTIGEN OF NORMAL

AND NEOPLASTIC HUMAN INTESTINE

D. M. GOLDENBERG and K. D. PANT

University of Kentucky
Lexington, Kentucky (USA)

The GW-39 tumor system consists of a human signet-ring cell carcinoma of the human colon which is serially transplantable in normal, unconditioned, adult golden hamsters either in the cheek pouch or intramuscularly (5). The microscopic morphology of GW-39, after staining with PAS, indicates that intracytoplasmic mucin is being produced by the tumor. Signet-ring cells are also prevelant. This tumor system has now been maintained by us in hamsters for about 10 yr, retaining human species-specific and organ-specific properties. For example, the few cells amenable to karyotyping have revealed a human pseudodiploid karyotype. As we reported earlier (3), one of the more interesting properties of GW-39 cells, which is consistent with their human origin, is their capacity to synthesize the carcinoembryonic antigen (CEA) of Gold and Freedman (2). Since this human colonic carcinoma retained a tumor-associated glycoprotein such as CEA even after long-term propagation in animal hosts, we reasoned that other human and perhaps organ-associated substances could likewise be detectable in this tumor model. The purpose of this paper is to describe the extraction and identification of an antigen, CSA, present in normal and neoplastic digestive tract tissue.

MATERIALS AND METHODS

Using a phenol-alcohol extraction procedure similar to that employed for blood group-specific glycoproteins (6), we were able to extract antigens distinct from CEA and other substances related to CSA. Of these, at least one appeared to be specific for human digestive tract mucosa, both normal and malignant (4).

The glycoprotein antigens were extracted in 88% phenol. The phenol was removed with ethanol and the antigens were separated by gel filtration. The phenol-alcohol prepared material from GW-39 tumors was used to prepare antisera in normal, adult hamsters. These antisera, following crossabsorption with normal human kidney, liver, spleen, and lung, as well as with similar hamster organs, reacted only with human colon and GW-39. The specific antigen in this system was called CSA.

In an attempt to characterize the molecular weight varieties of CSA, we obtained gel filtration profiles of phenol-alcohol extracts of GW-39 or normal human colon on Sephadex G-200 columns which were calibrated with radiolabelled CEA and other markers.

RESULTS AND DISCUSSION

As was previously reported (4), counterelectrophoresis patterns obtained by reacting hamster anti-CSA antiserum (GW-39 source) against extracts of human and hamster colon as well as GW-39 tumor revealed precipitin bands. However, no reaction was observed with human spleen, kidney, liver, and lung, or with either similar hamster organs or normal hamster serum when substituted for hamster antiserum. As also shown in that study (4), the antibody to CSA produced in hamsters did not show any reactivity in immunodiffusion to other tumor antigens tested, such as CEA, CCA-II, or CCA-III. No major blood group reactivity was detected for CSA.

Using our anti-CSA antiserum in counterelectrophoresis to detect antigen activity, the gel filtration profile of GW-39 CSA showed antigen activity in the 46,000 and 170,000 molecular weight ranges, thus indicating low molecular weight (LMW) and high molecular weight (HMW) varieties of CSA. A quite similar profile was obtained when normal human colon was chromatographed.

The LMW and HMW varieties of CSA were similar to their respective antigens in GW-39 and normal human colon, as shown in Fig. 1, where anti-CSA was reacted against HMW/CSA from normal human colon in well 1 and from GW-39 in well 2, and against LMW CSA from normal human colon in well 3 and from GW-39 in well 4. Although there was crossreactivity between HMW and LMW varieties of CSA from the same preparation, there seemed to be an antigenic distinction indicated by the spurring obtained when reacting the 2 varieties of CSA with the anti-CSA antibody. The immunoelectrophoretic patterns of HMW and LMW varieties of CSA, as compared to pure CEA, confirmed that there was a difference between these preparations. As shown in Fig. 2, the HMW is in the upper well, anti-CSA in the trough beneath, LMW in the middle well, and monospecific anti-CEA antibody in the lower trough, with pure CEA in the lowest well.

Both variants of CSA appeared to be more electronegative in charge than CEA.

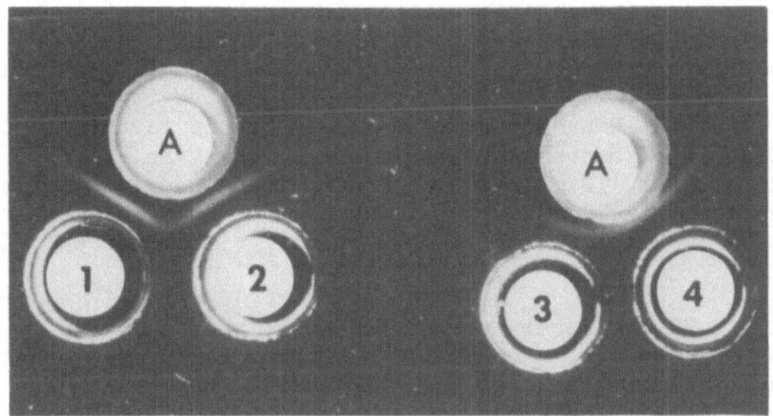

Fig. 1. Immunodiffusion patterns of high molecular weight (HMW) CSA from normal human colon (well 1) and GW-39 tumor (well 2) and low molecular weight (LMW) CSA from normal human colon (well 3) and GW-39 tumor (well 4) reacted against anti-CSA antiserum produced in hamsters (A), indicating reactions of complete identity.

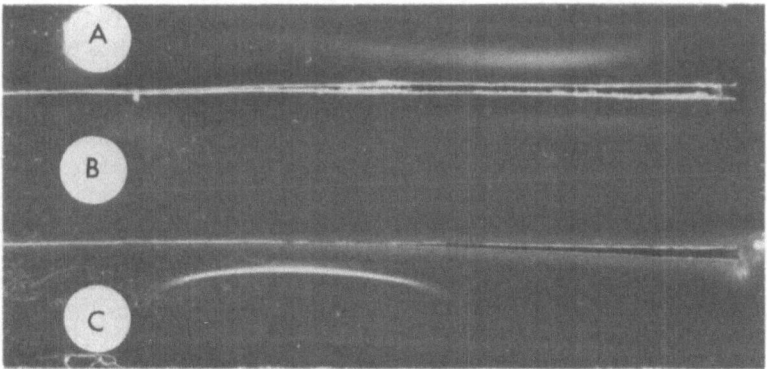

Fig. 2. Diagram of immunoelectrophoresis patterns obtained with high molecular (HMW) CSA in upper well, anti-CSA antiserum in the trough beneath, low molecular weight (LMW) CSA in the middle well, anti-CEA antiserum in the lower trough, and pure carcinoembryonic antigen (CEA) in the lowest well.

There seemed to be a difference in immunological specificity, however, between both variants of CSA. The LMW antigen did not appear to be as restricted to the digestive tract as the HMW antigen, since immunoreactivity to the former has been found in the mucoproteins of other organs, such as in normal cervix, in saliva, and in hamster colon. However, it has not been detected in any human cancers other than those of the digestive tract. On the other hand, the HMW variant of CSA seems to be restricted to the digestive tract mucosa, increasing in activity as one descends the digestive tract. Similarly, only normal and neoplastic human digestive tract tissues have been found to contain this HMW variant of CSA by immunodiffusion, thus supporting its organ-specific nature.

The cellular localization of CSA has been examined by means of indirect immunofluorescence of GW-39 tumor cells, and a bright continuous fluorescence on the cell surface was seen. The cell surface nature of this antigen encouraged us to consider whether we could detect any host immunoreactivity to CSA. Being present in both normal and neoplastic human colon, we thought that this might be the antigen described in cases of ulcerative colitis in which anti-colon antibodies have been demonstrated (1). To date, we have failed to document anti-CSA antibodies in the sera of patients with ulcerative colitis by immunodiffusion techniques. Likewise, we have been unable to consistently show anti-CSA antibodies or even circulating CSA in patients with colonic cancer. But we are aware that our testing methods are still too insensitive and are therefore attempting to develop a radioimmunoassay for CSA.

In a previous report (4), we demonstrated that in hamsters (Groups I and II), however, immunization with CSA led to failure of cheek pouch grafts to grow and tumor rejection by these animals correlated with their anti-CSA antibody titers. It was shown that the number of hamsters with tumors was inversely proportional to the precipitin band strength of the animals' serum against CSA. On the other hand, hamsters accepting the tumor grafts failed to show any increase in antibody levels. In the same study (4), hamsters (Groups III and IV) which received only Freund's adjuvant showed tumor takes and no serum antibody activity against CSA. Those results were of course far from suggesting that CSA may be of value in immunizing against human colonic cancer, since we were dealing with a xenogeneic model and the hamsters were recognizing a foreign antigen. It is for this reason that we are repeating these experiments with other control antigen preparations from human organs, just to ascertain whether the immunization is only specific for products of the GW-39 tumor or of normal human colon.

It is interesting, in retrospect, that we came upon the identification of organ-associated substances of the human digestive system by studying the antigenic components of a serially transplant-able human colonic carcinoma. Generally, it has been difficult to demonstrate the presence of any truly organ- or type-specific anti-gens in normal or neoplastic tissues, either in animals or in man. Hence, we are encouraged to believe that uniform and reproducible tumor systems of the kind described may reveal the presence of truly tumor-distinct antigenic markers. A discussion of the role of these substances in human cancer, or the possible response or lack of response of the RES to such substances, is obviously pre-mature until these various antigens have been better defined and characterized.

In summary, the extraction and identification of an antigen present in normal and neoplastic digestive tract tissue, CSA, is described. CSA was originally isolated from a human colonic car-cinoma serially propagated in hamsters and was found, after suitable absorptions, to contain antigenic substances crossreactive with similar material present in the human digestive tract. By means of gel filtration, 2 varieties of CSA have been identified, one of 46,000 (LMW) and the other of 170,000 (HMW). The HMW variety of CSA appeared to be truly distinctive of human digestive tract tis-sues, whereas the LMW form also had crossreactivity with mucopro-teins of normal cervix, saliva, and hamster colon. The digestive tract-specific CSA appeared to increase in activity descending from the esophagus to the rectum. CSA did not show any reactivity in immunodiffusion with other tumor-associated antigens tested, such as carcinoembryonic antigen, colon carcinoma antigen-II, and colon carcinoma antigen-III. Likewise, no major blood group reac-tivity was detected for CSA. Based upon preliminary immunization experiments of hamsters with CSA, it was found that this antigen could prevent subsequent challenge with GW-39 tumor cells, and that the number of hamsters with tumor "takes" was inversely proportion-al to the precipitin band strength of the hamster serum against CSA. Whether or not this finding is due to CSA or histocompatibil-ity antigens has not as yet been resolved.

ACKNOWLEDGEMENTS

This investigation was supported by Public Health Service Grant CA-15799 from the National Cancer Institute through the National Large Bowel Cancer Project.

REFERENCES

1. Broberger, O. and Perlmann, P., J. Exp. Med., 110 (1959) 657.
2. Gold, P. and Freedman, S.O., J. Exp. Med., 122 (1965) 467.
3. Goldenberg, D.M. and Hansen, H.J., Science, 175 (1972) 117.

4. Goldenberg, D.M., Pegram, C.A. and Vazquez, J.J., J. Immunol.,
    114 (1975) 1008.
5. Goldenberg, D.M., Witte, S. and Elster, K., Transplantation,
    4 (1966) 760.
6. Morgan, W.T.J. and King, H.K., Biochem. J., 37 (1943) 640.

TISSUE POLYPEPTIDE ANTIGEN (TPA) IN HUMAN CANCER DEFENSE RESPONSES

B. BJÖRKLUND, V. BJÖRKLUND, R. LUNDSTRÖM and G. EKLUND

National Bacteriological Laboratory, Central Hospital
and University of Stockholm, Stockholm (Sweden)

When cells multiply during embryonic development to regenerate damaged tissue, to form tumors or in certain cell culture conditions large amounts of new membrane protein has to be synthesized. This process involves activation of phase-specific genes which carry the code for the required amino acid sequences of the unit monomers.

The concept of phase variation is closely related to earlier thinking in the field of microbial antigens (13). For the student of cancer cells, it is of interest that bacteria in their logarithmic growth phase generally exhibit high nucleic acid protein ratio and loss of specialized functions in contrast to their resting phase when they have a low nucleic acid/protein ratio and carry out specialized functions such as producing polysaccharides or storing glycogen. Even though the mechanisms by which phase-specific genes are activated by the microenvironment are poorly understood, it is a reasonable conceptual approach to assume that mammalian cells during their growth phase undergo antigenic phase-specific variation which in principle resembles that of microorganisms (4). Herbert (13) pointed out that "there are few characteristics of microorganisms which are so directly and so markedly affected by the environment as their chemical composition". He also stated that the chemical composition of a cell is primarily dependent on the rate at which it is growing. Similar viewpoints have been expressed by Lacey (16) who listed 137 published instances of so-called "non-generic variation of surface antigens with environment". Anderson and Coggin (2) and Holleman and Palmer (14) have lucidly elaborated the concept of phase variation in relation to cancer and they coined the expression "phase-specific tumor-associated autoantigens".

A number of antigens have been described which may be regarded
as phase-specific. Most well-known are carcinoembryonic antigen or
CEA (12) and alphafetoprotein (1).

Another antigen of phase-specific nature has been described in
several reports by Bjorklund, et al. (5,6,7,9).This non-species-
specific antigen, which was designated tissue polypeptide antigen
or TPA, is present mainly in placenta, cancer cells in vivo and in
vitro and in regenerative conditions. Increased TPA levels can be
demonstrated in serum of most cancer patients. Successful surgery
or chemotherapy usually results in a lowering of serum TPA. Antibodies to TPA levels are present in man under certain conditions.

Some of the most pertinent findings on TPA will be described
here in conjunction with data on assay technique, occurrence, properties and clinical behavior of TPA. The observations will be
discussed in the light of phase variation.

MATERIALS AND METHODS

Assay Technique

For the determination of TPA a hemagglutination-inhibition
procedure was used based upon the ability of TPA in a specimen to
neutralize anti-TPA horse antibodies, which have been adjusted to
agglutinate 98-100% of sheep red blood cells labelled with TPA (8,
10,11).

The red cells, which were treated with tannic acid (20 μg/ml)
prior to labelling, were standardized to 160×10^6 cells/ml. About
6 U TPA/ml mixed with albumin was used for the labelling which was
carried out in the cold for 10 min.

Disposable microtitration plates with U-bottom wells were used
(Linbro IS-MRC-96). Testing was performed on minute quantities
(approx. 0.2 ml serum or 0.05 ml urine) and usually required 1-2
days, but could be performed in 6-7 hr. All serum samples were
absorbed overnight with a mixture of washed sheep red cells (packed)
and albumin-labelled red cells.

All plates were recorded photographically and readings were
made from Polaroid pictures by comparing the inhibition patterns
with those of identically titrated standards of TPA.

The reproducibility of the TPA-determination technique was
evaluated (Table I). The assay was reproducible, simple, TPA-
specific and suitable for large scale operation.

TABLE I

Deviations in the Determination of Polypeptide
Antigen Concentration (Two-step Titration Applied)

Deviation (No. of Steps)	Different recorders (Same technician) n = 253	Different technicians (Same recorder) n = 106
0	85%	93%
1	14%	7%
2	1%	0%
	100%	100%

RESULTS

Occurrence

The presence of TPA in various normal and malignant tissues
has been studied by different methods. More than 90% of the originally studied cancer tumors at various sites were found to contain substantial amounts of TPA. Later studies indicated that all
active tumors contained TPA and that earlier exceptions were due
to degradation during prolonged storage. Of particular interest
was the presence of TPA in all 75 placentas investigated (8).

TPA can be detected in the nutrient medium of cultures of
cancer cell lines such as HeLa, Detroit-6 and Hep-2. On the other
hand, WI-38 was never found to produce any detectable TPA. Fig.
1 illustrates the release of TPA in cultures of HeLa and KB cells
as compared to WI-38.

Gel filtration of human serum with Sephadex G-200 demonstrated
TPA close to albumin (Fig. 2).

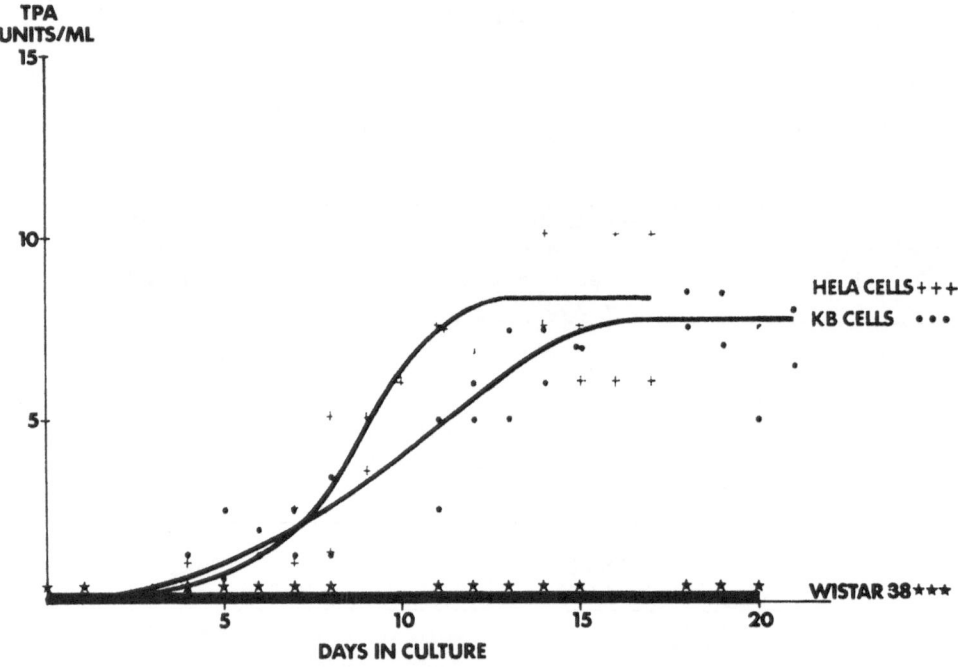

Fig. 1.   Release of TPA to medium in flask (Falcon 3012) cultures
of HeLa and Detroit-6 cell lines as compared with diploid WI-38
cell strain.   A suspension of 5 x 10^4 cells/ml was inoculated in
80% Eagle's MEM medium supplemented with 2 mM L-glutamine and 20%
serum.   Duplicate flasks were set for harvesting at day 0, 2, 5,
7, 9, 12, 14, 16, 19 and 26.   Medium without cells was used as
control.

        It is interesting that the antigenic specificity of TPA, re-
quired for complete hemagglutination-inhibition, has been found
also in a wide variety of other species.   Table II illustrates the
presence of TPA in random newborn calf, rat, moose, fallow-deer,
roe deer, rabbit, hare and wild boar.   Specific antibodies to TPA
were demonstrated in sera from donkey, moose and wapiti.   Table
III shows numbers and percentages of individuals with TPA in 7
different species.   It is noteworthy that horse, which was found to
be particularly suitable for immunization with TPA, in all 15 cases
showed the lowest levels of TPA.   In contrast, mice, hamsters and
rats exhibited high values of TPA in their serum.

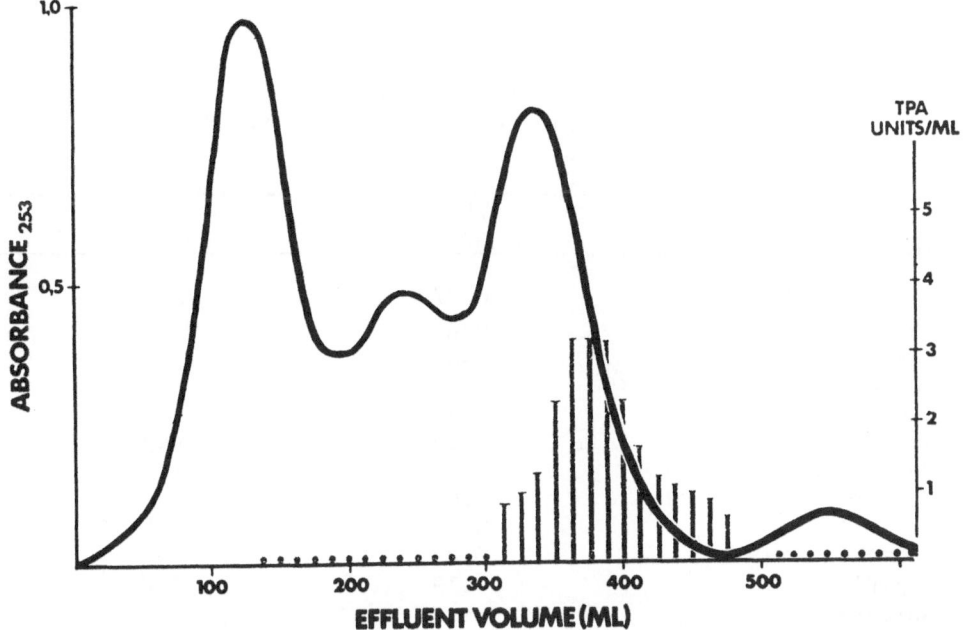

Fig. 2. Column fractionation of TPA-containing serum by gel filtration with Sephadex G-200.

Properties

The activity of TPA was attributed to a spectrum of aggregates of varying sizes representing a mixture of different oligomer states of a monomer. A state of equilibrium was observed. TPA was soluble at pH below 4 and it was irreversibly inactivated above the isoelectric point (pI=about 4.5). Tryptophan and cysteine were not present. TPA was neither inactivated by Koshland's reagent, performic acid nor by reduction or carboxymethylation. The specific antigenic activity seemed to be bound to a peptide without disulfide bonds of M_W approximately 22-23.000. Typical properties are absorption at 230 nm and fluorescence at 350 nm.

TABLE II

Occurrence of TPA and Anti-TPA in Serum
from Random Individuals of Various Species

Species	TPA U/ml	Anti-TPA
Newborn calf	0.37	neg
Horse	neg	
Goat	neg	
Sheep	neg	
Cat	neg	
Rat	1.50	neg
Monkey (Cercopidus)	0.09	
Monkey (Cynomolgus)	0.09	
Moose 1 - 73	0.37	1:4
Moose 2 - 49	0.09	neg
Wapiti	0.19	1:8
Reindeer	neg	
Fallow deer	0.75	neg
Roe deer	0.19	neg
Wisent	0.09	neg
Fox	neg	
Dog	neg	
Rabbit	1.50	neg
Hare 1	0.19	neg
Hare 2	3.00	neg
Pig	neg	
Wild boar	3.00	neg
Hen	0.09	neg
Donkey		1:8

TPA was found to be readily immunogenic. Circulating anti-bodies of the IgG type were formed and hemagglutination titers of 1:20.000-60.000 could easily be obtained in horses and most rabbits.

Studies on the incorporation of tritiated thymidine in TPA-immune horse lymphocytes in vitro have shown that the addition of TPA in autologous pre-immune serum causes significantly increased uptake of thymidine with a maximum on day 5.

TABLE III

Occurrence of Serum TPA in Various Species

Species	Number and Percentage of Individuals with TPA (U/ml)				Total Number of Individuals Tested
	<0.03 - 0.08	0.09 - 0.3	0.31 - 0.99	≥1.0	
Rabbit	13 (86.5)	1 (6.5)	1 (6.5)	0 (0)	15
Guinea pig	9 (60)	2 (13)	4 (26.5)	0 (0)	15
Mouse (ASN)	0 (0)	0 (0)	1 (5)	19 (95)	20
Golden Hamster	0 (0)	0 (0)	1 (6.5)	14 (93)	15
Monkey (Cercopidus)	5 (38.5)	5 (38.5)	1 (7.5)	2 (15.5)	13
Rat (Davby)	0 (0)	0 (0)	4 (44.5)	5 (55.5)	9
Horse	15 (100)	0 (0)	0 (0)	0 (0)	15

TABLE IV

The Percentage of Individuals (Serum Samples)
with Antigenemia within Different Groups in
the Total Study of 3,010 Analyses*

Group	% Positive Sera
1. Metastasized cancer	73 ± 10**
2. Untreated primary cancer without known metastases and cancer not radically treated	54 ± 10
3. Cancer radically treated	12 ± 7
4. Malignant disease other than carcinoma	26 ± 12
5. Other diseases	13 ± 2
6+7. Healthy adults and blood donors	2 ± 1
8. All children 10 yr and younger	6 ± 5
9. Umbilical cord blood	18 ± 10
10. Post partem mothers	17 ± 14
11. Pregnant women	3 ± 3
12. Cancer in situ	– –

*From Björklund, et al. (10)

**Standard error $= \sqrt{\dfrac{P\ (100 - P)}{n}}$

P = percent positive
n = number of cases

CLINICAL STUDIES

Initially, more than 3,000 patients were studied clinically
and in the laboratory by blind and double-blind techniques. The
patients were classified in groups as illustrated in Table IV
(from Bjorklund, et al., reference 10). The percentage of individuals or samples with elevated TPA in serum within the different
groups can be seen in this table. Values higher than or equal to
0.09 U TPA/ml of serum were considered to represent antigenemia,
while values lower than 0.09 U TPA/ml were considered negative.
This arbitrary border line was used throughout. The highest percentage of antigenemia was obtained for the group of cancer with
mestastases (73 ± 10%) and the lowest percentage for the group of
healthy persons and blood donors (2 ± 1%). For the 760 blood donors the percentage of antigenemia was 1 ± 1% (10).

Menendez-Botet and Schwartz (18) reported above normal serum TPA in 74% of 381 patients with cancer. In urine, elevated concentration of TPA was found in 63%. These figures agree well with the original Swedish studies.

Isacson, et al. (15) studied TPA in 109 patients with newly discovered verified cancer and in 50 controls. They found increased values of TPA in 78% of the serum samples. Nistor, et al. (19) found in a longitudinal study that recurrences of cancer could be picked up by elevated TPA in serum in 2 cases who were thought to be free of tumor.

Andren-Sandberg and Isacson (3) reported from a study of 300 patients, 80% of which had malignancy, that they had followed 14 patients with haematuria by cell cytology and TPA determinations. In 10 of the patients, cancer of the bladder was found and verified by histological examination after operation. Of these, 5 were diagnosed by urinary cytology and in 3 other cases the cytology indicated further control. The TPA values for urine were pathological in 9 of the 10 cancer cases. The remaining 4 cases with neither urinary infection nor malignancy had normal TPA and cytology in urine. The authors have extended their studies and confirmed the results.

From these and other studies of TPA it has become evident that there is a significant correlation between elevated TPA-levels in serum and/or urine and the occurrence of cancer.

The incidence of antigenemia in various types of infectious diseases was studied in 613 patients in connection with their hospital admissions (17). These patients fell into the following categories: hepatitis, 46; lower respiratory infections, 165; urinary infections, 104; viral infections, 105; skin infections, 34; gastrointestinal infections, 63; other bacterial infections, 78; viral meningitis, 18 (not included among the virus infections). The incidence of increased TPA in serum varied from 41% in hepatitis to 3% in skin infections. New tests were made in 107 random cases 3-5 weeks after admission. In all groups there was a significant decrease of TPA.

More recently, 173 hospitalized patients with infectious diseases and elevated serum TPA were retested at a follow-up 1-3 yr later (Table V). In 90% or 155 of the cases, TPA was within normal limits, in 9% (16 cases) there was a moderate elevation and in 1% (2 cases) TPA was high. None of the patients with elevated TPA was healthy. Separately, 202 hospitalized patients with infectious diseases without elevated TPA in their serum were given a repeat test a few weeks later. The TPA level remained normal in 97%. Only in 3% (7 cases) an increase was seen (Table VI). These re-

sults attest to the validity of the original concept that elevations of TPA, occurring in certain infectious diseases, are of a
temporary nature. This is in marked contrast to the findings in
progressive cancer.

One of the most difficult and delicate questions for the phy-
sician to answer concerns the prognosis of diagnosed cancer. In
an attempt to reveal the implications of TPA in this respect, 282
cancer patients were grouped according to their TPA values in serum and followed for 14 months with regard to survival. The
patients were grouped in the following 4 classes (TPA U/ml): a.)
<0.03 - 0.08 (n=155); b.) 0.09 - 0.3 (n=74); c.) 0.31 - 0.99 (n=
25);d.) ≥1 (n=28). As can be seen in Fig. 3, there is a remarkable
correlation between TPA and survival. The validity of the findings
is independent of type, stage, localization of cancer, and therapy. In the class with low TPA values, 66% were alive after 14
months. In the next class, comprising moderately elevated values,
45% were alive. The third class with stronger elevations of TPA
showed 37% survival after 7 months, and in the highest TPA class
only 23% were alive after 4 months.

Evidently, TPA values represent an important indicator of the
degree of lethal progress in cancerous disease even if individual
variations have to be properly considered.

TABLE V

TPA Values in 173 Hospitalized Patients with
Infectious Diseases which Caused Elevated TPA in Serum

TPA Serum levels (U/ml)	In Hospital		Check up after 1-3 yr	
	No.	%	No.	%
<0.03 - 0.08	–	–	155	(89.6)
0.09 - 0.3	108	(62.4)	16	(9.2)
0.31 - 0.99	48	(27.8)	2	(1.2)
≥1.0	17	(9.8)	–	–

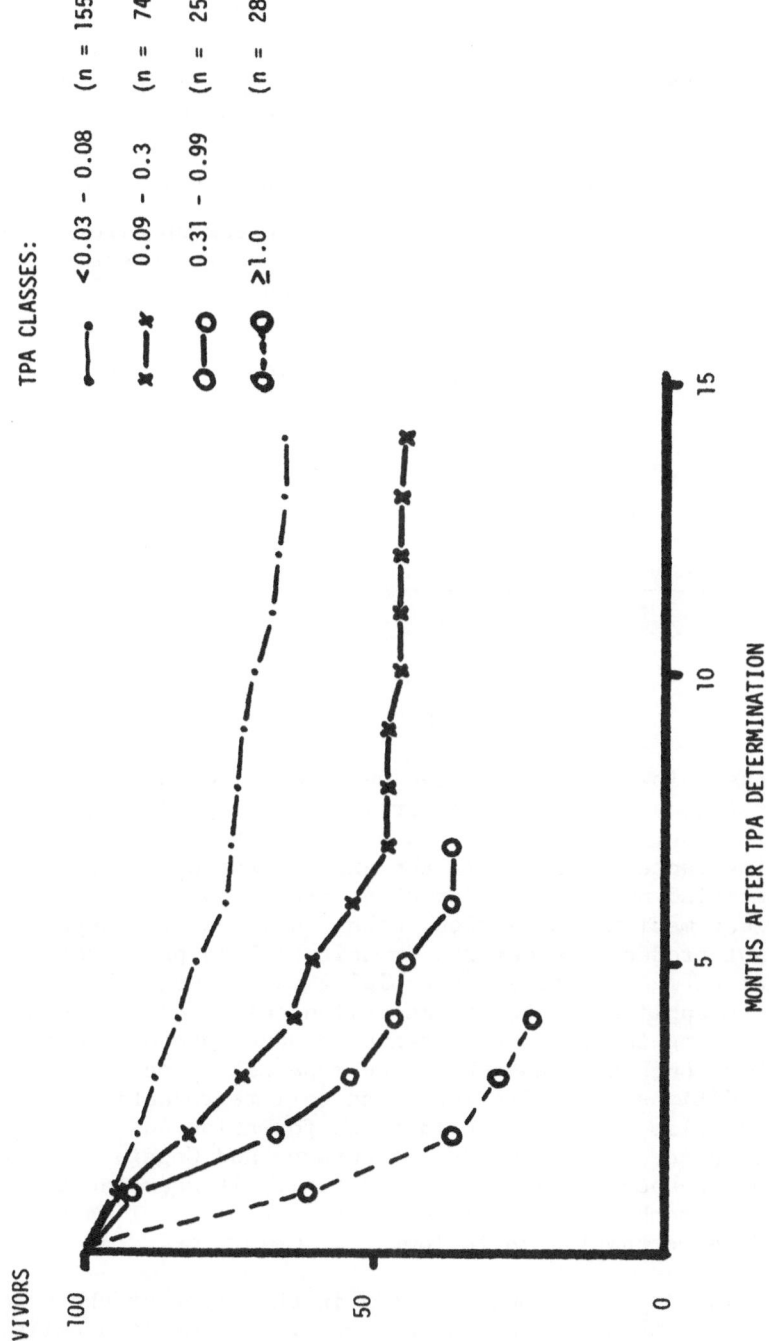

Fig. 3. Survival among 282 cancer patients grouped according to TPA in serum.

TABLE VI

TPA Values in 207 Hospitalized Patients with
Infectious Diseases without Elevated TPA in Their Serum

TPA Serum levels (U/ml)	In Hospital No.	%	Check up after a few weeks No.	%
< 0.03 – 0.08	207	(100.0)	200	(96.6)
0.09 – 0.3	–		6	(2.9)
0.31 – 0.99	–		1	(0.5)
≥ 1.0	–		–	

DISCUSSION

For bacteriologists it has been well-known for a long time
that in order to accomplish effective, immunity-producing vaccines or diagnostic antisera, it is pertinent to harvest the anti-
gen-producing bacterial cells in the exact phase of growth, i.e.
after a specified number of hr at a certain temperature in a specified nutrient medium. Otherwise, other antigens will appear
which may not produce the desired immunity. This phase.variation
of bacterial antigens represents a biological principle which
might well be applicable also to mammalian cells (4). If mammalian cells during their growth phase do produce phase-specific
antigens, then such antigens should be expected to occur in a variety of conditions involving increased cell reproduction such as
when damaged cells are replaced in tumor formation and in the development of placenta and embryo. Andersson and Coggin (2) have
described phase-specific antigens in detail. It seems as if a
conceptual approach of this type opens up a biological pathway
which could be worthwhile to follow.

The findings on TPA should be seen in the light of the hypothesis that growth-phase variation does occur in mammalian cells.
In agreement with this hypothesis are the findings that TPA is
produced in highly significant amounts by tumor cells of all sorts

of investigated locations and types in man and that TPA is being
made by each placenta examined. TPA is also released in regenerative, non-cancerous conditions and in atypical cell cultures in
vitro. Healthy tissues and healthy blood donors do not produce
elevated amounts of TPA, nor do cultures of diploid cells. Following removal of cancer tumors or convalescence after infectious
diseases, elevated TPA values usually return to normal. High TPA
values in cancer patients are strikingly correlated with decreased
survival and low values with prolonged survival. All these observations support the hypothesis but leave us with the question:
What changes of the microenvironment cause the growth-phase specific gene activation?

Whatever the best answer, it is fairly evident that human
cancer defense responses are inversely related to TPA. Whether
this is a casual relationship cannot be stated with certainty at
this stage.

The observations that TPA and anti-TPA exist in a wide variety of species point to possible regulatory mechanisms, which presumably were introduced at a very early stage during the evolution
of the species. Further exploration of this theme should be a fas-
cinating task.

Phase-specific reagents may perhaps offer improved possibilities in recognizing patients with internal pathological changes
in the direction of increased cell division.

ACKNOWLEDGEMENTS

Supported by the Swedish State, the Bonnier Group, the
Folksam and the Jameson Foundation. Skilled and valuable assis-
tance was given by Ms. K. Andersson, Ms. B. Gustafsson, Ms. V.
Lindh, Ms. A. Minnbergh, Mr. J. Paulsson, Mr. P. Redelius and
Mr. B. Wiklund.

REFERENCES

1. Abelev, G.I., Adv. Cancer Res., 14 (1971) 295.
2. Andersson, N.G. and Coggin, J.H., Jr., Proc. First Conf. Work-
    shop Embryonic and Fetal Antigens in Cancer, Conf.-710527, May
    24-26 (1971) 7.
3. Andren-Sandberg, A. and Isacson, S., VI Ann. Meeting Scand.
    Soc. Immunol., May 14-16 (1975).
4. Björklund, B., Conceptual Advances in Immunology and Oncology
    (Ed. R.W. Cumley, D.M. Aldridge, J. Haroz and J. McKay)(1963)503.
5. Björklund, B., Int. Arch. Allergy, 36 (1971) 191.

6. Björklund, B. and Björklund, V., Int. Arch. Allergy, 10 (1957)
    153.
7. Björklund, B., Björklund, V. and Hedlöf, I., J. Nat. Cancer
    Inst., 26 (1961) 533.
8. Björklund, B., Björklund, V., Wiklund, B., Lundstrom, R.,
    Ekdahl, P.H., Hagbard, L., Kaijser, K., Eklund, G. and Lüning,
    B., Immunological Techniques for Detection of Cancer (Ed. B.
    Björklund) Bonniers, Stockholm, Sweden (1973) 133.
9. Björklund, B., Lundblad, G. and Björklund, V., Arch. Allergy,
    12 (1958) 241.
10. Björklund, B., Lundström, R., Eklund, G. and Lüning, B., Im-
    munological Techniques for Detection of Cancer (Ed. B. Björklund)
    Bonniers, Stockholm, Sweden (1973) 237.
11. Björklund, B. and Paulsson, J.E., J. Immunol., 89 (1962) 759.
12. Gold, P. and Freedman, S.O., J. Exp. Med., 122 (1965) 467.
13. Herbert, D., Microbial Reaction to Environment (Ed. G.G. Meynell
    and H. Gooder) Cambridge University Press, London, England (1961)
    391.
14. Holleman, J.W. and Palmer, W.G., Proc. Second Conf. Embryonic
    and Fetal Antigens in Cancer, Conf.-720208, Feb. 14-16 (1972)
    117.
15. Isacson, S., Lindblad, C., Nistor, L. and Risholm, L., Abstr.
    XI Int. Cancer Congress in Florence, Italy, Oct. 24-26 (1974).
16. Lacey, B.W., Microbial Reaction to Environment (Ed. G.G. Meynell
    and H. Gooder) Cambridge University Press, London, England (1961)
    343.
17. Lundström, R., Björklund, B. and Eklund, G., Immunological
    Techniques for Detection of Cancer (Ed. B. Björklund) Bonniers,
    Stockholm, Sweden (1973) 243.
18. Menendez-Botet, C.J. and Schwartz, M.K., Abstr. 170th Nat.
    Meeting Amer. Chem. Soc. Div. Biol. Chem., August 24-29 (1975).
19. Nistor, L., Isacson, S., Lindström, K. and Risholm, L., Svensk
    Kirurgi,31, No. 2 (1974) 1.

AGE-DEPENDENT REGRESSION OF M-MSV TUMORS IN CBA/H MICE: REQUIREMENT FOR A MACROPHAGE-ADHERENT CELL POPULATION

O. STUTMAN

Memorial Sloan-Kettering Cancer Center
New York, New York (USA)

The effects of aging on immune responses in mice are complex and show some remarkable genetic influences (8,12,13,17,20,21,24, 25, 26,28). Certain strains will show profound impairment of both humoral as well as cellular immunity with increasing age, while other strains will have only moderate decline of immune functions (8,12,13,17,20,21,24,25,26,28). With the exception of the auto-immune-susceptible mouse strains which show a precocious decline of T-dependent functions and cell-mediated immunity (17,20,21,24, 25,28) it is accepted that cell-mediated immunity, measured both in vivo and in vitro, shows less impairment than humoral immunity, especially in long-lived non-autoimmune strains (8,12,13,17,20,21, 25). A good example of such a situation is the CBA/H subline, which is long-lived, does not show any detectable signs of "NZB-type" autoimmune disease (21,25), shows normal levels of T-dependent lymphocytes (20,21), shows intact thymus function (24,28) and a remarkable preservation of immune responses, especially cell-mediated immunity, even at advanced age, beyond the 10% survival range (17,21,25). However, during studies designed to test the susceptibility as well as the capacity to produce regressions of tumors induced by Moloney sarcoma virus (M-MSV) in mice of different strains and ages, it became apparent that the CBA/H mice showed a marked impairment for producing tumor regressions with age. Such defect appeared at approximately 600 days of age, at a time in which most immune functions, including humoral and cellular immunity, are still intact (21). An analysis of this defect is the subject of the present paper. By using several replacement procedures in the older mice, we could show that the actual defect in tumor regression is dependent on the supply of a macrophage-like cell population which seems to be deficient in the older CBA/H mice.

The immunological nature of the regression of M-MSV tumors in mice has been firmly established: immature animals (5,11) as well as animals immunosuppressed by irradiation (6), steroids (18), cyclophosphamide (7), anti-lymphocyte serum (10), thymectomy (2,4, 11) or congenital absence of the thymus (3, 23) are incapable of producing regression of the tumors. The immunological nature of these phenomena was also demonstrated by the transfer of resistance to tumor growth with cells (6) or serum (1,7,11) from animals that had regressing tumors. One study also indicated that older mice showed an impairment to produce regressing tumors with age (15). Some of the thymectomy experiments (2) as well as experiments with athymic-nude mice (23) actually indicated a strict thymus dependency of regression production, since thymus grafts could readily restore regression capacity in the deficient animals. This last aspect was observed regardless of the actual effector mechanism capable of destroying M-MSV tumor cells _in_ _vitro_ (9,14,16).

MATERIALS AND METHODS

CBA/H mice of both sexes were used in all the experiments. For details of origin of these strains and animal care see references 8,12,13,17,20,21,25 and 26. The survival of normal CBA/H under these conditions is: 50% survival of 960 $\pm$ 29 SE days for females and 940 $\pm$ 33 SE days for males (21). All animals are free of lactic-dehydrogenase-elevating virus (21,23) and were used at the ages indicated in the tables. No sex-dependent differences were observed, thus the results will be presented as combined data. Animals were injected intramuscularly in the left thigh with 0.1 ml of a 1:2 dilution of Moloney sarcoma virus (M-MSV) containing 5-9 x 10^7 focus forming units per ml when tested in secondary BALB/C mouse embryo cell cultures. The original stock of M-MSV was obtained from Electro-Nucleonics Laboratories (Bethesda, Md.) as M-MSV infected BALB-3T3 cells. For additional details on virus origin and virus preparation, see reference 23. Animals were inspected daily for tumor development after virus injection. The mean latent period for tumor development in CBA/H mice was 7.0 $\pm$ 1.6 SD days, when infected at 30 days of age and 10.2 $\pm$ 2.1 SD days when infected at 360 days, as observed previously (23). In the present experiments latent periods ranged from approximately 6-7 days to 12-14 days for all the age groups. The mean regression time, after tumor appearance in all the present experiments, when regressions were observed, was comparable to our published data (23) and ranged from 10.6 to 12.1 days with a SD range of 1.5 to 2.1. In all the experiments in which additional syngeneic cells were used for treatment, donor and host were sex-matched.

For details of thymus grafting and preparation of spleen cell
suspensions see references 20-24 and 28. For details of thymectomy
or thymectomy and lethal whole body irradiation see reference 22.
For details on nude athymic nu/nu-CBA/H see reference 23. In every
experiment, the treatment of the host was performed 30 days before
M-MSV infection. For details of the procedures for selective depletion of different subsets of lymphoid cells from the spleen
inocula, detailed in Table IV, see references 20 and 22.

Two basic experimental models were used: a) infection of CBA/H
mice of different ages with M-MSV, with subsequent determination
of tumor incidence and incidence of spontaneous regression of the
tumors (Table I) and b) infection of 900-day old CBA/H mice which
have been previously treated with either thymus grafts or spleen
cells from younger animals (Tables II and III) or with spleen cells
which have been specifically depleted of different subsets of lymphoid cells (Tables II and IV).

RESULTS AND DISCUSSION

Table I shows that, although the incidence of tumors produced
by M-MSV infection in CBA/H mice of different ages was comparable,
important age-related differences on the actual incidence of tumor
regressions were observed in these animals. At the younger age
studied (15 days), only 47% of the animals showed regressions, in
accordance with published data using newborn or young mice of
other strains (5,6,11). The young adult and adult mice (30 and
360 days of age) produced 100% regressions, while the older groups
ranging from 600 to 1080 days of age showed a marked reduction in
their capacity to produce regressions. This finding is comparable
to that obtained in another study (15). However, it was unexpected
since 600-day old and even older CBA/H mice have a normal humoral
and cellular immunity, determined by a variety of different assays
(17,20,21,24,24,28). A moderate deficit in antibody responses has
been described in 2-year old CBA mice of another subline (27) while
cell-mediated immunity measured in vivo and in vitro remained intact in 2-year old CBA/H mice (17,25,28) or even in older mice (20,
21,24).

Table II shows that when 900-day old CBA/H mice were treated
with spleen cells from 30-day old syngeneic sex-matched donors,
the incidence of tumor regressions was normalized (i.e. 75 to 100%
regressions, see lines 4 and 5, Table II). Similarly, spleen cells
from T-deprived young mice (neonatally thymectomized, thymectomized-
irradiated or athymic nude) could also restore the capacity to produce regressions in the 900-day old CBA/H mice (lines 6 to 8, Table
II). Conversely, and in accordance with the last statement, thymus
grafts or dispersed thymocytes were ineffective (lines 2 and 3,

Table II). This last set of results was somewhat expected, since
we already know that the T cell compartment as well as thymus
functions were intact in 900-day old CBA/H mice (20,21,24).

The age dependency of this phenomenon was also studied by the
administration of spleen cells from donors of different ages to
the 900-day old CBA/H mice, and Table III shows such results.
While newborn or 15-day old spleen cells were relatively ineffective in restoring the capacity to produce tumor regression, cells
from 30 to 400-day old donors were highly effective. Also, as
expected from the data in Table I, cells from 600 to 1200-day old
donors were totally ineffective.

Table IV shows the additional attempts to characterize the
cell in normal young spleen which was capable of restoring the
capacity to produce tumor regressions in the older mice. As expected from the lack of effect when thymocytes or thymus grafts
were used (see Table II), the removal of T cells by treatment
with anti thy. 1.2 antiserum (theta) and C had no effect (Table IV,
line 1). The procedures that remove mainly B cells (lines 2 to 4,
Table IV) also had no detectable effect on the capacity to restore
tumor regression. On the other hand, all the procedures which
were capable of removing adherent and/or macrophage-like cells
(lines 5 to 8, Table IV) also removed the capacity in the older
CBA/H mice. These results were interpreted as indicative of a
possible macrophage defect in the older CBA/H mice as the basis
for their inability to produce spontaneous regression of M-MSV
induced sarcomas.

The studies on macrophage function in young and old CBA/H
mice (these studies are still in progress) indicate that clearance of colloidal particles or opsonized bacteria are either
normal or increased in the older animals and that the function
of peritoneal macrophages, measured as capacity to ingest radiolabelled bacteria or to destroy adherent target cells, are also
preserved. It is apparent that a detailed analysis of macrophage function in these animals is mandatory.

The role of macrophages, as possible mediators of regression
of M-MSV induced tumors has been indicated in several publications
(10,14,23) and tumors growing progressively in either normal (14)
or athymic nude mice (23) have a decreased number of macrophages,
when compared to tumors that will regress in normal animals. Even
in the in vivo neutralization assays (i.e. mixtures of immune
lymphoid and tumor cells injected into normal animals), which require specific lymphocytes (19,29,30) for the expression of tumor destruction, a host cell of possible macrophage nature is
necessary for effective tumor neutralization.

TABLE I

Effect of Age on Regression of M-MSV Induced Sarcomas in CBA/H Mice

Age at Infection* (days)	Tumor Incidence per Number of Mice	Number of Regressions per total Tumors
15	21/21 (100%)	10/21 (47%)
30	16/17 (94%)	16/16 (100%)
360	12/20 (60%)	12/12 (100%)
600	17/22 (77%)	3/17 (18%)
900	18/20 (90%)	1/18 (5%)
1080	12/12 (100%)	0/12

*Injected intramuscularly in the left thigh with 0.1 ml of 1:2 diluted M-MSV containing 5-9 x 10^7 focus-forming units per ml when tested in secondary BALB/C mouse embryo cell cultures.

TABLE II

Effect of Thymus or Spleen on Inability to Produce Regression of M-MSV Sarcomas in 900-day Old CBA/H Mice

Treatment*	Tumor Incidence	Number of Regressions per Total Tumors
None	23/24 (96%)	3/23 (13%)
Thymus graft IP	14/14 (100%)	1/14 (7%)
100 x 10^6 thymocytes, IP	9/10 (90%)	1/9 (11%)
100 x 10^6 spleen cells, IP	12/12 (100%)	9/12 (75%)
500 x 10^6 spleen cells, IP	11/12 (92%)	11/11 (100%)
100 x 10^6 spleen, neonatally tx, IP	10/10 (100%)	9/10 (90%)
100 x 10^6 spleen, tx + x-rays, IP	10/10 (100%)	6/10 (60%)
100 x 10^6 spleen, nu/nu-CBA/H, IP	10/10 (100%)	7/10 (70%)

*900-day old CBA/H mice, treated 30 days before M-MSV injection (see Table I for M-MSV dose) with: a thymus graft implanted intraperitoneally (IP) from 30-day old CBA/H donors; dispersed thymocytes from 30-day old CBA/H donors; dispersed spleen cells from 30-day old CBA/H donors; spleen cells from 30-day old neonatally thymectomized CBA/H donors; spleen cells from 60-day old mice, thymectomized, irradiated with 980 R and injected with 10 x 10^6 CBA/H fetal liver (15-17 days of embryonic age) and the spleen cells obtained 30 days after treatment; spleen cells from 30-day old nu/nu-CBA/H donors.

TABLE III

Effect of Spleen Age on Capacity to Produce Regression
of M–MSV Sarcomas in 900–day Old CBA/H Mice

Age of Spleen Donor* (days)	Incidence of Regressions per Total Tumors
Newborn	2/12 (17%)
15	5/12 (42%)
30	12/12 (100%)
120	10/10 (100%)
360	10/10 (100%)
400	9/9 (100%)
600	3/12 (25%)
900	1/10 (10%)
1200	1/10 (10%)

*900–day old CBA/H mice treated at 30 days before M–MSV infection
(see Table I for M–MSV dose) with 100×10^6 spleen cells IP from
syngeneic CBA/H donors of different ages.

TABLE IV

Effect of Removal of Different Cell Types from 30–day Old
Spleen on Capacity to Produce Regression of M–MSV Induced
Sarcomas in 900–day Old CBA/H Mice

Treatment of Spleen Cells*	Incidence of Regressions per Total Tumors
Thy.1.2 and C (T cells)	9/10 (90%)
Anti-Ig and C (B cells)	9/10 (90%)
EAC rosettes (B sub-population)	9/10 (90%)
Ig-Anti Ig columns (B cells, cells with Fc receptor)	5/10 (50%)
Nylon wool, low serum (B cells, monocyte, macrophages)	3/10 (30%)
Nylon wool, high serum (monocyte, macrophages)	2/10 (20%)
Adherence to plastic (adherent cells, macrophages)	3/10 (30%)
Iron + Magnet (macrophages)	1/10 (10%)

*900–day old CBA/H mice treated 30 days before M–MSV infection with
100×10^6 spleen cells, IP, from 30–day old CBA/H donors, pretreated
in vitro as indicated (for details on techniques see references
20 and 22). In parentheses, the main cell type which is depleted
by each procedure.

In summary, CBA/H mice show an age-dependent cellular defect, probably involving macrophage cells, which is expressed as inability to produce regression of tumors induced by M-MSV. This defect occurs at an age when both humoral and cell-mediated immunity are still intact. These results also suggest that although the in vivo regression of M-MSV tumors has a strict thymus dependence (2,23) a macrophage-like cell seems required during the effector phases of the response.

ACKNOWLEDGEMENT

This work was supported by USPHS Grants CA-08748, CA-16599 and CA-17404.

REFERENCES

1. Bubenik, J. and Turano, A., Folia Biol.,14 (1968) 433.
2. Collavo, D., Colombatti, A., Chieco-Bianchi, L. and Davies, A.J.S., Nature, 249 (1974) 169.
3. DeClerq, E., J. Nat. Cancer Inst., 54 (1975) 473.
4. East, J. and Harvey, J.J., Int. J. Cancer, 3 (1968) 614.
5. Fefer, A., McCoy, J.L. and Glynn, J.P., Cancer Res., 27 (1967) 1626.
6. Fefer, A., McCoy, J.L. and Glynn, J.P., Cancer Res., 27 (1967) 2207.
7. Fefer, A., Cancer Res., 29 (1969) 2177.
8. Gerbase-deLima, M., Meredith, P. and Walford, R.L., Fed. Proc., 34 (1975) 159.
9. Lamon, E.W., Skurzak, H.M., Klein, E. and Wigzell, H., J. Exp. Med., 136 (1972) 1072.
10. Law, L.W., Ting, R.C. and Allison, A.C. , Nature, 220 (1968) 611.
11. Law, L.W., Ting, R.C. and Stanton, M.F., J. Nat. Cancer Inst., 40 (1968) 1101.
12. Makinodan, T. and Adler, W.H., Fed. Proc., 34 (1975) 153.
13. Nordin, A.A. and Makinodan, T., Fed. Proc.,33 (1974) 2033.
14. Owen, J.J.T and Seeger, R.C., Brit. J. Cancer 28, Suppl. I , (1973) 26.
15. Pazmino, N.H. and Yuhas, J.M., Cancer Res.,33 (1973) 2668.
16. Plata, F., Gomard, E., Leclerc, J.C. and Levy, J.P., J. Immunol. 111 (1973) 667.
17. Rodey, G.E., Good, R.A. and Yunis, E.J., Clin. Exp. Immunol., 9 (1971) 305.
18. Schachat, D.A., Fefer, A. and Moloney, J.B., Cancer Res., 28 (1968) 517.
19. Simes, R.J., Kearney, R. and Nelson, D.S., Immunology,29 (1975) 343.
20. Stutman, O., J. Immunol.,109 (1972) 602.

21. Stutman, O., Fed. Proc., 33 (1974) 2028.
22. Stutman, O., Ann. N. Y. Acad. Sci., 249 (1975) 89.
23. Stutman, O., Nature, 253 (1975) 142.
24. Stutman, O. and Good, R.A., Ser. Haematol., 7 (1974) 505.
25. Teague, P.O., Yunis, E.J., Rodey, G., Fish, A.J., Stutman, O. and Good, R.A., Lab. Invest., 22 (1970) 121.
26. Walford, R.L., Fed. Proc., 33 (1974) 2020.
27. Wigzell, H. and Stjernsward, J., J. Nat. Cancer Inst., 50 (1973) 701.
28. Yunis, E.J., Gernandez, G., Smith, J., Stutman, O. and Good, R.A., Microenvironmental Aspects of Immunity (Ed. B.D. Jankovic and K. Isakovic) Plenum Press, New York (1975) 301.
29. Zarling, J.M. and Tevethia, S.S., J. Nat. Cancer Inst., 50 (1973) 149.
30. Zbar, B., Wepsic, T.H., Rapp, H.J., Stewart, L.C. and Borsos, T., J. Nat. Cancer Inst., 44 (1970) 701.

HOST PROTECTION BY CELL-MEDIATED AND BY HUMORAL IMMUNITY IN MALIGNANCY

T. J. LINNA, C. HU and K. M. LAM

Temple University School of Medicine
Philadelphia, Pennsylvania (USA)

There is a complex and incomplete understanding of relationship between the immune system and malignancy. It is clear from
many experimental systems that both cell-mediated immunity and the
antibody-forming system can inhibit tumor development. There are
also clinical data supporting such a host-protective function for
immunity (3,6,12). However, the biological role of the immunological "surveillance" mechanisms in the development of spontaneous
tumors remains to be determined (19). The surveillance concept was
originally formulated with cell-mediated immunity in mind (24),
however, it has become increasingly clear in the past years that
the antibody forming system can also exert a powerful host-protective influence in vivo (9,15) and in vitro (8,17). The potential
of the mononuclear phagocyte system alone, or in conjunction with
other immunological defense mechanisms in tumor protection, is also
well documented in these proceedings and elsewhere (5,10).

This communication will review some recent findings from our
laboratory on contributions by cell-mediated immunity and by the
antibody-forming system to host defense against tumors. We have
used the chicken as the experimental animal because of the relatively clear delineation between thymus-dependent, cell-mediated
immunity, and bursa-dependent, humoral immunity, in this species
(4) which offers the possibility to study the contributions of
these 2 parts of the immune system to tumor development.

Tumor-frequency, mortality and size were compared between animals with impaired immunological functions and normal controls.
Specific immunological impairment was accomplished by surgical thymectomy, by surgical bursectomy (18), or by cyclophosphamide bur-

sectomy in the newly hatched period (14). We have used 3 tumors
of widely different origin and behavior in these studies, namely
a) avian reticuloendotheliosis, induced by a RNA tumor virus outside the avian leukosis-sarcoma complex, b) XC cells, a Rous sarcoma derived rat tumor cell line, and c) a chemical carcinogen induced transplantable tumor line.

We will demonstrate here that both cell-mediated immunity and
the antibody-forming system have a host-protective effect in these
malignancies, with the contributions of these 2 parts of the immune
defense system varying with the tumor used and with the experimental conditions. These similar findings in 3 widely different tumor
systems encourage us to postulate that both cell-mediated and humoral immunity have a host-protective function in malignancy. In
no case have we observed a tumor-enhancing function for the antibody-forming system. In the case of the reticuloendothelial tumor,
studied more extensively by us than the other tumors, immune serum
can be used to cure established tumors.

MATERIALS AND METHODS

Animals and Treatments

Hy-Line WC line chickens were used in all these studies. These
animals are isogenic for the major histocompatibility locus. Surgical thymectomies or bursectomies were performed on the day of
hatching, using the standard techniques (18). Cyclophosphamide
bursectomies were accomplished by intraperitoneal injection of 4 mg
of cyclophosphamide during each of the first 3 days of life, including the day of hatching (14).

Tumor Induction

Reticuloendothelial tumors growing locally in the wing web
were induced by administration of avian reticuloendotheliosis (RE)
virus, strain T, (20,23) into the wing web. This treatment causes
local tumor development. Depending on the tumor virus dose and on
the immune status and age of the animals, they can either grow progressively, ending in death with systemic tumor development, or
the tumor regresses with eventual survival and solid immunity to
further high doses of RE virus challenge. RE virus was used to induce wing web tumors in surgically bursectomized animals and their
respective controls on the third day of life, and in cyclophosphamide bursectomized animals and their controls, when these animals
were 1 month old.

XC Cell-Induced Tumors

They were induced by injection of 5×10^6 live (trypan blue-excluding) XC cells into the wing webs of thymectomized, of bursectomized and of control animals when 3 days old. XC cell administration to chickens causes development of local, sarcomatous tumors, and tumor spreads to distant sites (13,21,22). Karyotype determinations, kindly performed by Dr. W. Weber, University of Pennsylvania, show that the developing tumors consist of chick cells, while the XC cells have rat chromosome markers. Thus, the Rous sarcoma genome present in the XC cells is able to induce production of infectious Rous sarcoma virus in chickens, and those viruses infect surrounding cells, starting the tumor process.

Chemical Carcinogen-Induced Tumor

It was obtained by intramuscular injection of 1% (w/v) benzo-(a)pyrene in trioctanoin into a thymectomized WC chicken. The tumor was harvested 5 months later. It has the morphology of a fibrosarcoma, and is transplantable with a 100% take frequency in chickens sharing the major histocompatibility locus of the original donor. The take frequency is much lower when the major histocompatibility locus is crossed, even when much higher doses of tumor cells are used. Live tumor cells are necessary for tumor formation. In the study reported here, 1×10^3 live cells of the benzo(a)pyrene-induced fibrosarcoma were given subcutaneously to 1 day old bursectomized, thymectomized or control chickens.

Immune Serum Components

Immune serum was produced in WC chickens by inoculating them with 0.1 LD_{50} of RE virus when 3 days old, boosting the animals with 5 LD_{50} when 1 month old, and exsanguinating them from the heart 6 weeks later. Part of the serum was extensively absorbed with RE virus to remove viral specific antibodies. Immune and normal gammaglobulin was obtained from immune and normal serum by precipitating 3 times with ammonium sulphate at 35% saturation.

We had chosen to start tumor therapy on the sixth day after RE virus inoculation, when about 60% of the test chickens had measurable wing web tumors. Nine injections of immune serum or its components, or of normal serum or its components, were administered to the recipients 0.5 ml each time. A non-treated group was also included in each study.

Assessment of Tumor Development

The animals were observed daily for the presence of local tumor and for mortality. Local tumors were measured regularly with
a caliper. The presence of systemic tumors was determined at autopsy in dead or moribund animals, with histological confirmation
if necessary. The few animals not dying with tumors were excluded
from the evaluations. Differences in tumor frequency and mortality
between experimental and control groups were evaluated statistically
with Fisher's exact test using four-fold tables. Differences in
tumor size were evaluated with student's t-test or with analysis of
variance, and with the multivariate exact test for 2 independent
populations (16).

RESULTS

Thymectomy Studies

Complete surgical thymectomy is technically virtually impossible in the chicken. Even when performed in the newly hatched
period, it does not completely abolish thymus-dependent immunological functions. When RE virus was administered to thymectomized
and control chickens, tumor progression and tumor mortality rate
were significantly higher in thymectomized than in control chickens,
indicating that the thymus has a host-protective effect in this
malignancy. These data have been published in more detail elsewhere (15,25). In the case of XC cell-induced tumors, tumor frequency and mortality were significantly higher in thymectomized
than in normal animals. The mean tumor size was also much larger
in thymectomized than in control animals (Fig. 1). When tumor frequency and growth was compared between thymectomized and control
animals inoculated with the benzo(a)pyrene-induced tumor line, tumors appeared earlier in the thymectomized animals, but no significant difference between experimental and control animals could be
detected towards the end of the study. The addition of sublethal
irradiation to the surgical thymectomy in the newly hatched period,
accomplishing a more profound deficiency of cell-mediated immunity,
is necessary before any conclusions can be drawn from the lack of
effect of thymectomy on development of the transplanted tumors.

Bursectomy Studies

When RE virus is used to induce local tumors in bursectomized
and control animals, a significant increase in progressive tumor
growth and in tumor mortality can be observed in the bursectomized
group. This holds true both for surgically and cyclophosphamide

bursectomized animals. A more detailed account of these data can be found elsewhere (15,25). In XC cell-injected animals, significantly larger tumors, as evaluated by analysis of variance, could be observed in the surgically bursectomized group than the controls (Fig. 1). However, no clear difference could be observed between bursectomized and control animals in tumor frequency or mortality. When development of the chemical carcinogen-induced tumor line was evaluated in surgically bursectomized and control birds, there was earlier tumor appearance in the bursectomy group, but no difference in tumor frequency towards the end of the study. However, the bursectomized birds also developed significantly larger tumors than the controls in this tumor model (Fig. 2).

Since data from cyclophosphamide bursectomy experiments, which would more profoundly cripple humoral immunological functions, are not yet available in these last 2 tumor models, it can not be stated whether the function of the antibody-forming system is mainly to regulate tumor size, or whether a more stringent ablation of antibody-forming capacity, i.e. cyclophamide bursectomy, would also influence tumor frequency.

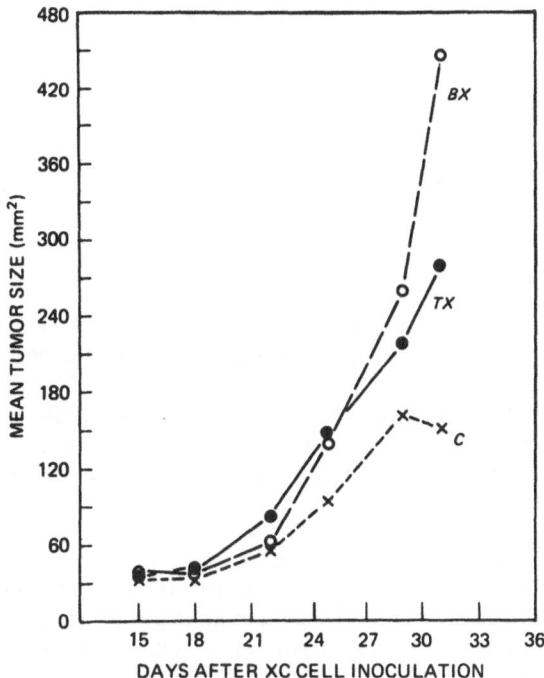

Fig. 1. Mean tumor sizes of thymectomized, bursectomized, and control chickens inoculated with XC cells.

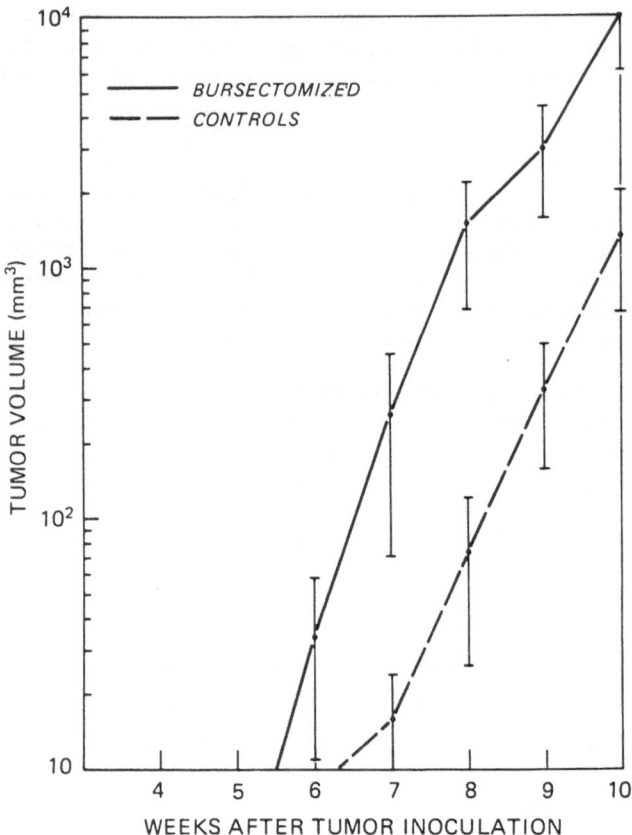

Fig. 2.   Mean BP-1 tumor volume in bursectomized and control chickens.

Tumor Immunotherapy Studies

Only the reticuloendothelial tumors have been studied by us
with regard to the capacity of immune serum to accomplish tumor
cure. With the protocol used by us, immune serum has been remark-
ably effective in this respect.

The results can be summarized as follows: using the above outlined protocol, immune serum is capable of accomplishing tumor regression, and of decreasing tumor mortality to a statistically significant level. There is no observable contribution of the antibody-forming system of the host to this effect, since the immune
serum works as well in tumor cure in cyclophosphamide-bursectomized,
tumor-bearing animals. Antiviral antibodies do not significantly

contribute to the curative effect, since immune serum is at least
as effective after extensive absorption of antiviral antibodies as
before absorption, but absorption with tumorous spleen cells abol-
ishes the curative effect (Table I). The serum has at least some
specificity, since it does not at all affect the development of tumors in chickens infected with a laboratory strain of Marek's disease virus. The gammaglobulin fraction of immune serum, but not
of normal serum, is effective in mediating tumor cure. For these
reasons, we find it reasonable to assume that the cure is afforded
by antibodies directed against transplantation-type antigens on
tumor cells.

DISCUSSION

Much data from experimental and clinical sources support the
concept of a surveillance function for the immunological system
(3,6,12). In the past, cell-mediated immunity has mainly been emphasized in this respect.

Our data obtained in the chicken, where the effects of cell-
mediated immunity and of the antibody-forming system can be better
separated than in other in vivo models, clearly support this function for cell-mediated immunity in 2 of the 3 tumors studied by us.
In the third - a very weakly immunogenic carcinogen-induced tumor -
no significant effect of thymectomy can be demonstrated. Experiments using more profoundly immunodeficient animals are necessary
before any conclusions can be drawn from these negative data.

The role of the antibody-forming system in tumor development
has been controversial, with much emphasis on the tumor-enhancing
ability of serum factors, presumed to be antibodies or antigen-
antibody complexes. It has been demonstrated long ago that antibodies, with capacity to enhance skin or tumor allografts, can be
obtained with certain experimental protocols (11). However, the
biological significance of such antibodies in tumor development is
unclear. Recent experiments, mainly from the laboratories of Alexander (1) and Baldwin (2), indicate that the "blocking factor" in
tumor-bearer serum is tumor antigen, or antigen-antibody complexes
in antigen excess. In the 3 tumor models of widely different genesis studied by us, bursectomy and consequently impairment of antibody producing capacity, results in a lowered ability of the host
in tumor defense, i.e. a surveillance-like function also for humoral immunity. The immunotherapy studies further emphasize this
host-protective effect of the antibody-forming system. Similar
findings in other tumor models are available, both in the older
(7) and the recent literature (9).

TABLE I

Effect of RE Tumor Cell-Absorbed Immune Serum on the Development of RE Tumors

Treatment	Regression of local tumors	Survivors	P
Non-absorbed immune serum	17/18 (94%)	17/18 (94%)	
Normal cell-absorbed immune serum	14/17 (82%)	14/17 (82%)	
RE tumor cell-absorbed immune serum	12/20 (60%)	11/20 (55%)	<0.05*
Normal serum	12/19 (63%)	11/19 (58%)	
None	9/20 (45%)	8/20 (40%)	

*Comparison with non-absorbed immune serum

Our findings, taken together with the data that "blocking" of cytotoxic functions is accomplished by tumor antigen, or by antigen-antibody complexes in antigen excess, indicate that both humoral and cell-mediated immunity are important for host defense.

ACKNOWLEDGEMENTS

This work was supported by USPHS, NIH Grant No. CA-13347 from the National Cancer Institute, by Grant No. IN-88F from the American Cancer Society, and by a NIH General Research Support Grant to Temple University. We thank Miss Annsofi Holst and Mr. Robert Kalwinski for skillful technical assistance.

REFERENCES

1. Alexander, P., Cancer Res., 34 (1974) 2082.
2. Baldwin, R.W., Bowen, J.G. and Price, M.R., Brit. J. Cancer, 28 (1973) 16.
3. Burnet, F.M., Immunological Surveillance, Pergamon Press, Oxford, (1970).
4. Cooper, M.D., Peterson, R.D.A., South, M.A. and Good, R.A., J. Exp. Med., 123 (1966) 75.
5. Evans, R., Cox, H. and Alexander, P., Proc. Soc. Exp. Med., 143 (1973) 259.
6. Good, R.A., Proc. Nat. Acad. Sci., 69 (1972) 1026.
7. Gorer, P.A. and Amos, D.B., Cancer Res., 16 (1956) 338.
8. Hayami, M., Hellstrom, I. and Hellstrom, K.E., Int. J. Cancer, 12 (1973) 667.
9. Hersey, P., Brit. J. Cancer, 28 (1973) 11.
10. Hibbs, J.B., Jr., Lambert, L.H., Jr. and Remington, J.S., Nature, 235 (1972) 48.
11. Kaliss, N., Cancer Res., 18 (1958) 992.
12. Klein, G., Immunological surveillance against neoplasia, Academic Press, New York, (1973-1974) 71.
13. Lam, K.M. and Linna, T.J., Fed. Proc., 34 (1975) 852.
14. Linna, T.J., Frommel, D. and Good, R.A., Int. Arch. Allergy, 42 (1972) 20.
15. Linna, T.J., Hu, C. and Thompson, K.D., J. Nat. Cancer Inst., 53 (1974) 847.
16. Morrison, D.F., Multivariate Statistical Methods, McGraw-Hill, New York, (1967).
17. Perlman, P., Perlman, H. and Wigzell, H., Transplant Rev., 13 (1973) 91.
18. Peterson, R.D.A., Burmester, B.R., Frederickson, T.N., Purchase, H.G. and Good, R.A., J. Nat. Cancer Inst., 32 (1964) 1343.
19. Prehn, R.T., J. Reticuloendothel. Soc. 10 (1971) 1.
20. Robinson, F.R. and Twiehaus, M.J., Avian Dis., 18 (1974) 278.

21. Svoboda, J., Nature, 186 (1960) 980.
22. Svoboda, J., Chyle, P., Simkovic, D. and Hilgert, I., Folia
    Biol., 9 (1963) 77.
23. Theilen, G.H., Zeigel, R.F. and Twiehaus, M.J., J. Nat. Cancer
    Inst., 37 (1966) 731.
24. Thomas, L., Cellular and humoral aspects of hypersensitive
    states, (Ed. H.S. Lawrence), Hoeber-Harper, New York, (1959)
    529.
25. Thompson, K.D. and Linna, T.J., Nature, 245 (1973) 10.

IN VITRO TUMOR GROWTH INHIBITION BY SYNGENEIC LYMPHOCYTES AND/OR MACROPHAGES

I. J. FIDLER and D. E. PETERSON

NCI-Frederick Cancer Research Center
Frederick, Maryland (USA)

We (3) have recently confirmed the hypothesis by Prehn (13) that the normal immune response to neoplasia might have a dual role. This hypothesis suggests that during the early development of cancer or with weakly immunogenic, transplantable tumors, the cell-mediated response might directly stimulate rather than inhibit tumor growth. Our studies (3) reported the results of the interaction of normal, sensitized and concanavalin A-stimulated syngeneic, allogeneic, and/or xenogeneic lymphocytes with the B16 melanoma, C57BL/6 and/or A mouse embryo cells in an in vitro colony inhibition-stimulation system. Specifically sensitized lymphocytes at ratios up to 1000:1 repeatedly and significantly enhanced the growth of the target cells. At higher lymphocyte ratios, target cell inhibition occurred.

In addition, we (4) also demonstrated that a low number of normal or sensitized syngeneic lymphocytes mixed in vitro with the B16 melanoma can increase the incidence of pulmonary metastasis in C57BL/6 mice given intravenous (i.v.) injections of the mixture. However, once a critical dose of immune cells was exceeded, cytotoxicity or inhibition of tumor metastasis was demonstrated. In subsequent studies (5), we reported that in vitro activated mouse macrophages injected i.v. into melanoma-bearing mice significantly reduced the number of established metastases.

These and other studies (8,13,14) have demonstrated an apparent dual role for immune cells in their interaction with syngeneic tumors. However, the mechanism responsible for the phenomenon has remained unclear. In this regard, several possibilities have emerged: a) there is one population of immune cells (lymphocyte)

389

producing lymphotoxins (lymphokines) which at low concentration
stimulates while at high concentration inhibits tumor growth (9);
b) there are 2 (or more) subclasses of lymphocytes which produce
stimulatory or inhibitory growth effects; and c) there are 2 distinct populations of immune cells, one principally responsible for
stimulation of tumor growth (lymphocyte) and the other responsible
for tumor inhibition (macrophage).

The present studies were performed to determine if a small
number of syngeneic macrophages within a purified lymphocyte preparation may be responsible for inhibition of tumor growth in vitro.

MATERIALS AND METHODS

Animals. Inbred C57BL/6 mice were obtained from the Jackson
Laboratories, Bar Harbor, Maine.

Tumor. The transplantable B16 melanoma, originating in
C57BL/6 mice, was obtained originally from The Jackson Laboratories,
Bar Harbor, Maine.

Immunizations of C57BL/6 Mice Against B16 Melanoma. C57BL/6
mice were given subcutaneous (s.c.) injections of 1×10^6 melanoma
tumor cells that had been irradiated to 15,000 rads and had been
mixed with complete Freund's adjuvant. Mice were given 1 injection
every 2 weeks over a six-week interval. These animals were challenged s.c. 2 weeks after the last injection with 1×10^5 viable
unirradiated B16 cells. Only mice rejecting the lethal challenge
of tumor cells were considered immunized.

B16 Melanoma Cultures. The transplantable B16 melanoma was
adapted to growth in tissue culture as described previously (3).
Stock cultures were maintained on glass in Eagle's minimum essential
medium, supplemented with 10% fetal calf serum, vitamin solution,
sodium pyruvate, nonessential amino acids, penicillin-streptomycin,
and L-glutamine (Grand Island Biological Co., Grand Island, New
York). This medium was designated as complete minimum essential
medium (CMEM). The cells were cultured in a humidified 37 C incubator containing 5% CO_2.

Preparation of Normal and Immunized Lymphocytes. Axillary,
cervical, and mesenteric lymph nodes and spleens were collected
aseptically from normal or immunized animals, placed in Hanks'
balanced salt solution (HBSS), and forced through a wire mesh
sieve. The resulting suspensions were filtered through a glass

wool column and centrifuged, and the cellular pellets were resuspended in CMEM. Viability was about 95% as determined by the
trypan blue dye exclusion test. These lymphocyte populations were
determined to have approximately 1% contamination with large mononuclear cells morphologically and functionally characterized as
macrophages.

Preparation of Peritoneal Macrophage Cultures. Macrophages
were obtained from normal C57BL/6 mice or from mice immunized
against the B16 melanoma. Mice were injected intraperitoneally
with 2.5 ml thioglycollate and killed 4 to 5 days later. Their
peritoneal exudate cells were harvested by repeated washing with
HBSS containing heparin (2 units/ml). The cells were then centrifuged, resuspended in CMEM, and plated in culture dishes (37 C, 5%
CO_2). At 4 to 5 days after thioglycollate, the peritoneal exudate
cells were composed of about 80% large mononuclear cells and about
20% small mononuclear cells, which morphologically resembled lymphocytes. The latter were completely removed from the cultures
by allowing cells to adhere to the plastic for 4 hr. At this time
the medium with all nonadherent cells was poured off and replaced
with fresh CMEM. The remaining cells might have contained some
granulocytes, but most of these died after 2 days in culture and
were removed with changes of medium. The remaining adherent cells
(3 to 4 days after initial plating) had a typical macrophage morphology and practically all cells demonstrated phagocytosis of
carbon particles. Macrophages were harvested with the aid of a
rubber policeman for the subsequent in vitro assays.

Tumor Growth Stimulation-Inhibition Assay. The basic technique has been described previously (2,3) in great detail. The
colony inhibition-stimulation test and its various microculture
modifications necessitate long and tedious microscopic observations
and counting of viable target cells to assess the degree of cytotoxicity accurately. The search for a more suitable technique has
been considerably advanced with the advent of radioactive labelling
of target cells and subsequent monitoring of isotope uptake or
release, indicating target cell damage. The details and virtues
of the [125]IUDR-labelling procedure for studies of tumor cells have
been previously reported from our laboratory (2).

The B16 melanoma tumor cells were harvested by short trypsinization (0.25%, 1 min). Cells were centrifuged and resuspended in
CMEM. Lymphocyte and/or macrophage suspensions were prepared as
described above. Following viability tests, the ratio of live
tumor cells and immune cells was adjusted as follows:

Tumor Cell	Lymphocyte	Macrophage
1	100	---
1	10,000	---
1	---	10
1	---	100
1	---	200
1	10,000	10
1	10,000	100

The above tumor-immune cell mixtures were rotated at room temperature for 90 min and then plated into 60 x 15 mm tissue culture dishes in triplicate (approximately 2.5 x 10^3 tumor cells per dish). On day 2 of the assay, 1 ml of CMEM was added to the cultures. On day 5 after plating, all cultures were re-fed with CMEM containing 0.3 µCi/ml ^{125}IUDR (New England Nuclear). Twenty-four hr later, the cultures were washed with HBSS to remove all unbound radioiodine and then were lysed with 0.5N NaOH. The lysate and 2 subsequent washes were monitored for radioactivity in a sodium iodide crystal counter. Each sample was counted 3 times, 5 min each. Statistical analysis was carried out by Student's t test.

RESULTS

Several in vitro experiments were performed with the B16 melanoma serving as target cells. The experiments varied with respect to the immune status of the donor. The results were similar and are summarized in Figures 1-3.

As shown in Fig. 1, cells from normal mice produced the following results: a) lymphocytes from normal C57BL/6 mice did not significantly affect tumor growth in vitro; b) macrophages from the same normal donor did not significantly affect tumor growth; and c) when lymphocytes (10,000:1) and macrophages (100:1) were mixed with the target (immune cell-tumor cell) slight tumor inhibition was observed ($P < 0.05$).

As illustrated in Fig. 2, the results differed when studies were carried out with lymphoid cells from immunized animals as follows: a) lymphocytes at 10,000:1, but not at 100:1, significantly inhibited tumor growth ($P < 0.01$); b) macrophages from immunized mice at 100:1 ratio also inhibited tumor growth ($P < 0.05$); and c) the most pronounced tumor growth inhibition was observed when both 10,000:1 lymphocyte and 100:1 macrophages were added to the tumor cultures ($P < 0.01$).

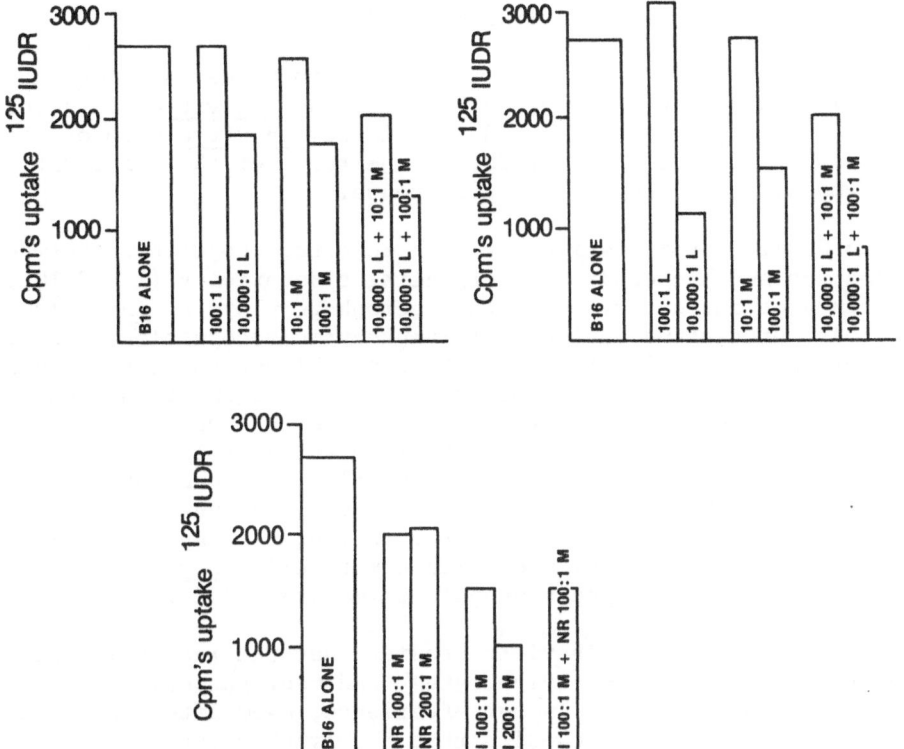

Figs. 1-3. In vitro reactivity of syngeneic lymphocytes and/or
macrophages with B16 melanoma. Fig. 1 - normal C57BL/6 (upper
left); Fig. 2 - immunized C57BL/6 (upper right); Fig. 3 - normal
and/or immunized C57BL/6 (lower).

        Thus it appeared that similar degrees of inhibition were ob-
served with either 10,000:1 lymphocyte to tumor or 100:1 macrophage to tumor ratios. In subsequent studies, only macrophages
from either normal or immunized animals were employed. The results
shown in Fig. 3 indicate that: a) macrophages from normal mice did
not produce significant inhibition; b) macrophages from immunized
mice were inhibitory to the B16 tumor and the degree of inhibition
was proportional to the number of the macrophages (P< 0.01); and
c) the level of inhibition observed with macrophages from immunized
mice was not altered by the addition of macrophages from normal
mice.

DISCUSSION

The interaction of lymphocytes and macrophages in effecting
syngeneic tumor cytotoxicity, both in vivo and in vitro, is becoming
recognized. In the present study, a six-day in vitro cytotoxicity
assay was utilized to further define the roles of the different
host lymphoid elements in such a response.

To date, many investigators have reported cooperation between
lymphocytes and macrophages in effecting cytotoxicity (1,6,7,10,11,
12). In some antigen systems, the macrophage plays an inductive
cell role, processing antigen for lymphocytic recognition, and/or
effector cell role with resultant target cell cytotoxicity. This
latter phenomenon might be thought of as activation of the phagocytic effector cell by the lymphocyte, either through direct cell-
to-cell interaction, or through the elaboration of lymphocytic
mediators. Macrophages from normal animals, although not apparently
cytotoxic against the syngeneic B16 melanoma, can be made cytotoxic
by supernatants derived from cultures of syngeneic spleen cells
sensitized in vivo and cultured with tumor cells in vitro, or by
incubation with sensitized syngeneic thymocytes (1,6,11).

It is important to in vitro studies that the respective cell
populations (tumor cells, lymphocytes, and macrophages) be as
homogeneous as possible. In our experience, even after purification,
morphologic examination revealed that the lymphocytes contained up
to 1% large, phagocytic, glass-adherent, mononuclear cells. This
level of contamination, however, was not observed with the populations of macrophages where initially only an occasional lymphocyte
or granulocyte was observed in the cultures. This result is inherent to the procedure of isolating the peritoneal cell populations
which were: a) washed 4 hr after initial plating to remove practically all nonglass-adherent cells; and then b) cultured for at
least 3 days before being employed in the assay. This effectively
eliminated the granulocytic population to levels below visual detection.

Thus, it is possible that the cytotoxicity observed in the
syngeneic lymphocyte-tumor culture could have been due to a small
number of macrophages within this culture. Indeed, the levels of
inhibition seen with 200:1 immune macrophages were similar to those
observed with 10,000:1 lymphocytes plus 100:1 macrophages. These
findings might not be consistent for all tumor systems. Specifically, it is entirely possible that host reactivity to syngeneic tumors is in many respects different from that of an allogeneic tumor. Nevertheless, it is felt that the syngeneic studies reported
above offer further insight into the complex host-tumor relationship.

In summary, we demonstrated previously that low ratios of syngeneic lymphocytes to tumor cells enhanced tumor growth in vitro and in vivo while high ratios of the same lymphocytes were inhibitory. Studies were performed to determine if a small number of syngeneic macrophages within a purified lymphocyte preparation may have been responsible for the observed inhibition. Effects on tumor growth by syngeneic lymphocytes, macrophages or both were tested in vitro. Various numbers of lymphocytes alone, macrophages alone or both were added to B16 cells and cultured. Tumor cell proliferation was assayed by monitoring the uptake of ^{125}I-iododeoxyuridine. Lymphocytes from immunized, but not normal mice, inhibited tumor growth in vitro. Normal mouse macrophages had no effect on the tumor, while macrophages from immunized mice significantly inhibited tumor growth. Furthermore, it was determined that contaminant macrophages in lymphocyte populations contributed to in vitro cytotoxicity in this system.

ACKNOWLEDGEMENTS

This research was sponsored by the National Cancer Institute under Contract No. N01-CO-25423, MOD. 27 with Litton Bionetics, Inc. and DE 281-03 from the NIDR.

REFERENCES

1. Evans, R. and Alexander, P., Transplantation, 12 (1971) 227.
2. Fidler, I.J., Immunological Comm., 2 (1973) 483.
3. Fidler, I.J., J. Nat. Cancer Inst., 50 (1973) 1307.
4. Fidler, I.J., Cancer Res., 34 (1974) 491.
5. Fidler, I.J., Cancer Res., 34 (1974) 1074.
6. Fidler, I.J., J. Nat. Cancer Inst.(1975) in press.
7. Gottlieb, A.A. and Waldman, S.R., Macrophages and Cellular Immunity (Ed. A. Laskin and H. Lechevalier) CRC Press, Cleveland, Ohio (1972) 13.
8. Jeejeebhoy, H.F., Int. J. Cancer, 13 (1974) 665.
9. Kolb, W.P. and Granger, G.A., Cell. Immunol., 1 (1970) 122.
10. Lipsky, P.E. and Rosenthal, A.S., JEM, 141 (1975) 138.
11. Lohmann-Matthes, M.L., Ziegler, F.G. and Fischer, H., Europ. J. Immunol., 3 (1973) 56.
12. North, R.J., Infect. Immun., 10 (1974) 66.
13. Prehn, R.T., Science, 176 (1972) 170.
14. Shearer, W.T., Philpott, G.W. and Parker, C.W., Cell. Immunol., 17 (1975) 447.

INHIBITION OF GROWTH AND DISSEMINATION OF SHAY MYELOGENOUS LEUKEMIC TUMOR IN RATS BY GLUCAN AND GLUCAN ACTIVATED MACROPHAGES

N. R. DI LUZIO, R. MCNAMEE, E. JONES, S. LASSOFF, W.
SEAR and E. O. HOFFMANN
Tulane University School of Medicine
New Orleans, Louisiana (USA)

Among the various agents which have been employed in an attempt to delineate the physiological and immunological role of the reticuloendothelial system (RES) has been zymosan. Zymosan was the name given by Pillemer and Ecker (33) to a yeast fraction having the property of inactivating the third component of complement. This cell wall preparation, derived from Saccharomyces cerevisiae, was initially demonstrated to be an effective reticuloendothelial (RE) stimulant by Benacerraf and Sebestyen (4). Zymosan was reported by Northcote and Horne (31) to be composed of protein, lipid and complex polysaccharides, namely mannan and glucan. Di Carlo and Fiore (11) reported that the composition of the yeast cell wall was approximately 58% glucan and 18% mannan. Glucan comprises the inner cell wall with mannan comprising the outer cell wall (30). Since the observations of Benacerraf and Sebestyen (4), zymosan has been uniformly demonstrated to produce marked stimulation of the RES. These events include increased phagocytosis, increased rate of intracellular degradation of phagocytized particles, increased resistance to infection, increased properdin levels, enhanced humoral immunity and inhibition or regression of various tumors (9, 12, 28, 32, 38).

In view of the profound biological activities produced by the administration of zymosan, studies were initiated in our laboratory in the late 1950's to isolate and identify the component of zymosan which possessed the specific ability to initiate activation and proliferation of the RES and thereby increase a variety of host defense mechanisms. In an extensive evaluation of various components present in zymosan, we reported in 1961 that the active RE stimulant in zymosan was glucan (34). Glucan has been characterized as a

polyglucose or neutral polysaccharide consisting of a chain of
gluco-pyranose units united by a β-(1,3) glucosidic linkage (23).
In addition to the major β-(1,3) D-glucan component, a minor β-(1,6)
D-glucan component has also been reported (2). The latter component
of yeast glucan has been found to be inactive relative to macrophage
activation and tumor inhibition (our unpublished observations).

On an experimental basis (13), the simultaneous subcutaneous
administration of glucan with Shay myelogenous leukemia cells was
found to significantly inhibit growth of the tumor. In clinical
studies involving 3 types of metastatic lesions, the intralesional
administration of glucan produced, in all cases studied, a rather
prompt and striking reduction in the size of the lesion (29). Regression of the glucan injected tumor was associated with selective
necrosis of the malignant cells and a pronounced monocytic infiltrate.

Since glucan appeared to offer distinct advantages (13) over
all other currently employed forms of immunotherapy which employ
viable organisms, such as BCG or C. parvum (Table I), the present
study was undertaken to extend our experimental observations on the
utilization of this unique polyglucose in the prevention and treatment of the Shay myelogenous leukemia tumor. In view of our recent
studies which appeared to denote the importance of the number and
function of host macrophages in resistance in neoplasia, as well
as tumor macrophages number and function in regard to growth and
dissemination of the malignant cells (14), additional studies were
conducted in which varying concentrations of glucan-activated peritoneal macrophages were added to tumor cells at the time of transplantation. The influence of altered tumor-macrophage cell ratios
on growth and dissemination of the tumor was ascertained. These
composite studies denote that glucan activation of macrophages,
either pre- or post-tumor cell transplant, significantly modified
the course of the leukemia. Additionally, the growth as well as
dissemination of tumor cells appears to be regulated by the number
of macrophages which exist within the primary tumor site.

                              RESULTS

The Shay tumor (36) was specifically selected as the tumor of
choice in view of our previous studies on the role of recognition
factor in controlling macrophage surveillance of malignant cells
(15-17). The characteristics of this uniquely virulent tumor are
listed in Table II and are essentially derived from the studies
of Lapis and Benedeczky (26) and Handler and Handler (21), as well
as our studies.

TABLE I

Clinical Impressions Relative to the
Comparative Toxicity – BCG and Glucan*

Patient Response	BCG	Glucan
Ulceration	+	none
Hypersensitivity	+	none
hypotension	+	none
chills	+	none
fever	+	none
nausea	+	none
vomiting	+	none
Malaise	+	1/23
Liver dysfunction	+	? (unknown) –in animals none
Hepatic granulomas	+	?
Facilitation of tumor growth	on occasion	none
Bacteremia	on occasion	none
Fatalities	on occasion	none
Problems in employment of immunostimulant:	viability strain	defined chemically
Possible Metabolites:	unknown	glucose (?)

*Glucan toxicity data are obviously much more limited both in time
and patient population and therefore must be considered preliminary
in nature. Above expressions with glucan based upon the preliminary published and unpublished observations of Mansell, et al.
(29).

TABLE II

Major Characteristics of Shay Myelogenous Leukemia Rat Tumor

---

1. Induced in Wistar rats by prolonged methylcholanthrene administration (1951).

2. Homogenous polygonal myeloblastic cells - loosely arrayed.

3. No intercellular attachments.

4. Large nuclei-varied shape and size.

5. Little cytoplasm and intracellular material - few mitochondria.

6. Rich homogeneously distributed free ribosomes, little RER.

7. Presence of pseudopodia.

8. Presence of virus-like particles of C type.

9. Few - if any - pigment granules (veroperoxidase).

10. Active pinocytosis.

11. Extremely rapid growth - any site.

12. Rapid dissemination to other organs (bone marrow, spleen, lung,
    liver, spinal cord, brain, thymus, lymph nodes, muscle).

TABLE III

Influence of Subcutaneous Implant of Glucan on
Growth of Shay Myelogenous Leukemia Cells*

	No.	Body Weight	Tumor Weight (g)
Saline	14	129 ± 8	12.0 ± 2.2
Glucan	18	132 ± 8	2.3 ± 0.5

*Tumor weight was determined 10 days post transplantation. Values
are expressed as mean $\pm$ standard error.

The influence of the simultaneous administration of yeast glucan (34) on the growth of the 20 x 10^6 subcutaneously implanted Shay tumor cells is presented in Table III. In agreement with previous observations (13), the administration of glucan (20 mg) significantly decreased tumor growth. Since glucan, when injected subcutaneously in rats, produced a profound monocytic infiltrate due to its chemotactic nature, the inhibition of tumor growth probably reflects an alteration in tumor macrophage ratio in favor of the latter cell. Additionally, since glucan is a profound activator of macrophages, the activation of macrophages at the tumor site may also contribute to the inhibition of tumor growth.

To determine whether or not glucan may modify the development of leukemia and therefore survival to intravenously administered leukemic cells, rats were pretreated with either saline, the lipid emulsion which was employed as a suspending medium for glucan (29), or glucan. The administration of glucan was made 1, 4 and 6 days prior to the administration of tumor cells which was conducted on the eighth day. As can be noted in Table IV, the 28-day survival in saline and lipid controls approximate 10%. In contrast, a significant decrease in mortality was seen in the glucan-pretreated group in which only a third of the animals succumbed to the leukemic episode.

TABLE IV

Influence of Pretreatment with Glucan on Survival of Rats to the Transplantation of 1 x 10^6 Shay Myelogenous Leukemia Cells

	No.	28-day Survival
Saline	15	6%
Lipid	26	15%
Glucan-Lipid	25	64%

Glucan in the amount of 5 mg prepared in lipid emulsion was given intravenously on days 1, 4 and 6 with tumor cells injected on day 8. No further injections of glucan were given following tumor cell administration.

In an effort to determine whether glucan may be used as a
treatment modality following the administration of leukemic cells,
rats were transplanted with 5 x 10^6 Shay myelogenous leukemic cells
intravenously. One and 3 days following transplantation glucan
was administered in 5% glucose intravenously in the amount of 50
and 25 mg, respectively. Control animals were injected with isotonic glucose. As can be noted (Table V), a significant enhancement in survival was observed in the post-treatment group. Additionally, studies have indicated that the development of leukemia
following leukemic cell transplant was identical in the glucan-
and dextrose-treated group. However, leukocyte levels rapidly
normalized in the glucan group denoting rejection of the leukemic
cell transplant. No signs of leukemia were present in the glucan-
treated group.

TABLE V

Enhanced Survival of Glucan-Treated Leukemia Rats*

	No.	28-day Survival
Control	20	15%
Glucan	16	88%

*Long-Evans rats (300 g $\pm$ 11) were given 5 x 10^6 Shay myelogenous
leukemia cells intravenously. One and 3 days following transplantation, 50 and 25 mg of glucan were given intravenously. Control rats were injected with isotonic dextrose and survival ascertained.

In an effort to further establish the importance of tumor
macrophages to the growth of the primary tumor as well as to its
dissemination, tumor cells were administered to normal rats either
alone or in the presence of varying concentrations of glucan-activated peritoneal macrophages. As can be noted in Table VI, in 2
experiments that were conducted in which tumor weights in the control groups varied by a factor of 3, a significant inhibition of
tumor weight was observed, particularly when macrophage-tumor cell
ratios approximated 1:1. In addition to the inhibition of tumor
growth as reflected by the weight of the tumor, distinct histological changes were noted in liver, lung and spleen, as well as
in the tumor of the animals bearing tumor implants with varying
macrophage populations.

TABLE VI

Influence of Addition of Glucan-Induced Peritoneal Macrophages
on Growth of Shay Acute Myelogenous Leukemia in Rats

Group**	Tumor Cells (x 10^6)	Peritoneal Macrophages*** Added (x 10^6)	Tumor Weight (g) Exp 1	Tumor Weight (g) Exp 2
Control	20	–	5.0 ±0.66	18.8 ±1.73
Control	40	–	–	18.4 ±1.73
Experimental	20	2	2.9 ±0.71*	14.6 ±1.87
Experimental	20	10	2.3 ±0.63*	13.0 ±2.65
Experimental	20	20	2.3 ±1.10*	12.5 ±1.00*

*p 0.5
**Each group had 5-7 rats in experiment 1 and 5-6 rats in experi-
ment 2, except for the control group (injected with 40 x 10^6 tumor
cells) which comprised 16 animals. Tumor weights were ascertained
on day 10 in experiment 1 and day 12 in experiment 2.
***Peritoneal macrophages were obtained 4 days following the intra-
peritoneal administration of 10 ml of a 0.5% suspension of glucan.
Tumor cells and glucan activated macrophages were subcutaneously
administered.

    In the control group which received tumor cells alone, liver
showed a heavy infiltrate of tumor cells which compressed and displaced parenchymal cells (Fig. 1). Tumor masses were extremely
prominent around portal spaces. The septi of the alveoli were
heavily infiltrated with tumor cells and occasionally blood vessels showed the presence of tumor emboli (Fig. 2). A heavy infiltrate of tumor cells, particularly in the red pulp, was observed in the spleen. Megakaryocytes were present in extremely
small numbers. In the tumor itself, which was characterized by
the presence of myeloblastic cells, limited areas of necrosis

and hemorrhage were observed. A relatively small number of mononuclear cells were present.

In the group that received tumor cells:macrophage implants at
the ratio of 10:1, the infiltrate of tumor cells in the liver was
not as prominent. Occasional macrophages which appeared to have
phagocytized tumor cells were present. The tumor infiltrate of
lung was quite extensive as was that of spleen. The tumor presented
irregularities in necrosis and hemorrhage and a mild infiltrate of
vacuolated macrophages.

In the animals which received tumor cells:macrophages in the
ratio of 2:1, the liver showed a significant decrease in tumor cell
infiltrates and indeed in some livers only a few tumor cells were
present. Likewise, the infiltrate of tumor cells in the lung was
characterized as mild to absent. Tumor infiltration of the spleen
was graded extensive to mild.

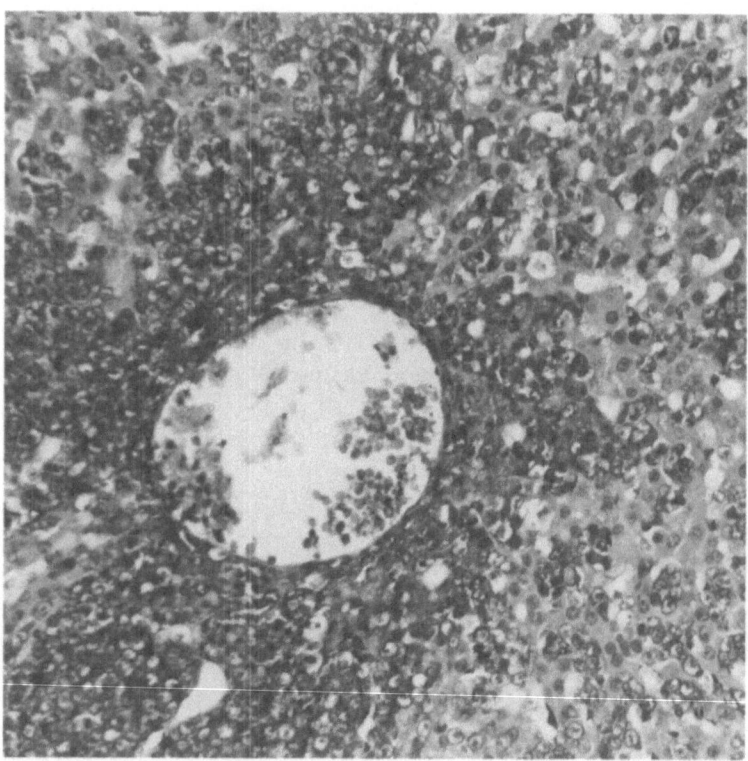

Fig. 1. Liver from a control rat injected with tumor cells and
saline shows a heavy infiltrate of tumor cells especially aggregated around a branch of the portal vein where hepatocytes have
been completely replaced by tumor cells. (H and E 250x)

(Figures 1-5 have been reduced 10% for reproduction).

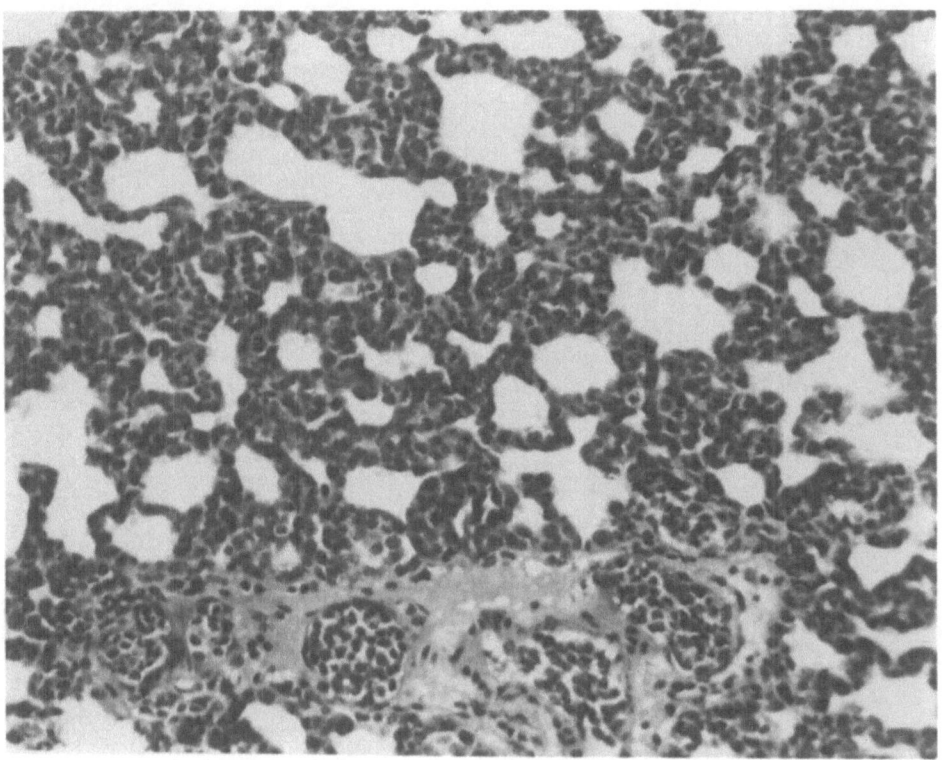

Fig. 2. Lung of a control animal injected with tumor cells and saline shows tumor infiltrates in the alveolar septi. Large aggregates of tumor cells appear to occlude blood vessels. (H and E 250x)

The most prominent histological findings were observed in those animals which received macrophages and tumor cells in the ratio of 1:1. In this group the presence of tumor cells in liver was judged to be slight to absent (Fig. 3). When tumor cells were present, they appeared to be undergoing destruction by the mononuclear population. The lung appeared normal (Fig. 4), and the spleen, relative to degree of tumor infiltrate, was characterized by mild to absent. The tumor presented more extensive areas of necrosis and hemorrhage and had a moderate infiltrate of vacuolated mononuclear cells (Fig. 5). These findings denote that by increasing tumor macrophage population to enhance the ratio of macrophages to tumor cells, a decreased dissemination of malignant cells occurred in association with a more extensive necrosis of the primary lesion.

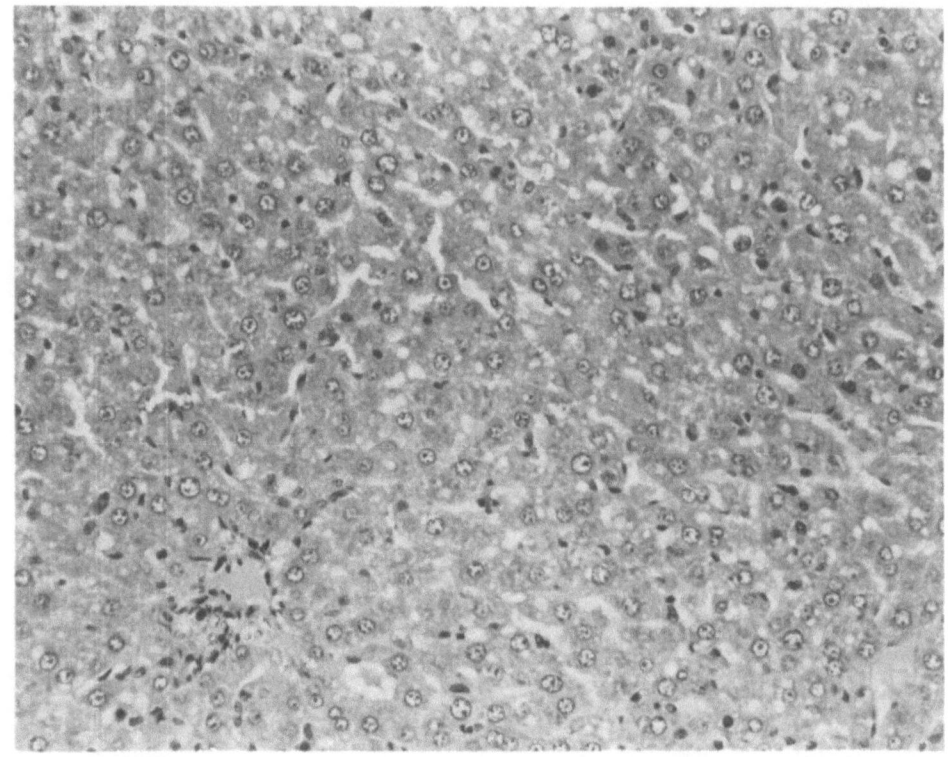

Fig. 3. Liver of an animal that received tumor cells and glucan-
stimulated macrophages (1:1 ratio) possesses essentially normal
structure with prominent Kupffer cells and an absence of disseminated tumor cells. (H and E 250x)

DISCUSSION

There is little question that in addition to the killer ability of T lymphocytes, macrophages (particularly those which are
activated either specifically or non-specifically) possess the
ability to destroy malignant cells by both a phagocytic and an,
as yet undefined, contact lysis mechanism (1,5). Krahenbuhl and
Lambert (25), studying the in vitro interaction between mouse
macrophages and tumor target cells, observed that activated macrophages had an enhanced ability to inhibit DNA synthesis in L-929
tumor cells. Direct contact between macrophages and target cells
was required for cytostasis. Employing 4 different assays of cytotoxicity, a cytotoxic effect of the peroxidase-H_2O_2 halide anti-
microbial system was demonstrated by Clark, et al. (8) on mouse
ascites lymphoma cells. The definitive contribution of the perox-

idase system to the killing ability of macrophages as the non-phagocytic mechanism of tumor killing is as yet to be defined.

The importance of macrophages to inhibition of tumor growth was recently stressed by Haskill, et al. (22) who employed velocity sedimentation to fractionate enzymatically or mechanically dispersed tumor cell suspensions. They reported that effector cells within 2 rat sarcomas evaluated were activated macrophages and that the proliferation of macrophages in vitro was inhibited by the presence of tumor cells (22). Whether glucan-enhanced activation and proliferation of macrophages may overcome any inhibitory action of the tumor on macrophage effector cells remains to be established.

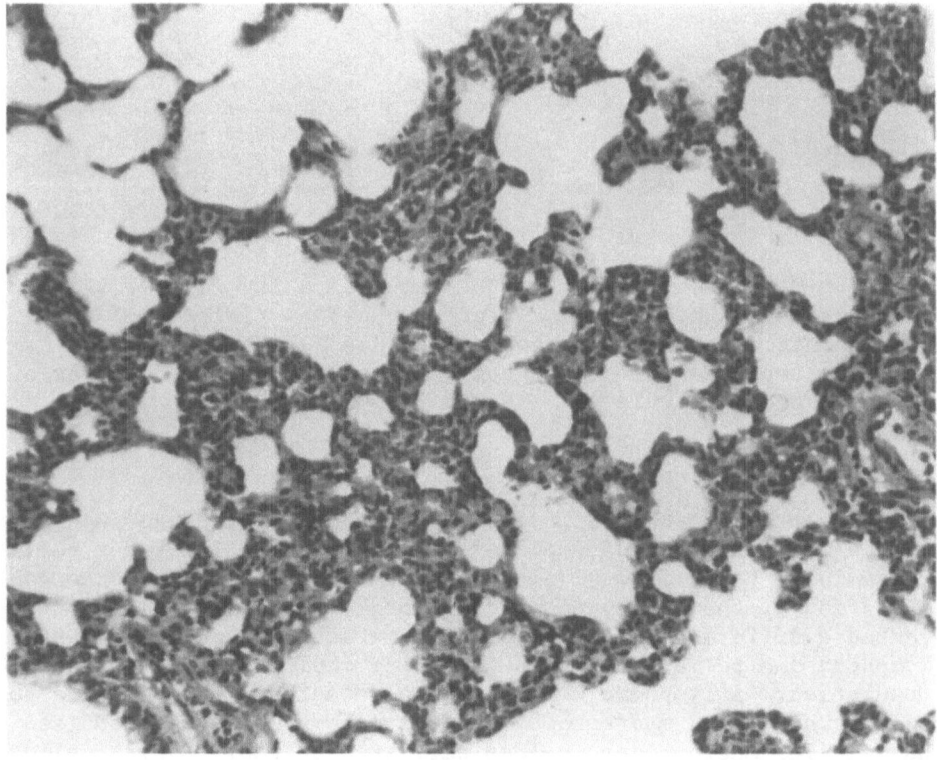

Fig. 4. Lung from an animal injected with tumor cells and glucan stimulated macrophages (1:1 ratio), presents normal structure. Blood vessels in the lower part of the photograph are free of tumor cells. (H and E 250x)

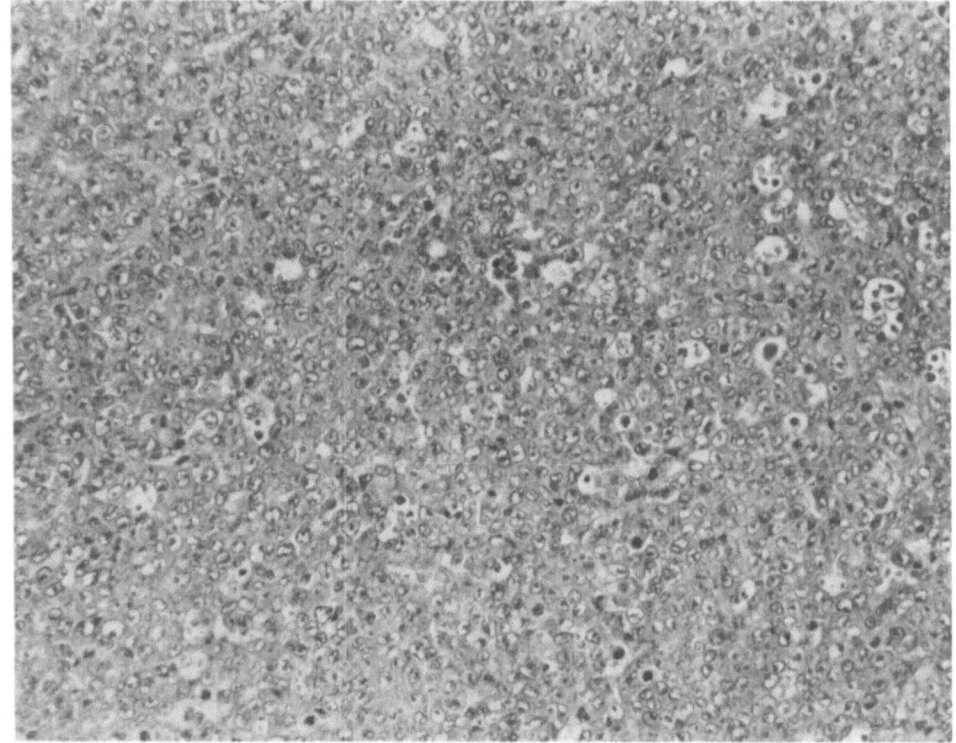

Fig. 5. Microphotograph from the tumor of a rat injected with tumor cells and glucan-stimulated macrophages (1:1 ratio). The section has a "starry sky" appearance due to the presence of numerous macrophages which have phagocytized tumor cells. Areas of necrosis are also evident. (H and E 250x)

The present studies indicate that tumor macrophage populations play a major role in determining the degree at which tumor cells disseminate to peripheral sites. This finding is in essence confirmation of the observations of Birbeck and Carter (6) who predicated a difference in behavior of a metastatic and non-metastatic lymphoma due to variations in their intratumor host macrophage population. A lymphoma which was non-metastasizing was found to possess activated macrophages while lymphoma which metastasized had smaller macrophages in less activated states. Additionally, the previous studies of Gershon, et al. (2) further add to the concept that failure of macrophage mobilization and activation intralesionally contributes significantly to the dissemination of tumor cells. They observed that sinus histiocytosis, which resulted in the region of a non-metastasizing lymphoma, was absent in lymph nodes draining a metastasizing lymphoma. The clinical importance of such observations may be denoted by the studies of Baum, et al. (3). These investigators studied macrophage phagocytic activity in patients with breast cancer. They reported that as long as the macrophage response of the host remained intact, as assayed by phagocytic response, the tumor remained localized to the breast and underlying muscle. In those individuals in which the tumor disseminated to lymph nodes and beyond, profound impairments were seen in macrophage function. These studies further contribute to an appreciation of the potential importance of macrophages as determinant cells in tumor cell dissemination. More recently, Keller (24) has demonstrated that activated peritoneal macrophages rapidly and effectively eliminate syngeneic tumor cells in vitro. Keller's studies indicated effector target cell ratios of 10:1 and 5:1 as well as a pronounced inhibition of thymidine incorporation of tumor cells at macrophage:target cell ratios of 1:1. The addition of silica to the incubation medium, which is selectively toxic to macrophages, eliminated the anti-tumor effect of the macrophage population. Eccles and Alexander (19) reported that a high macrophage content of the tumor was associated with decreased metastases while conversely mouse tumors which possess low macrophage content showed increased dissemination.

Eccles and Alexander (18) have indicated the existence of a monocytic defect rather than lymphocyte impairment in tumor induced "anergy". In their studies, immunological competence could be restored in rats possessing tumor implants by the administration of normal peritoneal macrophages, but not lymphocytes. From their work, they suggested that the resulting immunological anergy which develops as the tumor grows is due to the fact that the tumor competes for the available blood monocytes. They concluded that the state of anergy, in association with tumor development, may not be a failure of immunological recognition or reactivity, but due to the unavailability of monocytes to fulfill the expression of the immunological phenomenon. It is, therefore possible that by increasing the availability of macrophages through our mechanisms of glucan-induced macrophage proliferation and activation that immunological competence can be maintained.

It is evident on the basis of our experimental and clinical studies that glucan provides a unique means to initiate an enhancement of host resistance to neoplasia by macrophage activation and proliferation with a resulting significant increase in the level of host resistance. It is obvious that the ability to induce a functional macrophage population within the tumor mass, which has the ability to selectively kill neoplastic cells, would be of significant clinical importance. These macrophages would not only tend to function to eliminate neoplastic cells as such, but their in vivo events associated with macrophage antigen processing and

initiation of both T- and B-cell responsiveness would contribute to
a decrease in tumor cell dissemination. This concept is well supported by the reports of Vorbrodt, et al. (39-41). Vorbrodt, et
al. (41) have reported that during X-ray therapy of human skin cancer, macrophages were found to be actively engaged in tumor cell
destruction. Intimate contact between "invading macrophages" and
cancer cells with concomitant formation of cytoplasmic bridges and
fusion zones were frequently observed following radiation therapy.
In subsequent investigations to pursue this finding, Vorbrodt, et
al. (40) demonstrated that irradiation of macrophage and tumor cells
in culture was associated with macrophage activation, increase in
lysosomes and phagolysosomes as well as associated macrophage-ad-
herence to cancer cells and the presence of phagocytized cancer
cells. Vorbrodt, et al. (39) suggested that x-irradiation of the
tumor cells results in alterations in the tumor cell membrane which
promotes adhesion and contact between the macrophage and the tumor
cell and, in an as yet unknown fashion, leads to degeneration of
the tumor cell and phagocytosis of the cell by the macrophages.

The anti-tumor effect of yeast glucan is clearly evident by
the present experimental studies as well as our previous clinical
endeavors (29). Glucans obtained from a variety of other sources
have also been demonstrated to have anti-tumor activity in various
experimental systems (7,10,27,35,37). Sakai, et al. (35) have compared the anti-tumor action of certain glucans and attempted to relate anti-tumor action of chemical structure. Those glucans possessing a linear β-1,3 linked D-glucose structure were most effective. Glucans composed of alpha-configuration were ineffective.

In the studies of Sakai, et al. (35) the tumor employed was a
Sarcoma 180 in ascites form and injections of glucan were made intraperitoneally 24 hr post tumor cell transplantation and then
every other day for 20 days. Tumor growth was assayed on day 31.
The authors point out that with the exception of a tendency for a
slight weight loss in the glucan-treated group, no other ill effect
was noted.

Singh, et al. (37) also studied the anti-tumor action of a
scleroglucan derived from Sclerotium glucanicum. This glucan possesses a main chain of β-(1,3) D-glucopyranosyl units with every
third or fourth unit carrying a β-(1,6) D-glucopyranosyl group.
In agreement with the observations of Sakai, et al. (35), Singh,
et al. (37) indicated that scleroglucan did not show toxic effects
in test animals, which remained in excellent physical condition
throughout the testing period.

Chihara, et al. (7) have reported that lentinan, a water soluble extract of an edible mushroom, Lentinus edodes, possessed
strong anti-tumor activity when evaluated against Sarcoma 180.

They also noted in their report that no signs of toxicity were observed in lentinan-treated mice. The absence of overt toxicity
with glucan may be of future significance in their consideration
as immunostimulants.

The same investigators (7) also reported that the glucans,
which are very effective in modifying growth in experimental animals, were ineffective in modifying growth of tumor cell cultures
thus denoting that the action was host mediated and not directly
cytocidal. When our yeast glucan preparation was incubated with tumor cells <u>in vitro</u>, no direct cytotoxicity was manifested, confirm-
ing the observation of Chihara, et al. (7) that the anti-tumor effect of glucan is host mediated. They further observed that unlike
endotoxins, which cause cell necrosis also, glucan left no lesions
at the site of complete tumor regression, an event seen in our clinical studies.

Dennert and Tucker (10) have reported that lentinan, a linear
β- 1,3 glucan, obtained from the mushroom lentiusedodes possesses
pronounced anti-tumor activity. Anti-tumor effect was found in
normal, but not in neonatally thymectomized, mice denoting that an
intact T-cell system was a prerequisite for the demonstrated anti-
tumor effect of lentinan. It remains to be established whether
yeast glucan is also a T-cell adjuvant and whether its anti-tumor
activity is mediated through the proliferation and activation of
macrophages or an enhancement in T-cell populations and function.
Lentinan apparently does not exert its anti-tumor activity by macrophage stimulation, or by an enhancement in cell or humoral immunity (27), which appears to be the basis of our glucan-mediated
destruction of malignant lesions. Lentinan has been credited to
exert its effect by stimulation of histamine or serotonin (27).

There is little question in view of the diverse reports of
various investigators that glucans derived from various sources
but possessing a β-1,3 configuration have significant anti-tumor
activity against a variety of tumors in diverse animal species,
including man. It is obvious that glucan is a unique agent not
only relative to its ability to activate host defense mechanisms,
including those directed against malignant and non-malignant aber-
rant cells, but also as a chemically defined agent employable in
evaluation of the concept of immunotherapy and immunoprophylaxis.

These composite studies clearly establish the importance of
macrophages as host defense cells against the growth of leukemic
tumor cells as well as the importance of macrophages in determining
dissemination of malignant leukemic cell populations. It can well
be anticipated that through a further delineation of the role of
macrophages in tumor cell destruction as well as the importance of
macrophage populations of tumors to tumor growth and dissemination,

an appreciation of host defense mechanisms against neoplastic states will be forthcoming.

ACKNOWLEDGEMENT

This study was supported in part by USPHS Grant No. CA 13746.

REFERENCES

1. Amos, D.B., Ann. N. Y. Acad. Sci., 87 (1960) 273.
2. Bacon, J.S.D., Farmer, B.C., Jones, D. and Taylor, I.F., Biochem. J., 114 (1969) 557.
3. Baum, M., Sumner, D., Edwards, M.H. and Smythe, P., Brit. J. Surg., 60 (1973) 899.
4. Benacerraf, B. and Sebestyen, M.M., Fed. Proc., 16 (1957) 860.
5. Bennett, B., Old, L.J. and Boyse, E.A., Transplantation, 2 (1964) 183.
6. Birbeck, M.S.C. and Carter, R.L., Int. J. Cancer, 9 (1972) 249.
7. Chihara, G., Hamuro, J., Maeda, Y.Y., Arai, Y. and Fukuoka, F., Cancer Res., 30 (1970) 2776.
8. Clark, R.A., Klebanoff, S.J., Einstein, A.B. and Fefer, A., Blood, 45 (1975) 161.
9. Cutler, J.L., J. Immunol., 84 (1960) 416.
10. Dennert, G. and Tucker, D., J. Nat. Cancer Inst., 51 (1973) 1727.
11. Di Carlo, F.J. and Fiore, J.V., Science, 127 (1958) 756.
12. Diller, I.C. and Mankowski, T.Z., Extrait Acta Int. Contre Cancer, 16 (1960) 584.
13. Di Luzio, N.R., The Reticuloendothelial System, Williams and Wilkins Co., Int. Acad. Path. Monogr.,5 (1975) 49.
14. Di Luzio, N.R., Trace Components of Plasma: Isolation and Clinical Significance (Ed. G.A. Jamison and T.J. Greenwald) Alan Liss Publishing, Inc., New York, New York (1976) 1.
15. Di Luzio, N.R., Miller, E.F., McNamee, R. and Pisano, J.C., J. Reticuloendothel. Soc., 12 (1972) 314.
16. Di Luzio, N.R., McNamee, R., Olcay, I., Kitahama, A. and Miller, R.H., Proc. Soc. Exp. Biol. Med., 145 (1974) 311.
17. Di Luzio, N.R., Miller, E.F., McNamee, R. and Pisano, J.C., J. Reticuloendothel. Soc., 11 (1972) 186.
18. Eccles, S.A. and Alexander, P., Brit. J. Cancer, 30 (1974) 42.
19. Eccles, S.A. and Alexander, P., Nature, 250 (1974) 667.
20. Gershon, R.K., Carter, R.L. and Lane, N.J., Am. J. Pathol., 51 (1967) 1111.
21. Handler, E.E. and Handler, E.S., Regulation of Hematopoiesis, (Ed. A. S. Gordon) Appleton-Century-Crofts,47 (1970) 1273.
22. Haskill, J.S., Proctor, J.W. and Yamamura, Y., J. Nat. Cancer Inst., 54 (1975) 387.

23. Hassid, W.Z., Joslyn, M.A. and McGready, R.M., J. Amer.   Chem. Soc., 63 (1941) 295.
24. Keller, R., Brit. J. Exp. Path., 54 (1973) 298.
25. Krahenbuhl, J.L. and Lambert, L.H., Jr., J. Nat. Cancer. Inst., 54 (1975) 1433.
26. Lapis, K. and Benedeczky, I., Cancer, 27 (1967) 1544.
27. Maeda, Y.Y., Hamuro, J., Yamada, Y.O., Ishimura, K. and Chihara, G., Immunopotentiation (Ed. G. E. Wolstenholme and J. Knight) Elsevier, New York, New York (1973) 259.
28. Mankowski, Z.T., Diller, I.C. and Nickerson, W.J., Proc. Amer. Assoc. Cancer Res., 2 (1958) 324.
29. Mansell, P.W.A., Ichinose, H., Reed, R.J., Krementz, E.T., McNamee, R. and Di Luzio, N.R., J. Nat. Cancer Inst., 54 (1975) 571.
30. Mundkur, B., Exp. Cell Res., 20 (1960) 28.
31. Northcote, D.H. and Horne, R.W., Biochem. J., 51 (1952) 232.
32. Old. L.J., Clarke, D.A., Benacerraf, B. and Goldsmith, M., Ann. N. Y. Acad. Sci., 88 (1960) 264.
33. Pillemer, L. and Ecker, E.E., J. Biol. Chem., 137 (1941) 139.
34. Riggi, S.J. and Di Luzio, N.R., Amer. J. Physiol., 200 (1961) 297.
35. Sakai, S., Takada, S., Kamasuka, T., Momoki, Y. and Sugayama, J., Gann, 59 (1968) 507.
36. Shay, H., Gruenstein, M., Marx, H.E. and Glazer, L., J. Cancer Res., 11 (1951) 29.
37. Singh, P.P., Whistler, R.L., Tokuzen, R. and Nakahara, W., Carbohydrate Res., 37 (1974) 245.
38. Thiele, E.H., Proc. Soc. Exp. Biol. Med., 146 (1974) 1067.
39. Vorbrodt, A., Grabska, A., Krzyzowska-Gruca, S. and Gruca, S., Folia Histochem. Cytochem., 11 (1973) 185.
40. Vorbrodt, A., Grabska, A., Krzyzowska-Gruca, S. and Gruca, S., Folia Histochem. Cytochem., 11 (1973) 357.
41. Vorbrodt, A., Hliniak, A., Krzyzowska-Gruca, S. and Gruca, S., Acta Histochem., 42 (1972) 270.

MACROPHAGE PARTICIPATION IN A SPONTANEOUSLY REGRESSING SYNGENEIC TUMOR

R. B. LEVY*[1], R. L. ST. PIERRE[1] and S. D. WAKSAL[2]

Ohio State University[1],Columbus, Ohio (USA) and
NCI, National Institutes of Health[2],Bethesda, Maryland
(USA)

Investigations of experimentally induced (i.e. viral, chemical) neoplasms have begun to elucidate the significance of non-lymphoid cells in the interaction between host and tumor (1,2,3,4,5,6,7). Macrophages have been identified as effector cells in a number of tumor systems (1,2,4), and recent reports have suggested that these cells are capable of suppressing immune activity in tumor bearing hosts (3,5,6,7). We have been investigating the nature of tumor-immune system interactions occurring in the AKR mouse during the development and spontaneous regression of a syngeneic tumor. Studies carried out in our laboratory suggest that the role of the macrophage in host defense against this tumor may change during the periods of development and regression, and that failure to do so may result in irreversible tumor growth.

TUMOR DEVELOPMENT IN THE SYNGENEIC HOST

The tumor we have been examining arises in vitro, after the culturing of an undifferentiated fibrosarcoma (now in its 12th in vitro passage) of AKR origin (8). After 4 in vitro passages, an undifferentiated carcinoma-like cell predominates the plates. Intraperitoneal injection into AKR recipients of 2.5 x 10^4 or greater numbers of this latter cell type results in rapid proliferation, metastasis to liver, lung, and spleen, and death a short time later

*Present address: NCI, National Institutes of Health, Bethesda, Maryland (USA).

(Table I). Injection of 10^3 or fewer carcinoma-like cells does not
kill the host. Additional evidence of TSA was established from
subcutaneous injections into syngeneic recipients. Following inoculation of 5×10^6 cells, 2 distinct periods of tumor development
were identified (9). Tumor size increased, reaching a maximum
(12-15 mm) by day 8-9 (progressor phase), followed by spontaneous
regression by day 18-20 (regressor phase).

Paralleling tumor development in these animals is a marked
increase in spleen cell number. By the 10th day of tumor presence,
the spleens contained 1.5-2.0 x the number of mononuclear cells
found in normal, uninjected mice (Table II). This elevation in
cell number gradually receded as the tumor regressed, and returned
to near normal levels by day 18. Light and phase microscopic observations indicated a marked increase in the number of macrophages
in the spleens of animals bearing 10 day tumors. Macrophage numbers were not as evident in the spleens of day 18 animals.

TABLE I

Survival of AKR/J Mice Following Intraperitoneal
Injection of Carcinoma-Like Tumor Cells

Dose*	Death	Survival (Hrs)**
1.0×10^2	No	---
1.0×10^3	No	---
2.5×10^4	Yes	220
1.0×10^5	Yes	162
1.0×10^6	Yes	144
2.5×10^6	Yes	138
5.0×10^6	Yes	122

*All cells administered in 1.0 ml (total volume) of HBSS
**Each value is the average survival time of 4 injected mice

TABLE II

Total Spleen Cell Numbers Following Subcutaneous
Injection of 5×10^6 Tumor Cells

Time	Cell Number (10^8)
Normal (not injected)	0.65 - 0.80
Day 10	1.45 - 1.80
Day 18	0.75 - 0.90

IN VITRO RESPONSES OF SPLEEN CELLS FROM TUMOR-BEARING ANIMALS

We then observed the ability of spleen cells from mice with tumors at different stages of development to inhibit the tumor cell in vitro. Using a sensitive postlabelling assay (9) to monitor cell mediated cytotoxicity in vitro, 3 distinct periods of cytotoxic activity were identified. A representative experiment is shown in Table III. Spleen cells from progressor phases (day 5) were unable to demonstrate any inhibitory activity. Spleen cells obtained from animals at the onset of rejection manifested 35-40% tumor inhibition, and spleen cells obtained from regressor animals exhibited reduced, but significant, levels of inhibitory action. We find it noteworthy that a) the cytotoxic potential of these spleen cells paralleled the presence of increasing and decreasing macrophage numbers, and b) at the conclusion of each assay, only plates containing significant numbers of macrophages (adherent to the dishes) were found to have demonstrated significant levels of cytotoxic activity.

TABLE III

Cell Mediated Cytotoxicity by Spleen Cells from Tumor-Bearing (Immune) and Normal (Nonimmune) AKR Mice

Cells*	Day of Tumor Growth**	% Inhibition***
Immune	5	0 ± 1.4
Nonimmune		3 ± 0.8
Immune	10	34 ± 3.1
Nonimmune		5 ± 1.7
Immune	18	12 ± 2.1
Nonimmune		2 ± 1.2

*35 x 10 mm petri plates contained 3.74×10^6 spleen cells and 2.5×10^4 tumor cells (150:1)
**Days following subcutaneous injection of 5×10^6 tumor cells
***Cytotoxicity calculated $\frac{\text{Control} - \text{Experimental cpm}}{\text{Control (cpm)}}$ cpm X 100
Each value represents mean $\pm$ S.E. of 4 replicate plates

TABLE IV

Response of Spleen Cells from Tumor-Bearing
Hosts to PHA and Con A

Cells*	Mitogens**	CPM(3H)***	Stimulation Index****
Day 5	PHA	7,200	0.76
	Con A	30,560	3.22
	-----	9,500	----
Normal	PHA	29,500	2.50
	Con A	205,712	17.41
	-----	11, 812	----
Day 18	PHA	26,494	2.30
	Con A	146,215	12.70
	-----	11,541	----
Normal	PHA	33,230	2.45
	Con A	233,835	16.50
	-----	13,561	----

*8×10^5 spleens cells/well were cultured for 60 hr at 37 C, 95%
O_2- 5% CO_2 atmosphere. ^{3}H-thymidine (1.0 µCi/0.1 ml) was added to
each culture for the final 14 hr of incubation.
**PHA (Difco Laboratories, Detroit, Michigan) was added at 5.0 µg/
culture. Con A (Pharmacia, Fine Chemicals, Uppsala, Sweden) was
added at 0.5 µg/culture.
***Represents the mean cpm of quadruplicate cultures, harvested
with a M.A.S.H. and counter in a liquid scintillation counter.
****Stimulation index was computed by dividing the cpm of the stim-
ulated cultures by the cpm of the control cultures.

We then examined several additional in vitro responses of
spleen cells from tumor-bearing animals. Progressor phage (but
not regressor phase) spleen cells (Table IV) showed a marked reduction in response to both PHA and Con A. Previous work (9) has
shown that the ability of progressor spleen cells to generate
direct PFC after in vitro sensitization with SRBC, as well as their
ability to generate effector cells against allogeneic spleen cells
(C57BL/6) was also drastically depressed.

In an attempt to determine the cause of this depression, we
removed an adherent population of cells from the spleens of progressor animals and stimulated the nonadherent fraction with PHA
and Con A (Table V). Following the incubation of spleen cells on
petri dishes, the response of the nonadherent fraction was markedly
enhanced.

TABLE V

Response of Progressor Spleen Cells Following
Removal of an Adherent Population

| Cells* | Mitogen** | Treatment | |
		None	Adherence***
Progressor	PHA	7,000 ± 117	27,000 ± 167
	Con A	30,560 ± 207	97,400 ± 201
	----	10,033 ± 105	13,464 ± 153
Normal	PHA	33,230 ± 155	36,972 ± 201
	Con A	223,835 ± 324	180,700 ± 325
	----	13,561 ± 145	10,990 ± 96

*8 x 10^5 spleen cells/micro well cultured for 60 hr at 37 C 95% O_2-
5% CO_2. ^{3}H-thymidine (1.0 μc/Ci/0.1 ml) added to each culture for
final 14 hr incubation.
**PHA (same as Table IV), Con A (same as Table IV).
***75 x 10^6 spleen cells (freed of rbc) were incubated at 37 C 90%
O_2- 10% CO_2, in Dulbecco's Modified Eagle's Medium, for 75 min.
****Represents mean cpm of 4 replicate wells ± S.D.

SUPPRESSOR ACTIVITY IN SPLEENS OF PROGRESSOR PHASE ANIMALS

Previous work in our laboratory (9) had established the ability
of spleen cells from progressor phase (but not regressor phase) animals to inhibit the MLR responsiveness of normal AKR spleen cells
when added in low numbers to these cultures. We also found that
these cells could suppress the ability of normal AKR spleen cells
to generate cytotoxic effector cells in vitro against allogeneic
C57BL/6 spleen cells if administered during the sensitization period. In an attempt to characterize the nature of the population
responsible for the suppression observed, we passed spleen cells
from progressor animals over rayon ball columns (Table VI). Similar to our previous findings (Table V), we observed the reconstitution (even more so) of PHA and Con A responsiveness to near normal values following column passage. We also demonstrated (Table
VI) that progress spleen cells, including the adherent fraction,
could inhibit the mitogen response of normal spleen cells; but
that removal of the adherent population also removed the suppresive
capability.

TABLE VI

Effect of Rayon Column Treatment on the PHA
Responsiveness of Progressor Phase Spleen Cells

| Spleen Cells* | Uptake H^{3}TdR in cpm*** | | Stimulation Index (Exp/Control) |
	PHA	Control	
Tumor	5,000 ± 227	8,200 ± 185	0.6
Tumor-adh**	54,130 ± 385	10,175 ± 286	5.3
Normal	60,842 ± 262	11,115 ± 142	5.5
Normal-adh	55,000 ± 196	12,140 ± 197	4.5
Tumor + Normal	8,460 ± 155	6,200 ± 105	1.4
Tumor-adh + Normal	51,320 ± 201	10,845 ± 214	4.7
Normal-adh + Normal	69,620 ± 232	12,100 ± 187	5.8

*8 x 10^5 spleen cells cultured for 72 hr at 37 C, 95% O_2-5% CO_2
atm. ^{3}H-thymidine added for final 24 hr of incubation. Cultures
of "tumor" spleen cells and normal spleen cells contained 4 x 10^5
of each population.
**12 ml columns containing rayon balls (Parke-Davis, Detroit, Mich-
igan) were washed with incomplete media. Cells were added and incubated at room temperature. Cells were eluted with media + fetal
calf serum.
***cpm represent average of 4 replicate cultures ± S.D.

DUAL ROLE FOR THE MACROPHAGE?

Macrophages have been reported as suppressor cells in virus
infected mice (5,6,7) and rats (3). Since rayon treatment has been
shown to effectively remove phagocytic cells (5), our results suggest that macrophages may be mediating the suppression we have observed. Although we have not isolated the effector cell in this
system, our results suggest that the macrophage could be involved
with tumor cell destruction. The macrophage therefore appears to
function in 2 distinct activities, corresponding to the period of
tumor development. During the early period of tumor-immune system
interaction, macrophages would be responsible for inhibiting the
development of the immune response. During a later period of interaction, these cells would be involved in tumor destruction.
Such a dual role for the macrophage would raise a number of interesting questions concerning host defense against solid tumors,
including: a) is the suppressor macrophage generated by direct interaction with the tumor cell, or via control by T or B lymphocytes?

b) are 2 separate populations of macrophages involved during the periods of progression and regression, or is 1 population responsible for both functions? and c) what is the mechanism responsible for shifting the equilibrium from suppressive to effector operations, and if such a shift does not occur, what are the consequences to the host? This system provides an excellent model from which to isolate and characterize the role of macrophages in syngeneic tumor immunity.

ACKNOWLEDGEMENT

We would like to thank Judith Jaworek for preparation of the manuscript.

REFERENCES

1. Evans, R. and Alexander , P., Nature, 236 (1972) 168.
2. Evans, R. and Alexander, P., Immunology, 23 (1972) 615.
3. Glaser, M., Kirchner, H. and Herberman, R. B., Int. J. Cancer, in press (1975).
4. Keller, R., J. Exp. Med. 138 (1973) 625.
5. Kirchner, H., Chused, T.M., Herberman, R.B., Holden, H.T. and Lavrin, D.H., J. Exp. Med., 139 (1974) 1473.
6. Kirchner, H., Herberman, R.B., Glaser, M. and Lavrin, D.H., Cell. Immunol., 13 (1974) 32.
7. Kirchner, H., Muchmore, A.V., Chused, T.M., et al., J. Immunol., 114 (1975) 206.
8. Levy, R.B., St. Pierre, R.L., Barker, A.D. and Axler, D.A., J. Nat. Cancer Inst. (1975) in press.
9. Levy, R.B., Waksal, H.W., St. Pierre, R.L. and Waksal, S.D., J. Immunol. (1975) in press.

INTERACTION OF MACROPHAGES WITH TUMOR CELLS

C. C. STEWART[1], C. ADLES[1] and J. B. HIBBS, JR.[2]

Washington University School of Medicine[1], St. Louis, Missouri (USA) and V. A. Hospital[2] and University of Utah Medical Center[2],Salt Lake City, Utah (USA)

Macrophages obtained from animals injected with certain facultative intracellular bacteria and protozoa will kill or inhibit the growth of tumor cells in vitro (1,3). These activated macrophages are capable of selective cytolysis of neoplastic target cells under in vitro conditions in which non-neoplastic target cells are spared and grow to confluency. In contrast, macrophages obtained after injection of sterile inflammatory stimulants, such as thioglycollate or peptone, are ineffective in killing tumor cells. Furthermore, macrophages obtained from normal unstimulated mice are also not cytotoxic (2).

In the present studies, time-lapse cinemicrography was employed to study the interaction of macrophages with tumor cells. Macrophages were obtained from several sources and their cytostatic and cytotoxic activity was observed under various culture conditions.

MATERIAL AND METHODS

Adult C3H/anf (H-2^k) mice of both sexes or adult Balb/C (H-2^d) female mice were used as a source of peritoneal macrophages. Mice were injected intraperitoneally with either BCG (Trudeau Institute, Saranac Lake, New York; 2-6 x 10^8 viable organisms/ml or Paris strain BCG, 0.2 mg wet weight), thioglycollate or peptone. Cells were harvested from the peritoneum, usually after 14 days, by introducing and withdrawing 5 ml medium containing 5 units heparin/ml. Cells were washed once by centrifugation and resuspended in 10 ml medium. The medium used in this study was Alpha Minimal Essential Medium (Flow Laboratories, Rockville, Maryland) or Dulbecco's modified Eagles medium (Gibco, Grand Island, New York) supplemented

with 10% fetal calf serum,100 units penicillin/ml and 100 µg strepto-
mycin/ml.

Cellularity of the washed cells was determined using the pro-
nase-cetrimide procedure (6) and an electronic particle counter.
After adjusting the cell concentration, 0.1 ml of the suspension
was spotted near the center of a 35 mm culture dish (Falcon Plas-
tics, Oxnard, California) and incubated at 37 C for 45 min (10%
CO_2 in humidified air). The non-adherent cells were then removed
by adding 2 ml of medium to the dish and washing them away using
a transfer pipette. Two extra plates for each parameter were always prepared and used to determine the number of adherent cells
by removing them with cetrimide and counting (6).

EMT6 ($H-2^d$) tumor cells, and 3T12 ($H-2^d$) fibroblasts were
maintained as continuous lines of monolayer cultures and used as
target cells for these experiments. Non-neoplastic cells consisted
of embryo fibroblasts from either C3H or Balb/c mice in their fourth
to sixth passage. Monolayers were suspended with 0.25% trypsin
solution A (Gibco, Grand Island, New York), and adjusted to the appropriate concentration with medium; 3 ml were then pipetted onto
culture dishes. Control cultures consisted of target cells without macrophages. Cultures were incubated at 37 C (10% CO_2 in humidified air) for 72 hr, fixed in 10% formalin and stained with
methylene blue.

For time-lapse cinemicrography, 2 identical camera-microscope
units in a 37 C warm room were used so that macrophage-tumor cell
cultures and control cultures could be monitored simultaneously in
the same environment. Cultures were incubated in viewing chambers
having a 10% CO_2 in humidified air atmosphere. Cultures were viewed at a magnification of 100x with phase contrast optics. Kodak
Plus-X reversal film was used. The exposure time was 0.5 sec; the
interval time was usually 1 min/frame.

RESULTS

The effect of macrophages, from various sources, on tumor cells,
is shown in Table I. Since the yield of macrophages from the peritoneal cavity varies depending upon the source, the initial concentration of cells used was adjusted to produce as closely as possible the same density of adherent cells.

Since macrophages were plated in the center of the 35 mm culture dish and occupied an area of 0.78 cm^2 (10 mm diameter) while
tumor cells were added to the entire dish and occupied an area of
9.6 cm^2, the respective areas to which each had to attach was different. After counting the number of exudate cells which attached,

the cells/cm^2 were calculated and used to determine the ratio of macrophages to tumor cells. Tumor cells were plated at about 1.44 x 10^5 (1.50 x 10^4 cells/cm^2) or 4.8 x 10^4 (5 x 10^3 cells/cm^2) cells/ culture.

As shown in Table I, macrophages obtained from a peritoneal lavage of normal unstimulated mice were not cytostatic as measured by the time of the first division (column 6) at a ratio of 2.6 macrophages/tumor cell and were only slightly cytostatic at a ratio of 20 macrophages/tumor cell. Macrophages from these unstimulated peritoneal lavages survived poorly, with less than 75% viable cells remaining after 1 day of culture. Death was characterized by a rounding up of the cell followed by immotility and lysis. Dead cells were quickly phagocytized by viable macrophages. In contrast, macrophages induced by any one of the stimulating agents remained healthy with no evidence of death.

The motility and frequency of interaction of thioglycollate- or peptone-induced macrophages with tumor cells was greater than for normal macrophages. However, even at a very high macrophage density, performed to rule out a nutrient depletion effect, tumor cells proliferated and displaced attached macrophages. Motility of all cells decreased markedly after about 50 hr of culture, probably reflecting nutrient depletion. These peritoneal macrophages were only slightly more cytostatic than unstimulated macrophages.

Cinephotomicrographs A through D in Fig. 1 show BCG-activated macrophages (M) interacting with EMT6 tumor cells (T). While the motility, frequency and duration of interaction of BCG-activated macrophages with tumor cells were similar to those of thioglycollate- or peptone-induced macrophages, BCG-activated macrophages produced pronounced cytostasis and cytolysis of tumor cells (Table I). Tumor cells in contact with BCG-activated macrophages tended to form vacuoles. A time-lapse sequence of vacuole formation is shown in Fig. 1A, 1B and 1C representing 1 hr of elapsed time. Six macrophages are shown in contact with the tumor cells. Vacuoles increased in size as they moved from the periphery of the tumor cells toward the nucleus and then were absorbed (not shown) as new vacuoles formed. Also shown in this sequence is 1 tumor cell, TL, which in Fig. 1A rounded up and in Fig. 1B and Fig. 1C became detached and disintegrated. In Fig. 1C, a second tumor cell with macrophage attached began to undergo cytolysis. Four tumor cells undergoing cytolysis are shown after 40 hr of culture in Fig. 1D. Macrophages from "unstimulated" mice, shown in Fig. 1E and 1F, were less motile than cells obtained from stimulated mice. The large rounded cells in Fig. 1E are mitotic tumor cells, the small round and spread out cells are macrophages. In Fig. 1F, a confluent monolayer of tumor cells with "rounded up" macrophages on top of them is shown. There was no evidence of tumor cell cytotoxicity even though the macro-

phages had frequent contact with tumor cells.   Tumor cells proliferated during the 72 hr observation interval to form a confluent
monolayer, displacing the attached macrophages which rounded up on
top of the tumor cells.   These macrophages continued to be motile,
moving over the top of and underneath the attached tumor cells,
spreading out on the surface of the dish when space was found.

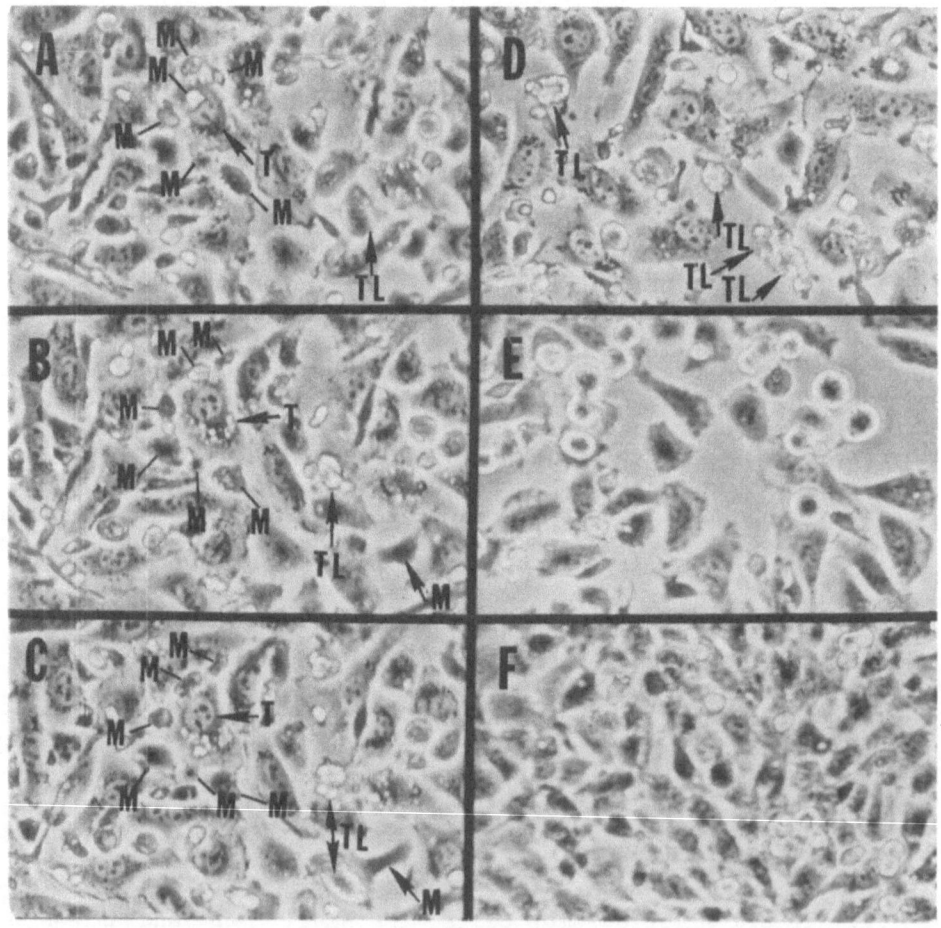

Fig. 1.   Time-lapse cinephotomicrographs of BCG-activated macrophages (M) interacting with EMT6 tumor cells (T).

TABLE I

Effect of Macrophages from Various Sources on Tumor Cells

Macrophage** Source	Cell Type	Cells/cm^2		Macrophage: target cell	Cytostasis hours***	Cytolysis
		macrophage	target			
Unstimulated	EMT6	40,000	15,000	2.6	0.10	none
	EMT6	100,000	5,000	20.0	7.9	none
Thioglycollate	EMT6	100,000	15,000	6.67	15.8	none
Peptone	EMT6	100,000	10,000	10.0	17.3	none
>10^6 viable BCG	EMT6	100,000	15,000	6.67	*	yes
<10^4 viable BCG	EMT6	100,000	15,000	6.67	27.9	none
>10^6 viable BCG	EMT6	100,000	5,000	20	*	yes
>10^6 viable BCG	EMT6	220,000	5,000	48	*	yes
>10^6 viable BCG	3T12	100,000	5,000	20	*	yes
>10^6 vialbe BCG	3T12	220,000	5,000	40	*	yes

*No divisions observed during the 72 hr observation period.
**Peritoneal exudate cells were obtained 14 days after injection of the indicated agent except for cells obtained from a normal "unstimulated" peritoneal lavage.
***Cytostasis is defined as the inability of surviving tumor cells to divide. The degree of cytostasis was measured by the length of time until the first division was observed.

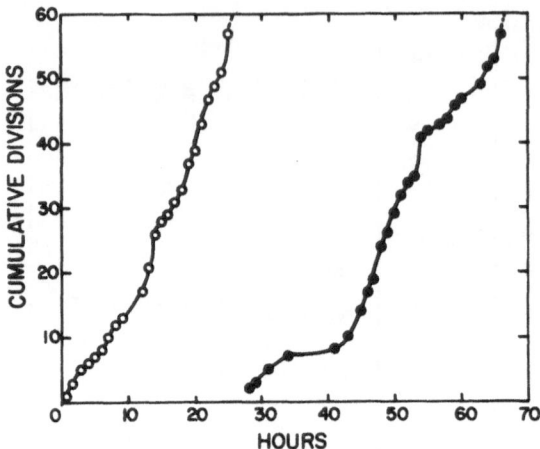

Fig. 2.   Macrophage induced cytostasis.

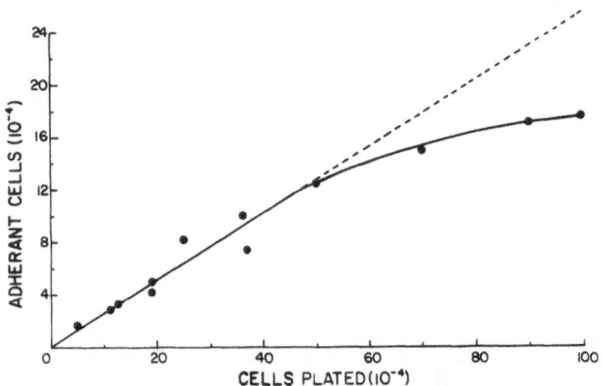

Fig. 3.   Macrophage confluence.

In these experiments, 3 kinds of macrophage interaction with
the tumor cells were noted; a) macrophages from all sources inter-
acted with the tumor cells making contact with them for varying
lengths of time and moving on top of, underneath and around the
target tumor cells, b) macrophages obtained from mice injected with
less than 1×10^4 viable BCG were cytostatic only, producing inhibition of tumor cell proliferation for about 28 hr.   The results
are shown in Fig. 2 where the cumulative number of divisions observed on film was plotted as a function of observation time.   Once
the first tumor cell divided, the remaining cells in the field be-

ing observed commenced proliferation shortly thereafter , and all
cells had divided at least 1 time by 40 hr. Control tumor cells
commenced proliferation after about 15 min. c) Cytolysis of tumor
cells began after 20 hr of culture when macrophages were obtained
from mice infected with 1 x 10^6 viable BCG, increasing in frequency
by about 40 hr. The cultures which exhibited cytolysis could further be divided into those in which both the tumor cells and macrophages died and those in which only the tumor cells died.

In order to investigate what conditions were necessary to produce cytolysis of both cell types, the macrophage and tumor cell
densities were varied at a constant ratio. The maximum number of
macrophages that would adhere to the dish was first determined by
plating increasingly higher concentrations of exudate cells and
counting adhering cells after washing away non-adherent ones. When
confluence was reached, there was no further increase in the number
of cells that attached.

As shown in Fig. 3, the number of adherent cells (from mice
injected 14 days earlier with 1 x 10^6 viable BCG) was linear up to
about 5 x 10^5 cells plated in 0.1 ml with 30% of the cells adherent.
As the concentration was further increased, the curve digressed
from linearity. We estimate 2.25 x 10^5 adherent cells/cm^2 produced
by plating 5 x 10^5 peritoneal cells in 0.1 ml reached greater than
95% confluence.

Starting with 2.25 x 10^5 exudate cells/cm^2 and 1.4 x 10^4 tumor
cells/cm^2, the effect of cell density at a constant macrophage to
tumor cell ratio of about 15 was measured. The results, shown in
Table II, indicate that the density of macrophages is important in
producing cytolysis.

TABLE II

Effect of Macrophage Density at a Constant Ratio

M*/cm^2 (1 x 10^3)	Target cells**/cm^2 (1 x 10^3)	Cytolysis	
		EMT6 Tumor cells	3T12 Fibroblasts
225	14	+	+
130	7.2	+	+
96	8.0	+	+
52	4.0	±	±
37	1.8	−	−
19	0.9	−	−

*BCG—activated macrophages
**EMT6 tumor cells or tumorigenic Balb/3T12 fibroblasts

At low cell densities, (less than 5.2×10^4 macrophages/cm^2)
tumor cells were not inhibited. At these low densities, few macrophages contacted tumor cells and those that did contacted them in-
frequently. This suggests that macrophage contact was an essential
component of the response. As the density was increased, only tumor cell cytolysis occurred at this ratio. While macrophages were
observed to phagocytize debris, no phagocytosis of viable or dying
tumor cells was observed.

The only conditions found to produce both tumor cell and macrophage cytolysis were when the ratio was less than 5 and the cell
density was $>1 \times 10^5$ cells/cm^2. Under these conditions macrophage
lysis also occurred simultaneously with tumor cell lysis between
39 and 50 hr of culture. While the former condition is not uniformly reproducible, the results suggest that a low ratio of macrophages to tumor cells cultured at high density produces sufficiently
poor conditions within the microenvironment for survival of both
cell types. That this is not dependent upon cell density alone is
reflected by the observation that macrophages cultured at $2.25 \times$
10^5 macrophages/cm^2 and 5×10^3 tumor cells/cm^2 produced only tumor
cell lysis. In addition, macrophages or tumor cells cultured separately, or tumor cells cultured at similar densities with non-BCG-
induced macrophages produced no detectable cytolysis of either cell
type.

Fig. 4 illustrates the importance of intimate cell contact for
the destruction of Balb/3T12 cells by BCG-activated macrophages.
Peritoneal cells from BCG-infected (Paris strain) or normal Balb/c
mice (6.0×10^5) in 0.1 ml of Dulbecco's medium were added to the
center of 35 mm Falcon plastic dishes and incubated at 37 C in 5%
CO_2 in air atmosphere for 30 min to allow for adherence of macrophages. Each dish was then washed with Hanks' balanced salt solution to remove nonadherent cells so that a central monolayer·of adherent cells was restricted to the size of the 0.1 ml drop in which
they were added to the dish. A rubber policeman was then used to
remove a narrow strip of adherent macrophages from the centrally
located monolayer. Several varieties of macrophage-free areas were
produced on different macrophage monolayers. Target cells (tumorigenic Balb/3T12 cells, 1×10^5) were added in 2 ml of complete
medium and attached evenly to the bottom of the dish. Cultures
were fixed with methanol and stained with Giemsa after incubation
for 60 hr.

The results underline the requirement for intimate activated
macrophage-target cell contact for cytotoxicity to occur. In all
cases, the Balb/3T12 cells had completely overgrown and formed a
multilayer over the normal macrophages. On the other hand, whenever the Balb/3T12 cells came into contact with BCG-activated
macrophages, there was a cytotoxic effect. Furthermore, it could

be seen that wherever a narrow strip of adherent macrophages was
removed from the dish with a rubber policeman prior to Balb/3T12
cell challenge, the macrophage-free area was overgrown with a monolayer of target cells, even though they were surrounded by viable,
fully cytotoxic BCG-activated macrophages. Microscopically, the
Balb/3T12 cells grew to the immediate edge of the macrophage-free
areas.

Although no time-lapse studies were performed using normal
fibroblasts, cultures of both C3H (allogeneic to EMT6) and Balb/c
(syngeneic to EMT6) embryo fibroblasts with BCG-activated macrophages were prepared and assayed 72 hr later. Unlike the central
clear zone observed with the tumor cells, fibroblast cultures contained no central clear zone and produced uniform monolayers similar to control cultures to which no macrophages had been added.

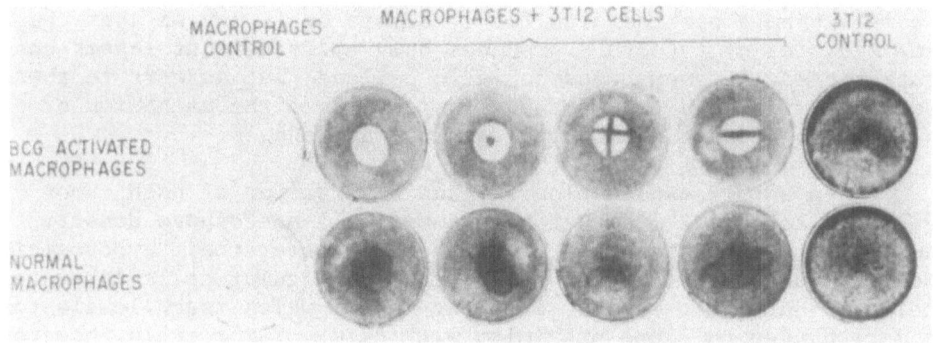

Fig. 4. Macrophage interaction with 3T12 cells.

DISCUSSION

In this study, time-lapse cinemicrography showed that macrophages from BCG-infected mice were capable of killing tumor cells,
while those from thioglycollate- or peptone-injected mice or from
normal mice were not. Similar results, also using time-lapse techniques, have been reported recently by Meltzer, et al. (5) who
quantitated translation movement, frequency and contact duration
of BCG-activated macrophages with neoplastic and non-neoplastic
established cell lines.

The question arises as to whether the nonspecific cytotoxicity
of BCG-activated macrophages was due to local nutrient depletion
because of a high cell density. The fact that 3T12 fibroblasts

form a monolayer on small areas of macrophage-free substrate completely surrounded by fully cytotoxic BCG-activated macrophages
suggests that nutrient depletion is not a factor (see Fig. 4).
In addition, the findings in Fig. 4 demonstrate the necessity for
intimate contact between BCG-activated macrophages and 3T12 target
cells for the expression of the cytotoxic effect. The results
shown in Table II also may reflect the requirement for activated
macrophage-tumor cell contact for the cytotoxic effect to occur.
Krahenbuhl and Lambert have reported similar findings (6).

In any system in which the measurement of host cell interaction with neoplastic cells is being studied, cytotoxicity of the
target cell should be distinguished from cytostasis. For example,
when only 1×10^4 viable BCG were injected, macrophages from these
mice were only cytostatic. This raises the question as to whether
macrophages obtained after BCG injection are a special population
of activated cells different from those induced by either thioglycollate or peptone, or whether they are the same cells activated
to a different degree or "armed" differently. Furthermore, cytostasis may be a prerequisite for cytolysis or it may be that the
same mediator of cytolysis produces only cytostasis at lesser concentrations. It seems reasonable to believe that answers to these
questions will lead to important insights into the mechanism of
tumor cell cytostasis and cytolysis by macrophages.

One possible explanation for the observation of both tumor
cell and macrophage lysis at high tumor cell-macrophage density
and low ratio is that after macrophages transfer their cytotoxicity-
inducing substance to tumor cells, the dying tumor cells liberate
their own autolytic enzymes into the medium which reach levels toxic
to macrophages in close proximity with them. Tumor cells, however,
not in contact with the macrophages (on the periphery) are not
killed due to the inactivation of these enzymes by the serum present in the medium or are diluted by the suspending medium as they
diffuse through it. Thus, the tumor cells not in contact with the
macrophages proliferate normally to produce the observed confluent
monolayer. When tumor cell density is reduced while macrophage
density is held constant, less autolytic enzymes are released as
a result of tumor cell lysis and, therefore fewer macrophages are
killed. Experiments are in progress to determine the validity of
this hypothesis.

ACKNOWLEDGEMENTS

Supported by NCI Grant No. CA13053 (CCS, CA), by the Veterans
Administration (JBH) and by NIH Grant No. CA15811 (JBH). Dr. Hibbs
thanks R. R. Taintor and C. C. Moore for their technical assistance.

REFERENCES

1. Hibbs, J.B., Jr., Science, 180 (1973) 868.
2. Hibbs, J.B., Jr., J. Nat. Cancer Inst., 53 (1974) 1487.
3. Hibbs, J.B., Jr., Lambert, L.H., Jr. and Remington, J.S.,
    Nature, 235 (1972) 48.
4. Krahenbuhl, J.L. and Lambert, L.H., Jr., J. Nat. Cancer Inst.,
    54 (1975) 1433.
5. Meltzer, M.S., Tucker, R.W. and Breuer, A.C., Cellular Immunol.,
    17 (1975) 30.
6. Stewart, C.C., J. Reticuloendothel. Soc., 14 (1973) 332.

Workshops

WORKSHOP: RES-RADIOBIOLOGY AND ENVIRONMENTAL FACTORS

K. FLEMMING[1], S. REICHARD[2] and M. ESCOBAR[3]
Institute of Biophysics and Radiobiology, University
of Freiburg[1], Freiburg, (West Germany), Medical College
of Georgia[2], Augusta, Georgia (USA) and Medical College
of Virginia, Virginia Commonwealth University[3], Richmond,
Virginia (USA)

This workshop was attended by those interested in the increasingly important area of radiobiological and environmental
effects on the RES. The 3 Chairmen briefly reviewed some experimental work which is presented here on radiation effects on macro-
phagic function, the interrelationship of the RES with endotoxins,
traumatic shock, radioprotection and aspects of nutritional effects. It is hoped that the initiation of this discussion will
be continued at future meetings with an expanded agenda of topics
in the environmental arena.

RADIATION EFFECTS ON CELLS OF THE RES

K. Flemming
University of Freiburg
Freiburg, (West Germany)

From the biological point of view, radiation as an environmental factor may be considered as a drug. Following high doses
of whole body X-radiation (WBR), toxic effects must be exerted on
all cells of the organism, including cells of the RES. According
to the literature, however, different effects have been reported
concerning phagocytosis (decreases, increases and no changes).

In experiments of our own, using the carbon clearance method
(8 or 16 mg carbon/100 g), a phagocytic increase was regularly
established in mice 4 to 10 days post WBR (250 to 1000 R). Considering the dependence of the increase on radiation dose, time

of assay and age of mice, it was concluded that we were dealing
with an abscopal effect of radiation, with increasing K values
to radiation death.

The phagocytic increase might be interpreted as being derived from bacterial endotoxins (lipopolysaccharides = LPS) which
enter the blood through the damaged intestinal mucosa.   This
hypothesis is supported by the following facts: a) the stimulation
of phagocytosis by LPS could also be demonstrated in irradiated
mice; b) the phagocytic increase after WBR was inhibited by oral
administration of non-resorbable antibiotics; c) Fine, et al.
ascribed responsibility for several pathophysiological changes to
the influx of bacterial LPS from the intestine into the blood (2);
and d) following WBR, a bacterial endotoxin has been found in the
blood of conventional but not of germ-free mice(12).

The question arises as to whether bacterial LPS from the
intestine could be responsible for the death of the irradiated
mice, too. We established that the phagocytic increase following
radiation was comparable to the effect of 0.5 µg to 1.0 µg of the
LPS from Salmonella abortus equi. This LPS dose has no fatal effect on non-irradiated mice (LD^{50} = 130 - 450 µg). However, 3 to
5 days after irradiation (at this time intestinal death is at its
maximum!) the sensitivity of mice towards endotoxin was increased
by a factor of 1000 (10). Therefore, 0.5 to 1.0 µg LPS originating
from the intestine could have caused the radiation death in our
experiments.

To determine whether macrophages were injured by radiation,
the response of the mice to phagocytosis stimulating drugs (LPS,
olive oil, triolein) was studied before and after irradiation.
The results showed that, following irradiation, the effectiveness
of those drugs was strongly reduced. From this one must conclude
that, notwithstanding the phagocytic increase described above, the
performance of the macrophages was actually depressed following
WBR. This was confirmed by further experiments in which the carbon dose for the carbon clearance test was increased to 50 mg/100 g.
Under these conditions, a decreased phagocytosis was established
from 1 to 10 days after radiation.

The syndrome following lethal WBR is in many aspects similar
to the traumatic and hemorrhagic shock, which is correlated with
a depressed activity of phagocytic cells. This phagocytic depression can not be attributed to reduced hepatic blood flow but may
be caused by a humoral substance being produced by the ischaemic
intestine (1). Moreover, shock has been implicated with endotoxins
of gram-negative intestinal bacteria (2). Therefore, the question
arises as to whether the phagocytic depression during shock could
be attributed to bacterial LPS.

Stimulation of phagocytosis by LPS is well-known, while occasional findings of depression were usually explained as side effects due to impurities. Recently, however, we have obtained a clear-cut depression of phagocytosis by LPS in mice. In these experiments, a highly purified LPS from S. abortus equi was used in sublethal doses ranging from 0.05 to 5.0 mg/kg body weight (Nothdurft and Flemming, unpublished). The degree and the duration of the phagocytic decrease was dose-dependent. With the largest LPS dose (5.0 mg/kg), a continuous depression from 4 hr to more than 48 hr was established, being followed by a steep and long-lasting increase (more than 4 days). In shock, even larger LPS doses might enter into the blood from the intestine. Therefore, one can understand that phagocytosis is deeply depressed for hours or even days under such pathological conditions.

I should like to comment briefly upon statements in the literature (4,11,14) which indicate that a high activity of the RES is correlated with an increased radioresistance. We have carried out numerous experiments with mice and rats in which both these parameters were compared. For this reason, the phagocytic activity of the RES was stimulated or depressed pharmacologically by a variety of substances like vegetable oils, fatty acid esters, drugs of plant origin, bacterial LPS and estrogens. When the phagocytic increase or decrease was at its maximum, the animals were irradiated and the radiosensitivity determined by means of 30 days survival rate. In many cases, a remarkable correspondence seemed to exist between both parameters at first sight. An analysis in more detail, however, revealed momentous non-conformities of the results. Thus these experiments did not yield any support for the hypothesis that RES function (as indicated by phagocytic activity) and radioresistance are casually connected.

One must emphasize, however, that such experiments are not well suited to elucidate the possible relationship between RES and chemical radioprotection. In fact, the concept that radioprotective drugs (3-6) act by affecting systemic biological reactions has obtained much support. Recently, it has been demonstrated that most of the variability in radioprotection is a product of host:drug interactions (13) in which hemopoietic tissues (stem cells, spleen, immunocompetent cells) are preferentially involved.

The physico-chemical basis of these interactions is still unknown. A decreased total metabolism (i.e., O_2 consumption) is perhaps one of the factors involved. An intimate connection between lowered metabolism and enhanced radioresistance could be observed after adaptation to heat or after psychotropic drugs (7, 8). The radioprotective effect of Solcoseryl (a protein-free blood extract), reported at this meeting, is also attributed to enzymatic and metabolic changes comparable to those occurring due to adaptation to heat. In this connection, I should like to point to the radioprotective effect of estrogens and chlorotrianisene. This pre-estrogen (see Workshop: RES-Pharmacological Aspects) causes a stagnation of animal growth due to the depression of the basic metabolic rate. Thus the radioprotective effect of these substances, too, might well be connected with the depression of metabolism.

REFERENCES

1. Blattberg, B. and Levy, M.N., Amer. J. Physiol., 204 (1963) 899.
2. Fine, J., Rutenburg, S. and Schweinburg, F.B., J. Exp. Med., 110 (1959) 547.
3. Flemming, K., Proc. 3rd. Int. Congr. Chemother., (Ed. H. P. Kuemmle and P. Preziosi), Georg Thieme-Stuttgart (1963) 1484.
4. Flemming, K. and Flemming, Ch., Naunyn-Schmiedebergs Arch. Exp. Path. Pharmak., 245 (1963) 129.
5. Flemming, K. and Flemming, Ch., Strahlentherapie, 124 (1964) 617.
6. Flemming, K. and Flemming Ch., Strahlentherapie, 125 (1964) 273.
7. Locker, A. and Bauer, D., Studio Biophysica, 28 (1971) 105.
8. Locker, A., Weish, P. and Krympholz, H., SGAE-Report SS-7/1971 (1971) 87.
9. Paterson, E., Brit. J. Radiol., 30 (1961) 577.
10. Smith, W.W., Alderman, J.M., Schneider, C. and Cornfield, J., Proc. Soc. Exp. Med., 113 (1963) 778.
11. Taplin, G.V., Finnegan, C., Noyes, P. and Sprangue, G., Amer. J. Roentgenol. Radium Ther. Nuclear Med., 71 (1954) 294.
12. Wilson, R., Barry, T.A. and Bealmear, P.M., Rad. Res., 41 (1970) 89.
13. Yuhas, J. M., Experientia (1975) In press.
14. Zweifach, B.W. and Thomas, L., J. Exp. Med., 106 (1957) 385.

PHYSIOLOGIC MECHANISMS IN RADIATION RESISTANCE

S. M. Reichard
Medical College of Georgia
Augusta, Georgia (USA)

The reticuloendothelial system constitutes a major defense system against bacterial agents and plays a vital role in the regeneration of the hemopoietic system. Exhaustion of certain cells of RE origin has been proposed as the prime cause of death after whole body irradiation (2). It is logical to assume that

the functional activity of the RES is an important factor in the
resistance against radiation effects.

The clearance of intravenously injected particulate material
from the blood stream is frequently used to measure the activity
of the RES assuming that this reflects the phagocytic activity of
the cells located largely in the spleen and liver. This measurement does not necessarily show changes within different RE organs,
and it is possible that disappearance rates may not be altered because damage to the RES at one site is balanced by RES regeneration
elsewhere. Thus, the uptake of colloidal agents by RE cells of
the spleen and liver at various times after receiving graded doses
of x-irradiation was investigated (5). Variation in the phagocytic
activity of these cells was observed with time and dose of irradiation and with the suspending medium for the colloidal particle.
For example, at 24 hr after exposure to 250 r of x-irradiation,
the uptake of colloidal thorium in the spleen of rats was significantly increased. With greater doses of x-irradiation, phagocytic
activity was increased further and the time at which this change
occurred was shortened. However, 10 days after exposure, the uptake returned to or fell below normal. Similar increases in activity have been observed in mice (1,10). When the thorium was
suspended in saline rather than dextrin (Thorotrast), its uptake
by the spleen was enhanced 3 to 5 times. No changes were seen in
livers of irradiated animals given thorium in dextrin, but amounts
of thorium in saline were greater than controls 24 to 72 hr following exposure, returning to normal or below at 10 days. This also
varied according to the strain of rats; in certain strains there
was a transient depression before the rise and a return to normal
levels or below at 20 days. The increased uptake seems to be due
to an increased activity of cells present or an increased proliferation of phagocytic cells since the accumulation by the whole organ is greater in spite of a decrease in organ weight.

The activity of peritoneal macrophages was also determined at
72 hr after 500 r. In this procedure, x-irradiated and sham-
irradiated rats were injected intraperitoneally with a suspension
of Pseudomonas aeruginosa containing approximately 10^6 cells in
0.25 ml. Peritoneal exudates were collected 2 hr later and 200
cells were examined for each animal. Activity was estimated by
recording the number of cells with no bacteria (nonactive) and
those containing bacteria (active). The percent of active phagocytes was greatly increased in the irradiated animals at a time
period corresponsing to the increased uptake of colloidal material.
This increased activity, however, may merely reflect the reduced
total number of leukocytes caused by the radiation. Smears of
peritoneal fluid from the x-irradiated animals also showed more
free bacteria, suggesting that the actual extent of phagocytosis
may be less than controls.

Ionizing radiation was also observed to sensitize rats to subsequent exposure to trauma, a severe stress (8). Survival decreased from 70% to 10% following exposure to trauma when animals had received nonlethal doses of whole body x-irradiation 10 days earlier. Stimulation of the phagocytic activity of the reticuloendothelial system with the administration of estrogens, denatured serum albumin or bacterial endotoxins prevented this potentiating effect of susceptibility of rats to trauma and prevented the protection afforded by stimulation. Protecting the RE elements of the spleen by lead shielding during irradiation similarly prevented the increased susceptibility observed following x-irradiation. In animals rendered resistant to traumatic shock prior to their exposure to ionizing radiation, no potentiation was observed. It was suggested that the increased susceptibility to traumatic shock may be caused by an altered function of the RES brought about by the ionizing radiation (7).

Protection against the lethal effects of radiation was obtained by several procedures that alter the reticuloendothelial system. The state of trauma resistance, produced in animals through repeated exposures to sublethal doses of shock and previously shown to be associated with the RES (3), reduced the mortality following 750 r x-irradiation from 80 to 25% (4). Impairing the function of the RES eliminated this acquired resistance to both traumatic shock and x-irradiation. Extracts prepared from the spleens and plasma of resistant animals and administered to normal animals were found to protect them against radiation lethality. Other procedures which stimulate phagocytic activity of the RES such as the prior administration of bacterial endotoxins (11,12) as well as injections of zymosan, a cell wall preparation of yeast, repeated injections of saccharated iron oxide and denatured human serum albumin aggregate also increased survival to x-irradiation (6). In all instances, impairment of the RES overcame this protection. A humoral mechanism seems to be involved, since protection can be passively transferred to otherwise unprotected rats. The RES, therefore, seems to be concerned with the elaboration or activation of an active factor capable of altering physiological adaptation to traumatic shock and ionizing radiation.

Survival of rats after whole body x-irradiation depends largely upon the restoration of the hemopoietic system (2). This probably occurs through the mobilization of undifferentiated progenitor cells of the RES that have survived radiation damage. The formation of endogenous spleen colonies in irradiated rats was used as a measure of the activity of this class of cells (9). A variety of erythropoietic stimuli were found to influence the number of spleen colonies. These include exposures to low oxygen, repeated bleedings and the administration of endotoxin, testosterone and anemic plasma extracts. These procedures enhanced survival following x-irradiation and increased hemopoietic recovery as measured by the number of spleen colonies, and by the uptake of ^{59}Fe and iododeoxyuridine by newly formed cells. Other procedures that undermine the rate of erythropoiesis such as hyperoxia, hypophysectomy and certain doses of prior radiation potentiated the lethal effects of subsequent whole body irradiation. The active "RES factor" capable of protecting against the lethal consequences of traumatic shock and x-irradiation may be concerned with the differentiation of primitive stem cells of the hemopoietic system.

It is clear that following x-irradiation, the sequestration of particulate material within the RE organs is increased. This may reflect heightened phagocytic activity; it may represent stimulated regeneration of RE cells after initial damage; or it may, in fact, reflect an impaired RES selectivity, with impaired function in other RE organs.

Perhaps the capacity for reticuloendothelial regeneration is itself a defense mechanism, and may represent a self-regulating process that stimulated greater activity of increased numbers of cells from mesenchymal precursors. This may account for the mobilization of cells for possible re-establishment of hemopoietic stem cells and to combat bacterial invasion. It may also be responsible for the elaboration of a humoral factor that protects against radiation lethality.

REFERENCES

1. Flemming, K., Flemming, C. and Nothdurft, W., J. Reticuloendothel. Soc., 7 (1970) 1.
2. Paterson, E., Brit. J. Radiol., 30 (1957) 577.
3. Reichard, S.M., Ann. N. Y. Acad. Sci., 88 (1960) 213.
4. Reichard, S.M., The Reticuloendothelial System: Morphology, Immunology and Regulation, Kyoto, Nissha, (1965) 359.
5. Reichard, S.M., Radiation Res., 31 (1967) 566.
6. Reichard. S.M., Radiology, 89 (1967) 501.
7. Reichard, S.M., Radiology, 91 (1968) 132.
8. Reichard, S.M. and Corrill, L., Fed. Proc., 27 (1968) 507.
9. Reichard, S.M. and Corrill, L.S., J. Reticuloendothel. Soc., 5 (1968) 584.
10. Šljivić, V.S., Brit. J. Exp. Path., 51 (1970) 130.
11. Smith, W.W., Alderman, I.M. and Gillespie, R.E., Amer. J. Physiol., 192 (1958) 549.
12. Zweifach, V.W., Kivy-Rosenberg, E. and Nagler, A.L., Amer. J. Physiol., 197 (1959) 1364.

IMMUNOLOGIC ASPECTS OF MALNUTRITION

M. R. Escobar
Medical College of Virginia
Virginia Commonwealth University
Richmond, Virginia (USA)

This workshop will be devoted to the effects of malnutrition on the immune response, infection and susceptibility to neoplasia. Other aspects of nutrition were discussed in another workshop during this Congress.

Although it has been well established epidemiologically that malnutrition interacting reciprocally and/or synergistically with infection is the greatest killer of infants and young children and the major cause of growth retardation (6), the manner in which nutritional status affects the immune response or host defense mechanisms is not clearly understood. In this connection, human experimentation has been mainly clinical, postmortem or retrospective. On the other hand, many investigators using animal models have provided often conflicting results. Their studies have dealt with both arms of the immune response, but they have differed regarding animal species; age; point in time during immunologic development at which malnutrition was instituted; the method, length and the severity of malnutrition; the type of superimposed infections, as well as the parameters selected to evaluate humoral and cell-mediated immune responses.

Several students working with me have investigated this problem in an attempt to confirm experimentally those human observations which were first made at least 100 years ago. During our earlier studies in search of an adequate animal model, one student (8) was able to demonstrate differences between host species not only as related to differences in the ontogeny of the immune response, but mainly in regard to differences at the cellular level. It was shown that human adenoviruses which were able to replicate in their primary host in vivo or in cell cultures of human cells in vitro failed to produce virions in an inbred strain of mice or in their cells grown in culture. This was subsequently explained by the finding that in cells derived from other species, the human adenoviruses undergo an abortive replicative cycle during which certain viral components are actually synthesized, but no capsid proteins or infectious progeny are produced. In later studies, we chose the guinea pig to study the effects of malnutrition on the immune response. There were several criteria which led us to make this choice: a) as in the human, nursing or suckling is not essential for survival of the newborn guinea pig; b) the newborn guinea pig is able to eat solid food; c) the size of the neonate is adequate for experimentation; and d) the similarity in the degree of immuno-

logic maturity at birth between the guinea pig and man, according
to physiologic age equivalence formulas devised by Solomon (7).
This choice of experimental animal allowed the creation of a model
significantly different from those used in previous studies.

The establishment of a "marasmic state" in the animals in
our studies was determined predominately by anthropometric measurements. A recent review of laboratory tests available for the
assessment of nutritional status concluded that no single biochemical procedure can satisfactorily evaluate protein and calorie
malnutrition in early or subclinical forms, and those indices
available are not capable of reliably distinguishing between the
2 extreme forms of marasmus and kwashiorkor (5).

Newborn guinea pigs were removed from their mother 24-48 hr
after birth. Animals in the experimental group were fed 2% of
their birth weight in grams of food (Purina Guinea Pig Chow) per
day. This amount was previously determined to be enough for survival, yet it was able to maintain them at their birth weight
(+10%). Immunologic studies were done at 10 and 42 days of age.
Histologic examinations of the thymus, bone marrow, lymph nodes
and spleen were also performed on these animals. Weights of all
animals were recorded daily. Constant surveillance on the infectious state of these animals was also carried out.

Most workers in this area have agreed that protein-calorie
malnutrition is associated with immunosuppression. However, because immunocompetence is a multifaceted system, it is not always
clear what is being affected by protein-calorie malnutrition.

In consonance with one of the aspects of our studies, another
student (1) investigated the effects of protein-calorie malnutrition
on the humoral immune response and found that humoral immunocompetence of the experimental animal group was unaltered or depressed
as determined by serum levels of immunoglobulins and complement,
antibody production, or morphologic alterations in lymphoid tissues after antigenic stimulation.

At 10 days of age, serum values for total protein were 4.0
and 3.8 g% in the experimental and control animal groups, respectively. At 42 days of age, the total protein serum levels had
reached 4.2 to 4.5 g% in both groups. The serum albumin was 3.1
and 2.8 g% in these groups, respectively. This finding of normal
serum albumin levels appears to indicate that the degree of malnutrition of the experimental animals was not severe enough to
cause elevation of plasma cortisol levels. At 42 days of age the
serum albumin levels were unchanged.

Total serum gamma globulin and IgG levels were determined

by radial immunodiffusion. No significant differences were found
between the experimental and control groups at 4 or 6 weeks of age.
Measurements of IgG and the third complement component (C'3)
revealed no difference between the 2 groups. Complement levels
were measured by radial immunodiffusion and by radial hemolysis
in gel and complement fixation. No significant differences were
observed between the 2 groups of animals.

While most of the major studies on the effects of malnutrition
on the immune system have shown that levels of immunoglobulins are
normal or slightly elevated in protein-calorie malnutrition, depressed levels of C'3 have been reported (2,10). Complement appears
in the serum of guinea pigs in late fetal life and increases during
postnatal development. It is possible that at 10 days of age C'3
levels were temporarily depressed but returned to normal following
nutritional repletion.

Macrophages from experimental and control animals in this
study were tested for their ability to phagocytize Salmonella
typhimurium. No difference was seen between the groups. This
finding differs from that of Cooper, et al. (3), who showed an
increased ability to ingest Listeria monocytogenes in macrophages
from chronically protein-depleted adult mice. Again, the normal
response seen at 42 days of age in this study might reflect recovery
from an altered phagocytic capacity during the period of malnutrition.

In constrast to the results of our studies regarding humoral
immunity and phagocytosis mentioned above, we were able to determine that the cell-mediated immune response of the experimental
animals was significantly impaired when compared to that of the
normal guinea pigs. Another student (4) demonstrated that neonatal malnutrition had no effect on the ability of lymphocytes to
respond to phytohemagglutinin or the percentage of rosette forming
cells present in the thymus or cervical lymph nodes at 6 weeks of
age. The ability to respond to DNCB 14 days after sensitization,
or at 42 days of age was, however, significantly depressed. Some
animals, negative at this first challenge, were able to respond
upon re-challenge 2 weeks later. It was postulated that malnutrition during the postnatal period interfered with either thymocyte seeding into peripheral lymph nodes, thus leading to a quantitative defect in the cell-mediated immune response, or the production of a nonspecific factor involved in the expression of delayed hypersensitivity.

Finally, determination of cyclic AMP levels in the plasma of
experimental and control animals revealed that cyclic AMP was reduced 6-8 times in the experimental group (8).

It should be emphasized that this study is still in a pilot

stage and that additional work is currently underway to expand the
data and confirm some of the results presented here. However, we
feel that this model is satisfactory for experimentation which may
help us to confirm epidemiologic observations made in humans.

Other work is in progress to determine possible differences
between the 2 animal groups in regard to their susceptibility
to cancer induction by cells from a nitrosamine-induced hepatoma.
The effect of BCG administered before, simultaneously or after
tumor transplants is also being investigated. The differential
susceptibility of the 2 groups to viral and fungal (Candida
albicans) infections will be studied since it appears that we
are dealing with impairment of the cell-mediated immune response
by malnutrition. Tests devised to assay for lymphokines are now
being developed for this model.

If one intends to reproduce more accurately the human manifestations of protein-calorie malnutrition, one has to consider
that food deprivation found in infants and young children goes
back for several generations. Therefore, malnutrition would
appear to exert its effects on the immune system, particularly
during gestation. Consequently, studies of postnatal malnutrition
may not be able to provide the answer to the problem. However,
experimental maternal deprivation, so as to determine its effects
on the fetus, may represent a technical feat, especially in the
guinea pig model.

REFERENCES

1. Berlinerman, D., Evaluation of Humoral Factors in Mal-
    nourished Newborn Guinea Pigs, M.S. Thesis (Pathology),
    Virginia Commonwealth University, Richmond, Virginia
    (1972) 1.
2. Chandra, R.K., J. Pediatrics, 81 (1972) 1194.
3. Cooper, W.C., Good, R.A. and Mariani, T., Amer. J. Clin.
    Nutr., 27 (1974) 647.
4. Israel, B.A., Persistent Depression of Cell-Mediated Immunity
    Following Neonatal Malnutrition in the Guinea Pig, M.S.
    Thesis (Pathology) Virginia Commonwealth University,
    Richmond, Virginia (1975) 1.
5. Sauberlich, H.E., Dowdy, R.P., Skalu, J.H., Laboratory Tests
    for the Assessment of Nutritional Status, CRC Press, Inc.,
    Ohio (1974) 1.
6. Smith, R.T., Biol. of Gestation, 2 (1968) 321.
7. Solomon, J.B., Frontiers of Biology (Ed. E. L. Tatum and A.
    Neuberger) North Holland Publishing Co., Amsterdam, 20
    (1971) 38.
8. Weeks, B.A., Escobar, M.R. and Dutz, W., Va. J. Sci., 26
    (1975) 98.

9.   Wilkinson, D., Factors Mediating the Resistance to Human
     Adenovirus in an Inbred Strain of Mice, M.S. Thesis (Pathology)
     Virginia Commonwealth University, Richmond, Virginia (1972) 1.
10. Work, T.H., Ann. Int. Med., 79 (1973) 701

METHODS IN BASIC LABORATORY IMMUNOLOGY - WORKSHOP SUMMARY

H. FRIEDMAN

Albert Einstein Medical Center, Philadelphia
Pennsylvania (USA)

As part of a post congress program for the International
Congress of the RES a "traveling" workshop devoted to laboratory
aspects of basic immunology was presented. By means of such a
workshop a forum was provided for a free and searching discussion
of current methodologies concerning the involvement of the RE
System in various aspects of immunity, including resistance to infectious agents, tumors, etc. Participants in this Workshop were
those attending the Congress who wished to participate in the
three day tour of the "Basque Country" of Northern Spain. Fifty
participants of the Congress, including spouses and family
participated in the three day tour, which included visits to
several exciting cities in Northern Spain. Each morning the
scientific participants had a "round table" discussion of various
aspects of laboratory immunology. In addition, one evening session
was held and many of the discussions continued on the bus and
during meals.

The major subject discussed was laboratory procedures for
detecting humoral and cellular components of the immune response
mechanism. In addition, some practical applications of these
laboratory techniques were discussed and described, including
newer approaches to a dental disease problem, i.e., periodontitis
which involves both cellular immunity and humoral hypersensitivity
reactions. Cellular immune responses in virus infections were also
discussed in detail as an example of how modern immunologic methodology can be applied to practical problems in microbiology and
infectious diseases.

In the first portion of the workshop sessions a general dis-

cussion of serologic procedures occupied most of the attention of
the participants. For example, basic consideration of the precipitin
reactions in liquid and gel media, agglutination procedures, including passive hemagglutination, latex and charcoal agglutination tests,
etc., as well as complement fixation procedures were briefly discussed and related to practical uses in various fields, including
those related to the RE system. Immunofluorescent and radioisotope
"tagging" procedures were also discussed briefly, as well as newer
methods using enzyme-linked immunoadsorbent methods to detect antigen-antibody reactions. The utilization of such tagged reagents,
i.e., either fluorechromes, radioisotopes, or enzymes, in a variety
of fields including endocrinology, virology, immunopathology, etc.,
was discussed, in some detail.

    Newer procedures for detecting antibody producing cells were
also described. The starting point of this discussion was the
now classic hemolytic antibody plaque assay in agar gel or cellu-
lose gum first described about a dozen years ago, as well as newer
modifications in which no supporting matrix is required, such as the
micro plaque assay of Cunningham using glass slides. Discussion of
the model system of hemolytic plaque assays was then expanded to
model systems based on analyses of antibody forming cells to a
variety of non-erythrocyte antigens using xenogeneic erythrocytes
sensitized with other antigens. For example, red blood cells
sensitized with bacterial antigens, proteins, polypeptides, polysaccharides or haptens have been widely used for enumerating antibody forming cells in a variety of experimental and clinical situations. In addition, direct plaque assays with bacteria as both the
immunogens and the target of the immunocyte were discussed. As an
example, E. coli was first utilized by Werner Braun and associates
approximately a decade ago for direct rather than indirect plaque
assays. Instead of coating E. coli antigens onto sheep erythrocytes, Braun and colleagues utilized living bacteria as a target
in agar gel containing splenocytes from mice immunized with E. coli
LPS. The release of bacteriolytic antibody to the E. coli into the
agar resulted in killing of the bacteria around specific immunocytes
when complement was added to the plates. Those bacteria which were
not killed grew into a visible "lawn" in the agar plate; the areas
of no growth, which were quite apparent as distinct "plaques,"
indicated the presence of specific anti E. coli antibody forming
cells. A similar direct vibriolytic plaque assay is also available
whereby cholera bacilli, first utilized over 100 years ago to
demonstrate complement dependent antibody mediated bacteriolysis,
are used. Incubation of vibrio bacilli in agar plates containing
spleen cells or other lymphoid cells from cholera immunized
individuals results in vibriolytic plaques which are readily distin-
guishable as areas of "no growth." Both high efficiency (19S)
and low efficiency (7S) plaques due to the release of various
immunoglobulin classes by the antibody forming cells are readily

detected. Similar to the passive hemolytic plaque assays, various
antigens, including haptens can be readily conjugated to bacteria
and antibody to such antigens can be detected by a bacteriolytic
plaque assay.

The applicability of antigen binding assays to detect "rosette"
forming cells, considered those immunocytes with specific receptors
on their surface for a variety of antigens, was also described.
For example, sheep erythrocytes coated with bacterial antigens,
serum proteins, polypeptides, haptens, etc., can be utilized to
detect antigen-binding rosette forming cells; such specific
antigen binding cells are readily detected even in normal nonimmune individuals, indicating the presence of immunocytes with
receptors which recognize such antigens. Furthermore, immunocytes
which recognize sheep erythrocytes are present in large numbers in
the spleen, lymphoid organs and peripheral blood of non-immunized
mice, as well as other animal species, including man. Such "background" erythrocytes are thought to represent immunocytes which
have either been stimulated by cross-reacting or similar antigens
in environment or, alternatively, genetically determined antigen-
recognizing cells important in the adoptive immune response. A
similar rosette assay with unsensitized sheep erythrocytes is also
used to detect T-lymphocytes in the peripheral blood and lymphoid
organs of humans. A variety of such rosette assay tests were described. Sheep erythrocytes sensitized with antibody and complement
can also be utilized to detect B-lymphocytes in the peripheral
blood of man. The number and distribution of such rosette-forming
B-lymphocytes correlate well with the number and distribution of
lymphocytes which have a high density of surface immunoglobulins
as detected by fluorescent antibody techniques.

A variety of laboratory procedures are now available which are
considered to be in vitro correlates of delayed hypersensitivity
reactions. The in vitro assay developed was the macrophage migration inhibition reaction whereby peritoneal macrophages were inhibited from normal migration from a capillary tube in the presence
of specifically sensitized lymphocytes plus the sensitizing antigen. Such migration inhibition in vitro correlates with delayed
skin reactions which is considered the standard indication of
cellular immunity. An indirect test for migration inhibitor factor
(MIF) which mediates macrophage migration inhibition has become a
standard in vitro assay in both basic and clinical immunology.
For this procedure lymphoid cells from individuals exhibiting cell-
mediated hypersensitivity to a specific antigen release MIF in
vitro when exposed to the specific antigen for 24 hours or longer
at 37 C. The cell-free supernatant of such antigen-treated lymphocytes has the ability of inhibiting the migration of normal macrophages in vitro. The active factor appears to be a relatively
small protein with a molecular weight of about 50,000 or so and is

not immunoglobulin in nature. Other related but probably distinct
factors which have a variety of other in vitro effects are present
in supernatants of antigen-stimulated lymphoid cells from sensitized
individuals. These include macrophage agglutinating or aggregating
factors, blastogenic stimulating factor, lymphotoxin, chemotactic fac
tor, permeability factors, etc. Interferon is also a lymphocyte-
released factor. These factors are now all considered to be lympo-
kinins and presumably released by T-cells only. However, it is now
clear that B-lymphocytes release many of these factors, including
MIF, when stimulated by antigen.

     T-lymphocytes have the ability to directly interact with
target cells and cause their cytolytic destruction. Such reactions are measured most readily in many laboratories by radioisotope
techniques in which the target cell is first labeled with an
isotope such as chromium or treated with tritiated thymidine and
the amount of radioactivity released into the supernatant after
incubation of the labeled target cells with washed sensitized
lymphoid cells determined. In many cases lymphoid cell populations
enriched in T-lymphocytes and depleted of B-lymphocytes are quite
active in such in vitro cytolysis. However, B-lymphocytes, including those from unsensitized individuals, can acquire in vitro cytolytic activity when sensitized with specific antibody. Such antibody dependent non-T-cell cytolysis can also be directly measured
by the release of a radioisotope from target cells in vitro.

     The enumeration of lymphoid cells involved in cell-mediated
reactions in vitro can also be determined by a viral plaque assay
since certain viruses appear to replicate preferentially in "activated" T-cells, including those involved in a T-cell mediated reaction to a variety of antigens. By enumerating virus plaques the
number of such lymphoid cells involved in cellular immunity can be
readily detected. In addition, the functional activity of lymphocytes, especially T-cells, can be assessed by radioisotope incorporation techniques. For example ^{3}H-thymidine is often utilized as
an indicator of cellular division. Lymphoid cells from individuals
sensitized to a variety of antigens can be stimulated by the
specific antigen in vitro that they undergo blastogenic transformation. The magnitude of the blastogenic response may be quantitated by determining the amount of radioactivity incorporated into
the nucleic acid moiety of the dividing cells. Certain plant
lectins, most notably phytohemagglutinin, Conconavalin A and pokeweed mitogen have been utilized successfully to assess the percentage of lymphocyte responsiveness. Furthermore, bacterial LPS has
been used as a mitogen to assess the blastogenic responsiveness of
B-lymphocytes in vitro. PPD, which was initially considered mainly a mitogen for T-cells, has also been utilized recently as a
mitogen for B lymphocytes.

Dr. Steven Mergenhagen, Director of the Laboratory of Microbiology and Immunology in the National Institute of Dental Research,
discussed the application of newer immunobiologic procedures to
biomedicine describing the immunologic aspects of periodontal
disease. He indicated that there is now a growing literature that
the cellular response to microbial antigens, including bacteria, in
plaque deposits on the tooth and at the gum line is intimately
involved in the progression of periodontal disease. Lymphoid cell
responsiveness to microbial antigens are altered in patients with
periodontal disease. For example, lymphoid cells, including those
in the peripheral blood, from patients with chronic severe periodontal disease show an increased blastogenic responsiveness to a
variety of bacterial and plaque extracts, including extracts from
the subjects own oral cavity. Furthermore, macrophages which are
known to infiltrate into the periodontal lesion are not only responsive to a variety of factors released by sensitized lymphoid
cells responding to the bacterial or plaque antigens, but also
appear to themselves release a variety of substances which have
marked influence on other tissue. For example, macrophages cultured
in vitro have now been shown to release soluble factors which result
in bone tissue resorption. Such resorption of bone is a major
characteristic of the periodontal disease. Therefore, the role of
lymphoid cells sensitized to bacterial and/or plaque antigens in
the progression of the disease, and the simultaneous involvement
of macrophages as a source of a bone resorbing factor explain many
of the features of oral pathology. Newer methodologies concerning
identification and isolation of bacterial plaque antigens will
undoubtedly also provide a better understanding of the stimulators
of the "unwanted" immunologic reactions in the oral cavity which
lead to bone loss and tooth resorption and/or exfoliation.

Dr. Joseph Bellanti, Professor of Pediatrics and Microbiology,
at Georgetown University Medical School, presented a broad over-
view of some of the newer information emerging concerning viral
immunopathology, especially the newer methods utilized to detect
immune responsiveness to virus infected and/or altered tissues and
cells. It had almost become axiomatic that virus immunology was
concerned with the serologic identification of virus and/or titration of antiviral serum antibody. Titration of a variety of antibodies to viruses in patients can be accomplished by many techniques, including virus neutralization, complement fixation,
hemagglutination inhibition, etc. A several fold rise in titer
to a specific virus in a patient with appropriate symptoms is
considered diagnostic for a virus infection. These procedures
assess only the changes in serum antibody titers to virus antigen. It is known, however, that in many situations in which B-
lymphocyte immunoresponsiveness is suppressed or altered, such as
in certain types of hypogammaglobulinemia, patients nevertheless
show a remarkable resistance to viral infection. It had been felt

that although such individuals may lack a normal antibody forming
mechanism, they show resistance to viruses because of interferon
production or that cell mediated immunity may be a more important
defense to certain viruses. Thus it seems important to measure
cellular immunity to virus antigens, rather than only humoral
(antibody) responses. Dr. Bellanti described a number of such
tests designed to assess cellular immune responses of patients with
viral infections. Peripheral blood leukocytes respond to either
crude or purified virus antigens by blastogenic transformation in
vitro. Furthermore, lymphocytes from patients with active viral
infection and/or following recovery from infection or active
immunization often acquire the ability to lyse virus-infected
target cells. Direct microscopic examination can obviously be
utilized to demonstrate such cytolytic reactivity of lymphocytes
against viral-infected cells. However, the most useful technique
is measurement of the release of a radioisotope marker from the
target virus-infected cells. For such procedures peripheral blood
lymphocytes from patients or experimental animals evincing cellular
immunity to a virus are incubated with appropriate target tissue
culture cells which had previously been infected with the specific
virus. The target cells are labeled by appropriate tagging with
an isotope such as ^{51}Cr. After incubation of the tagged cells
with blood leukocytes at 37 C the amount of chromium released into
the supernatant of the culture mixture is measured and the intensity of the lymphoid cell response to the virus antigen assessed.
One of the drawbacks for such procedures has been the need for
freshly infected target cells growing in a log phase. Recently
Dr. Bellanti and associates described a procedure for infecting
cells with virus and then freezing such cells at -70 C. Shortly
before a specific test is to be performed, the cells are thawed,
labeled with the radioisotope and then utilized as the target.
By this means a relatively large stock of virus infected cells can
be available in small aliquots for individual test procedures.

    Immunological tests using virus infected target cells have
also been adapted to serodiagnosis based on measurement of humoral
antibody.

    Similar leukocyte or antibody dependent cytolytic tests are
now being utilized with lymphoid cells as the "killer" cell
against target cells infected either in vivo or in vitro with
a variety for tumorigenic viruses. Furthermore, target cells of
bearing tumor-associated antigens are also being utilized in such
in vitro cytolytic reactions. It is presumed that the tumor
associated antigen on a cell surface may be due either to depression of certain fetal antigens or, alternatively, a result of a
tumorigenic virus (at least in the case of experimental animals).
Regardless of the exact nature of the antigen expression on
surfaces of the target cells, the use of radiolabeled markers per-

mits assessment of lymphoid cells capable of recognizing antigen
and subsequent lysis of the target cells.

It should suffice to state that during the course of this
workshop much discussion and indeed excitement was evoked among
the participants. One of the most consistent comments, however,
was the need for "simplified" direct assays for clinical laboratory
immunology which can be easily performed in a reproducible manner.
There was a strong feeling that those procedures which are essentially restricted by complexity or need for experience and
specialized equipment to only basic research laboratories have
little attraction for the general immunology laboratory. The
need for reproducible techniques is certainly evident and it
seems likely that the parallel development of both immunological
technologies and newer concepts is important. In this regard, it
seemed sufficient to emphasize that the 1970's are certainly
considered a golden era not only for conceptual advances in
immunology, but also for development of useful immunologic "tools."

LIST OF PARTICIPANTS

ABRAMOFF, P.
Marquette University
Milwaukee, Wisconsin USA

ACTON, J. D.
Bowman Gray School of
Medicine
Winston-Salem, North Carolina
USA

AKAZAKI, K.
Aichi Cancer Center
Research Institute
Chikusa-Ku, Nagoya 464, Japan

ASTRUC, J. A.
Medical College of Virginia
Virginia Commonwealth
University
Richmond, Virginia USA

AVILA, J. L.
Instituto Nacional de
Dermatologia
Caracas 101, Venezuela

BABNIK, J.
Josef Stefan Institute
University of Ljubljana
Jamova 39, Ljubljana,
Yugoslavia

BAIRD, L. G.
Medical College of Virginia
Virginia Commonwealth
University
Richmond, Virginia USA

BATTIFORA, H. A.
Northwestern Memorial
Hospital
Chicago, Illinois USA

BATTISTO, J. R.
Cleveland Clinic Foundation
Cleveland, Ohio USA

BELLANTI, J. A.
Georgetown University Medical
Center
Washington, D. C. USA

BERGHEM, L.
National Defense Research
Institute
Sundbyberg 4, Sweden

BIANO, G.
Harvard Medical School
Boston, Massachusetts USA

BIGLEY, R.
University of Oregon Health
Sciences Center
Portland, Oregon USA

BJÖRKLUND, B.
National Bacteriological
Laboratory, Central Hospital
and University of Stockholm
Stockholm, Sweden

BLIZNAKOV, E. G.
New England Institute
Ridgefield, Connecticut USA

BOLOS PI, C.
University of Barcelona
Barcelona, Spain

BORTIN, M. M.
Mount Sinai Medical Center and
Medical College of Wisconsin
Milwaukee, Wisconsin USA

BOUTHILLIER, Y.
Fondation Curie-Institut du
Radium
Paris, France

CARR, I.
University of Sheffield and
Weston Park Hospital
Sheffield, United Kingdom

CASTRO SERRANO, R.
University of Barcelona
Barcelona, Spain

CEGLOWSKI, W. S.
Pennsylvania State University
University Park, Pennsylvania
USA

CHAI, C. K.
The Jackson Laboratory
Bar Harbor, Maine USA

CHEDID, L. A.
Institut Pasteur
75 Paris, France

CLARK, J. L.
708 Thornby Drive
Wilmington, Delaware USA

COOPER, M. D.
Comprehensive Cancer Center
University of Alabama in
Birmingham
Birmingham, Alabama USA

DAEMS, W. Th.
Laboratory for Electron
Microscopy
Leiden, The Netherlands

DI CARLO, F. J.
Warner-Lambert Research
Institute
Morris Plains, New Jersey USA

DIENER, E.
University of Alberta
Edmonton 7, Alberta, Canada

DI LUZIO, N. R.
Tulane University School of
Medicine
New Orleans, Louisiana USA

DUTZ, W.
Medical College of Virginia
Virginia Commonwealth
University
Richmond, Virginia USA

ELSBACH, P.
New York University School
of Medicine
New York, New York USA

ESCOBAR, M. R.
Medical College of Virginia
Virginia Commonwealth
University
Richmond, Virginia USA

ESMANN, V.
Marselisborg Hospital
8000 Aarhus C, Denmark

EVANS, R.
Chester Beatty Research
Institute
Belmont, Sutton, Surrey,
Great Britain

EVERETT, N. B.
University of Washington School
of Medicine
Seattle, Washington USA

FAGUET, G. B.
Medical College of Georgia
Augusta, Georgia USA

FARBER, P. A.
Albert Einstein Medical Center
and Temple University School
of Dentistry and Medicine
Philadelphia, Pennsylvania USA

FINKE, J. H.
Cleveland Clinic Foundation
Cleveland, Ohio USA

FLEMMING, C.
Institute of Biophysics and
Radiobiology
University of Freiburg
78 Freiburg i.Br., Albertstr.
23, West Germany

FLEMMING, K. B. P.
Institute of Biophysics and
Radiobiology
University of Freiburg
78 Freiburg i.Br., Albertstr.
23, West Germany

FÖRSTER, O.
Institute of General and
Experimental Pathology
University of Vienna
Vienna, Austria

FRIEDMAN, H.
Albert Einstein Medical
Center
Philadelphia, Pennsylvania USA

GALLILY, R.
Hebrew University
Hadassah Medical School
Jerusalem, Israel

GALVIN, M. J., JR.
Medical College of Georgia
Augusta, Georgia USA

GANDER, G. W.
Medical College of Virginia
Virginia Commonwealth
University
Richmond, Virginia USA

GARCIA, J.
Instituto Municipal de
Investigaciones Medicas
Barcelona, Spain

GARZON, P.
Facultad de Medicina
Universidad de Gaudalajara
Jalisco, Mexico

GEE, J. B. L.
Yale University School of
Medicine
New Haven, Connecticut USA

GERHARDT, N. B.
University of Oregon Health
Sciences Center
Portland, Oregon USA

GERSHON, R. K.
Yale University School of
Medicine
New Haven, Connecticut USA

GIACOMO, M. De
Istituto di Anestesiologia e
Rianimazione
Universita Cattolica del
Sacro Cuore
Rome, Italy

GILLISSEN, G.
Abteilung fur Med. Mikrobiologie
der Med. Facultat der Techn.
Hochschule
51 Aachen, West Germany

GLAUMANN, H.
Sabbatsberg's Hospital
Stockholm, Sweden

GOLDENBERG, D. M.
University of Kentucky
Lexington, Kentucky USA

GOOD, R. A.
Memorial Sloan-Kettering Cancer
Center
New York, New York USA

GORDON, D.
Center for Disease Control
Atlanta, Georgia USA

GRAS RIERA, J.
Urgel 253
Barcelona, Spain

GROSS, R. L.
Massachusetts Institute of
Technology
Cambridge, Massachusetts USA

GUSDON, J. P.
Bowman Gray School of Medicine
Winston-Salem, North Carolina
USA

HALIE, N. R.
University Hospital, State
University
Groningen, The Netherlands

HARRIS, N. S.
Shriners Burns Institute and
University of Texas Medical
Branch
Galveston, Texas USA

HARRIS, S.
The Children's Hospital of
Philadelphia and School of
Medicine, University of
Pennsylvania
Philadelphia, Pennsylvania USA

HARRIS, T. N.
The Children's Hospital of
Philadelphia and School of
Medicine, University of
Pennsylvania
Philadelphia, Pennsylvania USA

HAURANI, F. I.
Cardeza Foundation for
Hematologic Research
Thomas Jefferson University
Philadelphia, Pennsylvania USA

HEINE, J. W.
Baltimore City Hospitals
Baltimore, Maryland USA

HEISE, E. R.
Bowman Gray School of Medicine
Winston-Salem, North Carolina
USA

HINCHMAN, S.
Cleveland Clinic Foundation
Cleveland, Ohio USA

HOEBEKE, J.
Janssen Pharmaceutica
2340 Beerse, Belgium

HOFFMANN, E. O.
Tulane University School of
Medicine
New Orleans, Louisiana USA

HOSSAINI, A. A.
Medical College of Virginia
Virginia Commonwealth
University
Richmond, Virginia USA

HUYBRECHTS-GODIN, G.
Institute of Cellular and
Molecular Pathology
Brussels, Belgium

INO, S.
Fukushima Institute of Health
15-4 Mitouchi, Hokida
Fukushima 960, Japan

JACQUES, P. J.
International Institute for
Cellular and Molecular Pathology
Universite Catholique de
Louvain en Woluwe and Institute
Pasteur du Brabant
Brussels, Belgium

JAEGER, K. H.
Institute for Theoretical
Physics
Vienna, Austria

JAP, P. H. K.
University of Nymegen
Nymegen, The Netherlands

KAPLAN, A. M.
Medical College of Virginia
Virginia Commonwealth
University
Richmond, Virginia USA

KENYON, A. J.
Memorial Sloan-Kettering
Cancer Center
New York, New York USA

KIJISTRA, A.
University Hospital, University
of Leiden
Leiden, The Netherlands

KLEIN, M.
Temple University School
of Medicine
Philadelphia, Pennsylvania USA

KOHOUT-DUTZ, E.
Veterans Administration
Hospital
Richmond, Virginia USA

KOJIMA, M.
Fukushima Medical College
Fukushima, Japan

KOPITAR, M.
Josef Stefan Institute
University of Ljubljana
Jamova 39, Ljubljana, Yugoslavia

LANDUCCI, G.
University of Minnesota
Minneapolis, Minnesota USA

LA VIA, M. F.
Emory University School of
Medicine
Atlanta, Georgia USA

LÁZÁR, E.
Institute of Biology
University of Szeged School
of Medicine
Szeged, Hungary

LÁZÁR, G.
Institute of Pathophysiology
University of Szeged School
of Medicine
Szeged, Hungary

LEAKE, E. S.
Bowman Gray School of Medicine
Winston-Salem, North Carolina
USA

LEFKOWITZ, S. S.
Texas Tech University School
of Medicine
Lubbock, Texas USA

LESKOWITZ, S. S.
Tufts Medical School
Boston, Massachusetts USA

LINDEMAN, J.
University of Leiden
Leiden, The Netherlands

LINNA, T. J.
Temple University School of
Medicine
Philadelphia, Pennsylvania USA

LONG, A. P., JR.
Portsmouth General and Mary
View Hospitals
Portsmouth, Virginia USA

LOPEZ-ROMAN, A.
University of Navarra Faculty
of Medicine
Pamplona, Spain

MACHADO, E. A.
University of Tennessee Memorial
Research Center
Knoxville, Tennessee USA

MARIANI, T.
University of Minnesota
Minneapolis, Minnesota USA

MERGENHAGEN, S. E.
National Institute of Dental
Research
Bethesda, Maryland USA

MEURET, G.
Medical Clinic C
Kantonsspital
CH-9006 St. Gallen, Switzerland

MITCHELL, R. H.
Medical College of Virginia
Virginia Commonwealth
University
Richmond, Virginia USA

MODOLELL, M.
Max Planck Institute
71 Freiburg, Germany

MONTGOMERY, J. R.
School of Primary Medical Care
Huntsville, Alabama USA

MORRELL, R. M.
Baylor College of Medicine
Houston, Texas USA

MORRISON, D. C.
Scripps Clinic and Research
Foundation
La Jolla, California USA

MOUTON, D.
Fondation Curie-Institut du
Radium
Paris, France

MUSETESCU, M.
Institut Merieux Charbonniers
Marcy l'Etoile, France

NAHMIAS, A. J.
Center for Disease Control and
Emory University School of
Medicine
Atlanta, Georgia USA

NELSON, R. A.
University of Montreal
Montreal 101, Canada

NITULESCU, G.
Institute of General and
Experimental Pathology
University of Vienna
Vienna, Austria

NOMURA, N.
Tohoku University School of
Medicine
1-1 Seiryomachi, Sendai, Japan

NOTANI, G. W.
Memorial Sloan-Kettering Cancer
Center
New York, New York USA

OEHLING, A.
University of Navarra Faculty
of Medicine
Pamplona, Spain

OGLE, C. K.
University of Cincinnati
Cincinnati, Ohio USA

OGLE, J. D.
University of Cincinnati
Cincinnati, Ohio USA

OUCHI, E.
Tohoku University School of
Medicine
1-1 Seiryomachi, Sendai, Japan

PANT, K. D.
University of Kentucky
Lexington, Kentucky USA

PARANT, M.
Institute Pasteur Immunothe
Exptle
28 rue de Dr. Roux
Paris, France

PARKER, C. W.
Washington University School
of Medicine
St. Louis, Missouri USA

PATRIARCA, P.
Istituto di Patologia, Generale
Universita Trieste
Trieste, Italy

PIERCE, G.
University of Kansas Medical
Center
Kansas City, Kansas USA

PONZIO, N. M.
New York University School
of Medicine
New York, New York USA

POULIK, M. D.
William Beaumont Hospital and
Wayne State School of Medicine
Detroit, Michigan USA

PUSZTAI, Z.
Oberarztin der Abteilung
Medizinische Mikrobiologie
der Techn. Hochschule
Aachen, West Germany

REICHARD, S. M.
Medical College of Georgia
Augusta, Georgia USA

RITTS, R. E., JR.
Mayo Graduate School of
Medicine, University of
Minnesota
Rochester, Minnesota USA

RODRIGUEZ, G. E.
Medical College of Virginia
Virginia Commonwealth
University
Richmond, Virginia USA

ROLLEY, R. T.
Johns Hopkins University School
of Medicine
Baltimore, Maryland USA

ROOS, D.
Central Laboratory
Netherlands Red Cross Blood
Transfusion Service
Amsterdam, The Netherlands

ROSE, N. R.
Wayne State University School
of Medicine
Detroit, Michigan USA

ROSSI, F.
Istituto di Patologia, Generale
Universita Trieste
Trieste, Italy

SATO, T.
Tohoku School of Medicine
1-1 Seirymachi, Sendai, Japan

SBARRA, A. J.
St. Margaret's Hospital and
Tufts University School of
Medicine
Boston, Massachusetts USA

SCHELL-FREDERICK, E.
Institut de Recherche
Interdisciplinaire
Free University of Brussels
Brussels, Belgium

SCHILDT, B. E.
University of Linkoping
Regional Hospital
S-581 85 Linkoping, Sweden

SCHORN, H.
Euratom-C.E.A./S.P.T.E.
92 Fonteayaux-Roses, France

SEIJI, K.
Tohoku School of Medicine
1-1 Seirymachi, Sendai, Japan

SIGEL, M. M.
University of Miami School
of Medicine
Miami, Florida USA

SORANZO, M. R.
Istituto di Patologia
Generale Universita Trieste
Trieste, Italy

STENDAHL, O.
Linkoping University School
of Medicine
S-581 85 Linkoping, Sweden

STEWART, C. C.
Washington University School
of Medicine
St. Louis, Missouri USA

STINNETT, J. D.
University of Cincinnati
Medical Center
Cincinnati, Ohio USA

STUART, A.
University of Edinburgh Medical
School
Edinburgh, Scotland

STUTMAN, O.
Memorial Sloan-Kettering
Cancer Center
New York, New York USA

SUGIUCHI, I.
Tohoku School of Medicine
1-1 Seirymachi, Sendai, Japan

SYROP, H.
Medical College of Virginia
Virginia Commonwealth
University
Richmond, Virginia USA

TAKEHISA, Y.
Tokushima University School
of Medicine
Tokushima, Japan

TANAKA, O.
Tokushima University School
of Medicine
Tokushima, Japan

THORBECKE, G. J.
New York University School
of Medicine
New York, New York USA

TORO, I.
University of Budapest
Budapest, Hungary

TRUEHEART, R. E.
St. Francis Hospital
Evanston, Illinois USA

TUSET OLLER, N.
Sants 13
Barcelona, Spain

VACHER, J. M.
Roussel-Volaf Laboratories
93230 Romainville, France

VAZQUEZ, J. J.
University of Navarra Faculty
of Medicine
Pamplona, Spain

VICTOR, T. A.
Evanston Hospital
Evanston, Illinois USA

VILDÉ, F.
Universite Paris 5, Hopital
Boucicaut
Paris, France

VILDÉ, J. L.
Universite Paris 7, Hopital
Claude Bernard
Paris, France

WATABE, S.
Tohoku University School
of Medicine
1-1 Seiryomachi, Sendai, Japan

WEEKS, B. A.
Old Dominion University
Norfolk, Virginia USA

WILKINS, D. J.
Centre Europeen de Recherches
Mauvernay
63201 Riom, France

WILLIS, J. I.
The Ohio State University
Columbus, Ohio USA

WILSON, W. R.
Mayo Clinic
Rochester, Minnesota USA

WIRTZ, P.
University of Nymegen
Nymegen, The Netherlands

WISSÉ, E.
Laboratory for Electron
Microscopy, University of
Leiden, The Netherlands

YU, C.
UCLA School of Medicine
Los Angeles, California USA

YU, D. T. Y.
UCLA School of Medicine
Los Angeles, California USA

YUNIS, E. J.
University of Minnesota
Minneapolis, Minnesota USA

INDEX

A cell, 17-22, 137, 138
A_1 protein, 140, 142
ADCC, *see* Antibody Cytotoxicity
Adenosine deaminase deficiency, 143, 161, 163
Adenosine monophosphate, cyclic, 113, 116, 446
Adenovirus, 444
Agammaglobulinemia, Bruton's X-linked, 158-159, 206
AKP, *see* Mouse strains
AKR, *see* Mouse strains
Alloantibody, 257, 261, 263
Alloantigen and allograft rejection, 232
Allogeneic cell, 254
    tissue, 257
Allograft
    enhancement, 16
    in mouse, 256-257
    rejection, 231, 232, 256, 293-299
    retention, 254-256
    tolerance, 295
Allotransplantation, 231-251, 294
Allotype, defined, 8
Amphetamine sulfate, 270, 273
Amphotericin B, 333
Amyloid, 129, 130, 283-288
Amyloidosis, 130, 283-285, 288
Anaphylactic shock, 191
Anergy, 196, 198, 409
    in skin test, 189, 191
    tumor-induced, 409
Ankylosing spondylitis, 244

Antibody
    antigen complex, 71
    cytotoxicity, antibody-mediated, 132
        cell-mediated, 217
        -dependent cell-mediated (ADCC), 218
    plaque-forming cell (PFC), 342
    role of, 122
    synthesis, an orderly process, 53
        persistent, 69-70
        shift from M to G, 53-64
        titer and degree of disease, 212
    see Immunoglobulins
Antigen, 139
    antibody complex, 71
    capping, 22
    effect on localization of B cells, 65-75
    -induced suppression of delayed hypersensitivity, 189
    microbial, 453
    plasma cell, 199, 205
    recognition, 136
    θ (Theta), 111
Anti-graft strain alloantibody, 256
Antiserum, 107
Arthritis, 244
Arthus reaction, 210
Ascitic fluid globulin, 257
Astrocyte, 133, 134
Ataxia telangiectasia (AT), 122
    immunodeficiency in, 164-166
    in malignancy, 165